Undergraduate Lecture Notes in Physics

For further volumes:
http://www.springer.com/series/8917

Undergraduate Lecture Notes in Physics (ULNP) publishes authoritative texts covering topics throughout pure and applied physics. Each title in the series is suitable as a basis for undergraduate instruction, typically containing practice problems, worked examples, chapter summaries, and suggestions for further reading.

ULNP titles must provide at least one of the following:

- An exceptionally clear and concise treatment of a standard undergraduate subject.
- A solid undergraduate-level introduction to a graduate, advanced, or non-standard subject.
- A novel perspective or an unusual approach to teaching a subject.

ULNP especially encourages new, original, and idiosyncratic approaches to physics teaching at the undergraduate level.

The purpose of ULNP is to provide intriguing, absorbing books that will continue to be the reader's preferred reference throughout their academic career.

Series Editors

Valerio Faraoni

Special Relativity

Valerio Faraoni
Physics Department
Bishop's University
Sherbrooke, QC
Canada

ISSN 2192-4791 ISSN 2192-4805 (electronic)
ISBN 978-3-319-01106-6 ISBN 978-3-319-01107-3 (eBook)
DOI 10.1007/978-3-319-01107-3
Springer Cham Heidelberg New York Dordrecht London

Library of Congress Control Number: 2013942663

Printed on acid-free paper

Springer is part of Springer Science+Business Media (www.springer.com)

To my parents

Preface

Special Relativity arises from the need for a textbook at intermediate level, especially in North-American universities. There are nowadays several technological applications of Special Relativity in everyday life, including the GPS system, PET scanners, and other medical instruments. There are plenty of college-level, deliberately watered-down introductions to Special Relativity. These introductions typically serve a Modern Physics course and devote two chapters to the subject. Such short introductions are inadequate for physicists or scientists requiring a modern university course. What is worse, and few teachers will deny that, fundamental aspects of Special Relativity are usually not understood by the students and often not presented by the teachers. The four-dimensional world view is usually introduced only in a very colloquial way while the mathematical formalism of four-vectors and four-tensors, which is necessary to understand and formalize the idea of spacetime, is ignored. Yet, the latter is not so complicated and is accessible to students who have taken a linear algebra course provided that room is made in the relativity course for it. For most students the Modern Physics course is the only introduction to Special Relativity that they receive in their entire curriculum and, as a result, physics and mathematics students (the very ones who *must* know it) miss this fundamental subject. This knowledge gap is very serious since Special Relativity is, without doubt, one of the great intellectual achievements of mankind, exactly the kind of stuff that a student gets into physics for. From a teacher's logistical point of view, it is also relatively simple in comparison with, e.g., General Relativity or particle physics. Yet, too often the science student is robbed of this part of his or her education by systems too bent on entertainment and student counts. An essential part of the physics curriculum is sacrificed for fear of mathematics.

For students carrying on to a course on General Relativity, perhaps in graduate studies, this problem quickly becomes irrelevant but such students are a minority and physics departments in many small universities do not offer a General Relativity course or offer it infrequently. This situation is too common and it is true that the mathematical background of second- or third-year undergraduates is limited and that the formalism of tensors necessary for the study of Special Relativity needs to be taught in the course. One problem of this rationalization of the curriculum is the lack of a proper textbook on Special Relativity at a level higher than college physics or current Modern Physics courses, yet accessible to undergraduates who are not prepared to tackle General Relativity textbooks or Special Relativity classics such as Wolfgang Rindler's *Introduction to Special Relativity*.

Another complication is that, in an elementary course in Special Relativity, typical undergraduates have not yet taken a serious course in electromagnetism and do not yet master the Maxwell equations. Therefore, the usual textbook approach using Maxwell's theory as a trampoline for Special Relativity ends up being ineffective for students at this level. In retrospect, although the historical line of thought and the traditional pedagogical approach was electromagnetism first and then Special Relativity, and it is true that Maxwell's theory receives its most elegant and revealing formulation in Special Relativity, the latter is not just electromagnetism pushed to the extreme (this point was stressed early on by Pauli). Although electromagnetism was a useful trampoline to discover relativity, Einstein's 1905 theory is about the unveiling of the four-dimensional nature of the world, spacetime and its Lorentzian geometry, the equivalence of mass and energy, the modified mechanics, and the fundamental symmetries of this theory. The Maxwell field is only one of the possible forms of mass-energy that can live in Minkowski spacetime. This point of view is obvious after one takes a course in General Relativity or particle physics, but it is at odds with the discover-Special-Relativity-through-electromagnetism approach of many textbooks. An axiomatic approach to Special Relativity based on the Principle of Relativity and the constancy of the speed of light seems the best option here—the key physical concepts of the theory (relativity of simultaneity, time dilation, and length contraction) can be derived easily from these two postulates, and this is the path originally followed by Einstein. A constructive approach would need to be based on electromagnetism and the average undergraduate at this point still views the Maxwell equations as rather abstract or has not seen them at all. It is possible, however, to introduce the Lorentz transformation as the transformation relating inertial frames which leaves electromagnetism invariant without considering explicitly the Maxwell equations. An example showing a simple electromagnetic phenomenon will do, and this is the avenue taken in this book. The Lorentz invariance of the full Maxwell equations can be checked later when the student is familiar with them.

The book begins at an elementary level exposing and discussing in Chap. 1 and Chap. 2 all the basic concepts normally contained in college-level expositions, including the Lorentz transformation. Then, in Chap. 3, it introduces the student to the four-dimensional world view, making clear that this is implied by the Lorentz transformations mixing time and space coordinates. This is as far as the best Modern Physics textbooks seem to get. In addition, we make use of spacetime diagrams already in this part of the book (as well as, of course, in the following parts) to visualize the relevant discussion. Following this introduction, in Chap. 4, is the part that is avoided in lower-level courses; the formalism of tensors. It is my experience, gained by teaching Special Relativity courses in Canada, that once the student is persuaded that space and time do mix and is motivated by the need to understand the four-dimensional world, and once time is made during the course to explain this part, this chapter goes surprisingly easily and the fear of tensors proves unfounded. Chapter 5 then introduces the essential concept of causality missed in the Modern Physics courses and details the application of the general formalism of tensors to Minkowski space. The following Chap. 6 discusses the relativistic mechanics of point particles, four-momentum and four-force, the equivalence between mass and energy, and some applications. Chapter 7 describes relativistic optics and, whenever possible, uses the similarities between the motion of massless and massive particles to facilitate understanding and memorization. An optional short Chap. 8 follows, in which measurements in Minkowski spacetime are discussed to dispel the widespread impression that physical observables are coordinate components of four-vectors or four-tensors (and therefore, that all measurements are based on coordinate-dependent components).

Matter in Minkowski spacetime is discussed in Chap. 9. Here the energy-momentum tensor of a continuous distribution of mass-energy and its covariant conservation are introduced, and various (optional) energy conditions are presented. Angular momentum is discussed briefly. This part is followed by a discussion of the scalar field (presented first, as this is the simplest field in theoretical physics), of perfect fluids, and of the Maxwell field (which is now presented as one of the possible energy distributions in Minkowski space, although as an important one describing one of only four fundamental interactions).

For pedagogical reasons, it is easier to work in Cartesian coordinates and all the material introduced thus far (with the exception of the tensor formalism of Chap. 4, which is quite general) is restricted to these coordinates. This restriction facilitates the introduction of the ideas and concepts of Special Relativity, but is too restrictive. At this point, Chap. 10 introduces general coordinates, covariant differentiation, and geodesics (the emphasis is on computational skills rather than rigour or proof) and reformulates the previous material in arbitrary coordinate systems using covariant formulas. This chapter could be skipped if short of time during a course.

The mathematics essential to study Special Relativity is relatively simple: some calculus and linear algebra. The real difficulties are conceptual, not mathematical. The beginning student is fighting his or her own physical sense derived by everyday low-speed intuition, which conflicts with the results of Special Relativity. Mathematics is a tool which facilitates understanding and, by banning it from the course, current low-level oversimplified courses preclude the understanding of the basic concepts of the theory. It is much better to allow room for it in the course, although stripping it down to the essential, and then use it rather than paraphrasing it with obscure and wordy discussions which invariably fail to convey the essence of relativity.

Every chapter is supplemented by a section containing practice problems. These exercises constitute an essential part of the textbook and the student is urged to try them. The solution to selected exercises, as well as the numerical answers to others, appears at the end of the book.

For pedagogical purposes, in this book we retain explicitly the speed of light c in the formulas, i.e., we do not set c to unity except in spacetime diagrams and we include several steps in the calculations to facilitate comprehension by the beginner. We avoid the obsolete notions of rest mass and dynamical mass,

Acknowledgments

It is a pleasure to thank all my students who provided feedback and comments over the years. Special thanks go to Andres Zambrano Moreno for typing most of the content of this book. Many thanks go also to Dr. Aldo Rampioni, Springer Editor for Theoretical and Mathematical Physics for his support and friendly assistance during the writing of this book.

Contents

Symbols

c	Speed of light in vacuo
τ	Proper time
l_0	Proper length
γ	Lorentz factor
β	v/c
λ	Affine parameter along a curve
$\{x^\mu\}$	Coordinate system
∂_μ	$\partial/\partial x^\mu$
$f_{,u}$	$\partial f/\partial x^\mu$
$\nabla_\mu f$	Gradient of f
A^μ	Contravariant vector
A_μ	Covariant vector
$(\mu\nu)$	Symmetrization of the indices μ and ν
$[\mu\nu]$	Antisymmetrization of the indices μ and ν
ds^2	Minkowski line element
$g_{\mu\nu}$	Metric tensor
$g^{\mu\nu}$	Inverse metric tensor
$\eta_{\mu\nu}$	Minkowski metric in Cartesian coordinates
$\eta^{\mu\nu}$	Inverse Minkowski metric in Cartesian coordinates
δ^μ_ν	Kronecker delta
$\varepsilon_{\mu\nu\alpha\beta}$	Levi-Civita symbol in Minkowski spacetime
u^μ	4-Velocity
a^μ	4-Acceleration
f^μ	4-Force
p^μ	4-Momentum
E	Energy
T	Kinetic energy
ν	Frequency
ω	Angular frequency
λ	Wavelength
$\mathbf{k}$	Wave vector
$T_{\mu\nu}$	Stress-energy tensor
ρ	Energy density

P	Pressure
J^{μ}	Energy current density
j^{μ}	Electric current density
w	Equation of state parameter
ϕ	Scalar field
$V(\phi)$	Scalar field potential energy density
$\mathbf{E}$	Electric field
$\mathbf{B}$	Magnetic field
A^{μ}	4-Potential
$F_{\mu\nu}$	Maxwell tensor
n^{μ}	Unit normal to a surface
$\mathrm{d}S^{\mu}$	Surface element
$h_{\mu\nu}$	3-Dimensional metric
∇_{μ}	Covariant derivative operator
$\Gamma^{\alpha}_{\mu\nu}$	Connection coefficients
$\equiv$	Equal by definition
$\doteq$	Equal in a special coordinate system or in a particular gauge

Chapter 1
Fundamentals of Special Relativity

Concepts that have proven useful in ordering things easily achieve such authority over us that we forget their earthly origins and accept them as unalterable givens.

—Albert Einstein

1.1 Introduction

The theory of Special Relativity formulated by Albert Einstein in 1905 [1, 2] is, without doubt, one of the great intellectual achievements of the twentieth century. Our everyday experience is about objects moving at speeds much smaller than the speed of light in vacuo, $c = 2.99792458 \times 10^8$ m/s. Newtonian mechanics was developed to describe phenomena at typical speeds $v \ll c$ and fails when speeds are not negligible in comparison with c. This situation is not infrequent; for example, it is relatively easy to accelerate electrons to speeds $v = 0.99\,c$ in accelerators. However, as their velocities become closer and closer to c, it becomes harder and harder to accelerate these electrons further. As we will see, it is believed that the speed of light c is a barrier that particles and physical signals cannot break.

The theory of Special Relativity was a one man's work which incorporated previous ideas and contributions (especially those of Poincaré), but in a way that nobody was able to fully grasp before Einstein.[1] To summarize, Special Relativity correctly predicts phenomena for velocities v which are not negligible in comparison with the speed of light, or for $v \to c$, and it has Newtonian mechanics as its low-velocity limit. Further, it agrees with Maxwell's theory of electromagnetism (which predicts electromagnetic waves travelling at speed c in vacuo) and describes physics in the

[1] See Ref. [3] for a brief outline of the ideas leading to the development of Special and General Relativity and, e.g., Ref. [4] for a discussion of the limitations of the ideas of Henri Poincaré and Hendrik-Antoon Lorentz in the development of Special Relativity.

V. Faraoni, *Special Relativity*, Undergraduate Lecture Notes in Physics,
DOI: 10.1007/978-3-319-01107-3_1, © Springer International Publishing Switzerland 2013

absence of gravity. The class of phenomena modelled by Special Relativity includes the dynamics of particles moving at speeds comparable with c, the slowing down of clocks and the contraction of lengths in moving reference frames, the Doppler effect for electromagnetic radiation, the equivalence of mass and energy illustrated by the "most famous formula of physics" $E = mc^2$, and the creation and annihilation of particle/antiparticle pairs. These phenomena have many applications, including particle accelerators, the Global Positioning System (GPS), commercial nuclear reactors, Positron Emission Tomography (PET) scanners used in hospitals, and synchrotron light imaging in industry. Special Relativity has now been tested by countless experiments and is still tested every day in particle accelerators. The merging of Special Relativity and Quantum Mechanics produced relativistic Quantum Field Theory, on which the standard model of particle physics is based. Nowadays it is impossible for physicists to conceive of a non-relativistic fundamental theory.[2]

In our exposition of Special Relativity we will omit gravity from the picture and we will assume that inertial frames extend to spatial infinity. Special Relativity is a theory for a world in which gravity is switched off; the relativistic theory of gravity, General Relativity, contains Special Relativity as the limit for vanishing gravitational fields, and was created by Einstein in 1915. We will not consider General Relativity and gravity in this book, except for some considerations at the end of it.

1.2 The Principle of Relativity

Physical phenomena are normally described with respect to a given reference frame, which consists of a set of spatial axes and a set of synchronized clocks with respect to which one measures positions and time intervals. Newton's second law is valid in all inertial frames, which are defined by Newton's first law. According to this law, an *inertial frame* is one in which a free body (i.e., a body not subject to forces) is not accelerating. Any system moving with constant velocity with respect to an inertial frame is also an inertial frame.

The *Principle of Newtonian* (or *Galilean*) *Relativity* states that

the laws of mechanics are invariant under change of inertial frame.

In other words, the outcome of a physical experiment performed in two different inertial frames must be the same; no experiment can detect a difference between the two frames. This statement applies not only to kinematics, but to any kind of (non-gravitational) physics. Therefore, *there is no absolute motion* through space, and *there is no preferred inertial frame.*

In pre-relativistic physics, the transformation from one inertial frame to another (*change of frame*) is given by the *Galilei transformation*. For simplicity, consider the motion of two inertial frames S and S' such that the x', y', and z' Cartesian axes

[2] Exceptions are certain theories of quantum gravity, but these are completely speculative and probing them does not seem feasible in the near future.

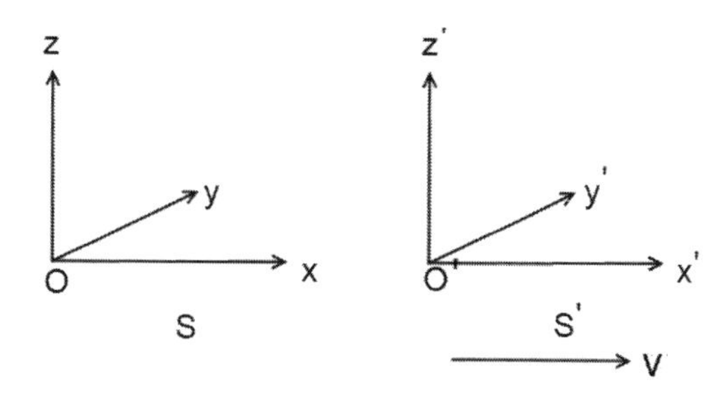

Fig. 1.1 Two inertial frames S and S' in standard configuration

of S' are parallel to the x, y, and z axes of S, respectively, and such that the system $\{x', y', z'\}$ moves along the x-axis with *constant* velocity v (*standard configuration*), as in Fig. 1.1. We adopt the convention that the velocity v is positive if the system S' is moving *away* from S.

Suppose that you synchronize clocks in the two inertial frames in such a way that the time indicated by a clock in S coincides with the time indicated by a clock in S' when the origins of S and S' superpose. An inertial observer refers his or her measurements to an inertial coordinate system and a clock.

The *Galilei transformation* relating two inertial frames S and S' in standard configuration is

$$
\begin{aligned}
x' &= x - vt, \\
y' &= y, \\
z' &= z, \\
t' &= t.
\end{aligned}
\tag{1.1}
$$

In Newtonian mechanics, time is *assumed* to be the same in the two frames and clocks run at the same rate in two inertial frames connected by a Galilei transformation, irrespective of their relative velocity v. The time interval between two events is the same for both observers. (As we will see shortly, in Special Relativity this assumption turns out to be incorrect, which becomes manifest when v is comparable to c. This fact is striking and counterintuitive.)

The inverse of the Galilei transformation is obtained by changing primed and unprimed quantities and by the substitution $v \longrightarrow -v$, producing

$$x = x' + vt,$$
$$y = y',$$
$$z = z', \tag{1.2}$$
$$t = t'.$$

This procedure is intuitive: if S sees S' moving away with velocity v, then S' sees S moving away with velocity $-v$, a clear application of the Principle of Relativity.

A fact that is familiar from experience is that, in Newtonian mechanics, 3-dimensional lengths are Galilei-invariant, i.e., if $\mathbf{x}_A$ and $\mathbf{x}_B$ are the positions of two points of space and $l_{(3)} \equiv |\mathbf{x}_A - \mathbf{x}_B|$, then

$$l'_{(3)} = l_{(3)} \tag{1.3}$$

(the formal proof of this statement is left to the reader as a simple exercise). Physically, Eq. (1.3) means that if a rod has length $l_{(3)}$ in one inertial frame, its length will be measured to be $l_{(3)}$ also in any other inertial frame related to the former by a Galilei transformation, a statement that matches everyday's experience. If x^i and $x^i + dx^i$ ($i = 1, 2, 3$) are the Cartesian coordinates of two infinitesimally close points of space, the *Euclidean line element*

$$dl^2_{(3)} = dx^2 + dy^2 + dz^2 = \sum_{i,j=1}^{3} \delta_{ij} dx^i dx^j \tag{1.4}$$

(where δ_{ij} is the Kronecker delta) is invariant under Galilei transformations.

The importance of the Galilei transformation in Newtonian mechanics is due to the fact that

Newton's laws are invariant under Galilei transformations

To wit, consider Newton's second law $\mathbf{F} = m\, d^2\mathbf{x}/dt^2$ in an inertial frame $\{t, x, y, z\}$. Is it $m\, d^2\mathbf{x}'/dt'^2 = \mathbf{F}$ also in another inertial frame $\{t', x', y', z'\}$? The answer is affirmative, for

$$\mathbf{x}' = (x - vt, y, z),$$

$$\frac{d\mathbf{x}'}{dt'} = \left(\frac{dx}{dt} - v, \frac{dy}{dt}, \frac{dz}{dt}\right)$$

and, since v is a constant,

$$m\frac{d^2\mathbf{x}'}{dt'^2} = m\left(\frac{d^2x}{dt^2}, \frac{d^2y}{dt^2}, \frac{d^2z}{dt^2}\right) = \mathbf{F}.$$

1.3 *Groups: The Galilei Group

Transformations acting on a physical system which leave it invariant are called *symmetries* of that system. For example, a ball is invariant under rotations about an axis passing through its centre (its symmetries). In theoretical physics, symmetries can be much more abstract than rotations, but the idea is still the same: certain mathematical operations are defined which leave the physical system (particle, rigid body, wave, field, etc.) invariant. Usually, the operations used in theoretical physics and acting on a physical (or mathematical) system have a mathematically well-defined structure. The simplest such algebraic structure is a *group*. Thinking of the symmetry transformations as elements of a set, the mathematical definition of a group follows.

Definition 1.1 Let G be a (non-empty) set and let $\circ$ denote an operation in G, i.e., a single-valued map

$$\circ\colon G \times G \longrightarrow G \tag{1.5}$$

which associates to a pair of elements $g_1, g_2 \in G$ another element of G which is denoted $g_1 \circ g_2$ (the *image* of the ordered pair (g_1, g_2) under the operation $\circ$). The important thing is that performing the operation on g_1 and g_2 one obtains as a result an object $g_1 \circ g_2$ *which is still an element of* G or, in more technical parlance, that G be *closed* with respect to $\circ$.

The pair $(G, \circ)$ is a *group* if

1. The operation is *associative*, i.e.,

$$(g_1 \circ g_2) \circ g_3 = g_1 \circ (g_2 \circ g_3) \quad \forall g_1, g_2, g_3 \in G. \tag{1.6}$$

2. There exists a *neutral element* e such that

$$g \circ e = e \circ g = g \quad \forall g \in G. \tag{1.7}$$

3. Any $g \in G$ has an *inverse*, i.e.,

$$\forall g \in G \ \ \exists g^{-1} \in G \ : \ g \circ g^{-1} = g^{-1} \circ g = e. \tag{1.8}$$

In general, the operation $\circ$ is not commutative, i.e., $g_1 \circ g_2 \neq g_2 \circ g_1$.

Definition 1.2 The group $(G, \circ)$ is *Abelian* or *commutative* if the operation $\circ$ is commutative, $g_1 \circ g_2 = g_2 \circ g_1 \quad \forall g_1, g_2 \in G$.

Definition 1.3 A *subgroup* of the group $(G, \circ)$ is a subset $S \subseteq G$ which is itself a group with respect to the operation $\circ$.

A subgroup of G must necessarily contain the neutral element e of G.

Example 1.1 The set of integers $\mathbb{N} = \{0, 1, 2, 3, \ldots, n, \ldots\}$ is *not* a group with respect to the addition $+$ because the inverse of any $n > 0$ (which would be $-n$) does not belong to $\mathbb{N}$.

Example 1.2 The set $\mathbb{Z} = \{0, \pm 1, \pm 2, \pm 3, \ldots, \pm n, \ldots\}$ is a group with respect to the addition $+$. It is not a group with respect to the multiplication.

Example 1.3 The set of real numbers $\mathbb{R}$ is a group with respect to addition.

Example 1.4 The set of real numbers without the zero $\mathbb{R} \setminus \{0\}$ is a group with respect to multiplication.

Example 1.5 The set of real functions $f : I \longrightarrow \mathbb{R}$, where I is an interval of $\mathbb{R}$, is a group with respect to the sum of functions defined as

$$(f + g)(x) \equiv f(x) + g(x) \qquad \forall x \in I. \tag{1.9}$$

* * *

Let us return to the physics related to the change of inertial frames:

the Galilei transformations form a commutative group with respect to the composition of transformations.

In fact, let g be a Galilei transformation that sends (x, y, z, t) into $(x' = x - vt,\ y' = y,\ z' = z,\ t' = t)$ and consider the set G of all Galilei transformations. *The composition of Galilei transformations is still a Galilei transformation:* if g_1 represents the map $(x, y, z, t) \longrightarrow (x - vt, y, z, t)$ and g_2 represents the map $(x, y, z, t) \longrightarrow (x - ut, y, z, t)$, we then have that $g_2 \circ g_1$ is

$$(x, y, z, t) \longrightarrow (x', y', z', t') = (x - vt, y, z, t) \longrightarrow (x'', y'', z'', t'')$$

$$= (x' - ut, y', z', t') = \Big(x - (v + u)\,t, y, z, t\Big)$$

and $g_1 \circ g_2$ is

$$(x, y, z, t) \longrightarrow (x', y', z', t') = (x - ut, y, z, t) \longrightarrow (x'', y'', z'', t'')$$

$$= (x' - ut, y', z', t') = \Big(x - (u + v)\,t, y, z, t\Big),$$

hence $g_1 \circ g_2 = g_2 \circ g_1$ is still a Galilei transformation and the operation $\circ$ is commutative.

Let us now check the *associativity:*

1. if $g_3 \in G$ is another Lorentz transformation with velocity w, $(g_1 \circ g_2) \circ g_3$ acting on (x, y, z, t) produces

$$\Big(x-(u+v)\,t,\,y,\,z,\,t\Big) \longrightarrow (x''', y''', z''', t''') = \Big(x-[(u+v)+w]\,t,\,y,\,z,\,t\Big)$$

while

$$g_1 \circ (g_2 \circ g_3)$$

produces $\Big(x - [v + (u + w)]\,t,\, y,\, z,\, t\Big)$, which is identical to the previous expression, hence

$$(g_1 \circ g_2) \circ g_3 = g_1 \circ (g_2 \circ g_3) \quad \forall\, g_1, g_2, g_3 \in G. \tag{1.10}$$

2. Let us check now for the existence of a neutral element (identity). This is obviously the Galilei transformation with zero velocity $v = 0$, or the map e described by $(x, y, z, t) \longrightarrow \left(x', y', z', t'\right) \equiv (x, y, z, t)$. It is $g \circ e = e \circ g = g \;\; \forall\, g \in G$.
3. Finally, let us check for the existence of an inverse. Let g be the Galilei transformation with constant velocity v. Then its inverse g^{-1} is the transformation with velocity $-v$. In fact, $g \circ g^{-1} = e$ is given by

$$(x, y, z, t) \longrightarrow (x - vt, y, z, t) \longrightarrow (x - vt - (-vt)\,, y, z, t) = (x, y, z, t)$$

so that $g \circ g^{-1} = g^{-1} \circ g = e$.

We conclude that the Galilei transformations form an Abelian group. Since each transformation of this group corresponds to one and only one value of the constant velocity v, it is said that the Galilei transformations form a *1-parameter group*. The statement that Newton's laws are invariant under Galilei transformations is rephrased by saying that the Galilei group is the symmetry group of Newtonian mechanics.

Group theory is an abstract branch of mathematics that plays a fundamental role in atomic, molecular, particle, and solid state physics, to the point that the search for new or more general symmetries of the equations describing a physical theory may lead to new discoveries, or may serve as a guide for the formulation of new theories. An important result due to the German mathematician Emmy Noether states that conserved quantities are associated with symmetries (see. e.g., Ref. [5]): this fact makes it very valuable to know the symmetries of a physical system. For the moment, we will content ourselves with the definitions given.

One can consider two consecutive Galilei transformations with velocity vectors $\mathbf{v}_1$ and $\mathbf{v}_2$ that are not parallel, i.e., not in standard configuration. These two transformations commute (i.e., the result does not depend on the order in which the two transformations are applied). The first transformation gives

$$(\mathbf{x}, t) \longrightarrow (\mathbf{x}' = \mathbf{x} - \mathbf{v}t, t)$$

and, for two such transformations with velocity vectors **v** and **u**, it is

$$\mathbf{x} \longrightarrow \mathbf{x}' = \mathbf{x} - \mathbf{v}\,t \longrightarrow \mathbf{x}'' = \mathbf{x}' - \mathbf{u}\,t$$

$$= \mathbf{x} - (\mathbf{u} + \mathbf{v})\,t = \mathbf{x} - (\mathbf{v} + \mathbf{u})\,t.$$

1.4 Galileian Law of Addition of Velocities

Consider two events separated by an infinitesimal distance $\mathrm{d}x$ and an infinitesimal time $\mathrm{d}t$ in two inertial frames S and S' in standard configuration. According to the Galilei transformation, we have $\mathrm{d}x' = \mathrm{d}x - v\mathrm{d}t$ (v is constant). The velocity of an object covering the space $\mathrm{d}x$ in the time $\mathrm{d}t$ in S is $u^x \equiv dx/dt$. The velocity of an object covering $\mathrm{d}x'$ in the time $\mathrm{d}t' = \mathrm{d}t$ in S' is $u^{x'} \equiv u = \frac{d(x - vt)}{dt} = \frac{dx}{dt} - v$, or

$$u^{x'} = u^x - v \tag{1.11}$$

(*Galileian law of addition of velocities*), which agrees with everyday's low-speed intuition. The relative velocity of two particles approaching each other along the same line with speeds v_1 and v_2 (and velocities v_1 and $-v_2$) is $u = v_1 - (-v_2) = v_1 + v_2$, i.e., "velocities add up". For example, two cars moving toward each other at 50 km/h each will have a relative velocity of 100 km/h. If Eq. (1.11) were correct for light signals, then the relative velocity of a light beam and an observer approaching the beam would be larger than c. However, in Special Relativity the Galilean transformation formula (1.11) is incorrect and leads to wrong results when applied to electromagnetic waves or to particles moving at relativistic speeds.

1.5 The Lesson from Electromagnetism

The Maxwell equations[3] formulated in 1864,

---

[3] Remember that, in Cartesian coordinates, the electric and magnetic fields have components $\mathbf{E} = (E^x, E^y, E^z)$ and $\mathbf{B} = (B^x, B^y, B^z)$, $\nabla \cdot \mathbf{E} \equiv \frac{\partial E^x}{\partial x} + \frac{\partial E^y}{\partial y} + \frac{\partial E^z}{\partial z}$, $\nabla \times \mathbf{B} \equiv \begin{vmatrix} \hat{i} & \hat{j} & \hat{k} \\ \partial_x & \partial_y & \partial_z \\ B^x & B^y & B^z \end{vmatrix}$, and $\frac{\partial \mathbf{E}}{\partial t} \equiv \left(\frac{\partial E^x}{\partial t}, \frac{\partial E^y}{\partial t}, \frac{\partial E^z}{\partial t} \right)$.

$$\nabla \cdot \mathbf{E} = 4\pi\rho, \tag{1.12}$$

$$\nabla \times \mathbf{E} = -\frac{1}{c}\frac{\partial \mathbf{B}}{\partial t}, \tag{1.13}$$

$$\nabla \cdot \mathbf{B} = 0, \tag{1.14}$$

$$\nabla \times \mathbf{B} = \frac{4\pi}{c}\mathbf{j} + \frac{1}{c}\frac{\partial \mathbf{E}}{\partial t} \tag{1.15}$$

(where ρ is the electric charge density, $\mathbf{j}$ is the current density, $\mathbf{E}$ is the electric field, $\mathbf{B}$ is the magnetic field and cgs Gaussian units are used), summarize the entire theory of electromagnetism, together with the law for the Lorentz force acting on a charged particle of charge q, mass m, and velocity $\mathbf{v}$ in electric and magnetic fields

$$\mathbf{F} = q\left(\mathbf{E} + \frac{\mathbf{v}}{c} \times \mathbf{B}\right). \tag{1.16}$$

It was noted early on that *the Maxwell equations are not invariant under Galilei transformations*, as one can deduce by checking these equations explicitly. Instead of doing this exercise consider, for illustration, a phenomenon ruled by the Maxwell equations, a spherical flash of electromagnetic waves irradiated by a point-like source and propagating radially outward with speed c. For a pulse emitted at time $t = 0$ at the origin (Fig. 1.2) one has, at the time $t > 0$, that $c^2t^2 = x^2 + y^2 + z^2$, or

$$-c^2t^2 + x^2 + y^2 + z^2 = 0 \tag{1.17}$$

in the $\{x, y, z, t\}$ inertial frame. But in the inertial frame $\{x', y', z', t'\}$ moving with constant velocity v with respect to the previous one, it is not $-c^2\left(t'\right)^2 + x'^2 + y'^2 + z'^2 = 0$. Instead we have, using Eq. (1.17),

$$-c^2t'^2 + x'^2 + y'^2 + z'^2 = vt\,(vt - 2x) \neq 0. \tag{1.18}$$

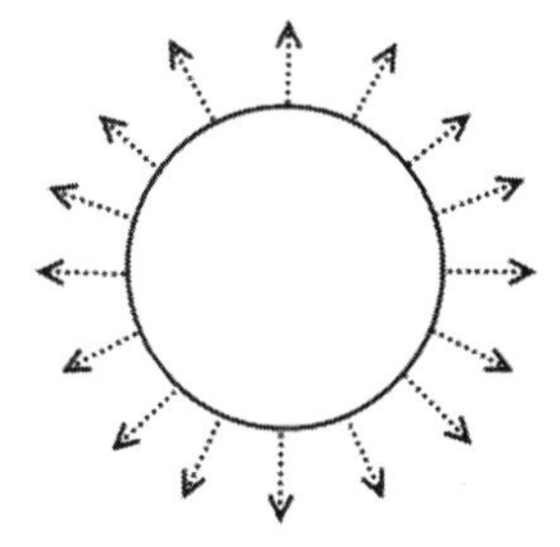

Fig. 1.2 A spherical flash of electromagnetic waves irradiated by a *point-like* source at the origin at time $t = 0$

In general, the Maxwell equations and the phenomena that they rule are not Galilei-invariant. Based on the success of Maxwell's theory and on the inconsistence of Newtonian mechanics which, with its forces dependent only on the position of particles, requires an instantaneous action at distance and an infinite propagation speed of interactions, Special Relativity rejects the Galilei transformation and replaces it with the transformation which leaves the Maxwell equations invariant under changes of inertial frame (this is the Lorentz transformation which we are going to see in Chap. 2). Newton's theory, which is invariant under Galilei transformations and therefore is incompatible with electromagnetism, is discarded (in the sense that it is only believed to be valid in the limit $|v| \ll c$), while Maxwell's theory is retained. After all, the latter correctly predicts electromagnetic waves which, propagating with speed c, are the most relativistic physical entities already well known in Einstein's time.

1.5.1 The Ether

Newtonian mechanics is incompatible with Maxwell's theory of electromagnetism. Consider, for example, two particles interacting at a distance. In Newtonian mechanics, a change in the position of one particle is felt immediately at the position of the second particle, implying an infinite speed of propagation of interactions. By contrast, the electromagnetic interaction between two charges propagates at finite speed c in Maxwell's theory. At the end of the 1800s it was proposed that electromagnetic waves are disturbances of, and propagate in, a medium called *luminiferous ether* and that c is the speed of these waves with respect to a frame in which the ether is at rest.[4] Then, according to Galilei's law of composition of velocities, in any other frame moving with speed v with respect to the ether, the speed of light would be $c \pm v$.

Since the earth orbits the sun with a linear speed $v \approx 30$ km/s, one could regard the earth as a moving frame with respect to the ether and would detect a velocity $c \pm v$ instead of c when the earth is at opposite positions in its orbit around the sun (Fig. 1.3). Various experiments, and especially the famous Michelson–Morley experiment, failed to detect changes in the speed of light and the motion of the earth with respect to the ether.

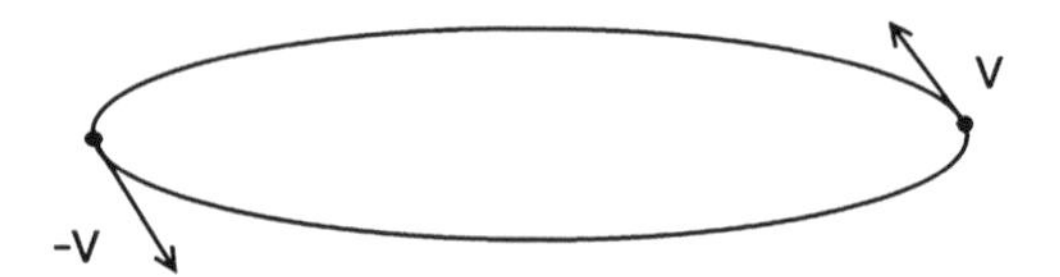

Fig. 1.3 The motion of the earth with respect to the ether at two opposite points of the orbit

[4] Today the concept of ether seems artificial even at first sight, but it was not so unnatural in the mechanistic views of the physicists of the 1800s.

1.5.2 The Michelson–Morley Experiment

The purpose of the 1881 Michelson–Morley experiment [6] was to measure the speed of the earth with respect to the ether. The result of this classic experiment is negative and contradicts the ether hypothesis.

The experimental apparatus, now known as the Michelson interferometer, is described in Fig. 1.4. It consists of two perpendicular arms of equal length L. Monochromatic light from a source passes through a beam splitter (a half-silvered mirror) at the origin and is sent trough each arm, the light in the two arms starting out in phase. In each arm, the light beam bounces off a mirror at the end of the arm and is reflected back to the origin, where it is collected and sent to a telescope. There, the light beams collected by both arms interfere and, if their phases differ, an interference pattern is observed. Assume, for simplicity, that the ether is moving in the direction of the horizontal arm of Fig. 1.4 and that the light moves with speed c with respect to the ether: then the travel times in the two arms will be different.

Non-relativistic analysis: consider arm 2 (horizontal) of the interferometer, oriented in the direction of the motion of the earth through the ether. This means that the ether is moving by at speed v and that the speed of light *with respect to the earth* is $(c - v)$ before reflection on the mirror M_2 and $(c + v)$ after reflection on M_2. The time of travel to the right is $\frac{L}{c-v}$ and the time of travel to the left is $\frac{L}{c+v}$. The total travel time for the round trip on the horizontal arm is

$$t_2 = \frac{L}{c-v} + \frac{L}{c+v} = \frac{2L}{c\left(1 - \frac{v^2}{c^2}\right)}. \quad (1.19)$$

Let us consider now arm one of the interferometer (the vertical one in Fig. 1.4) perpendicular to the ether wind: the speed of this light beam relative to the earth is

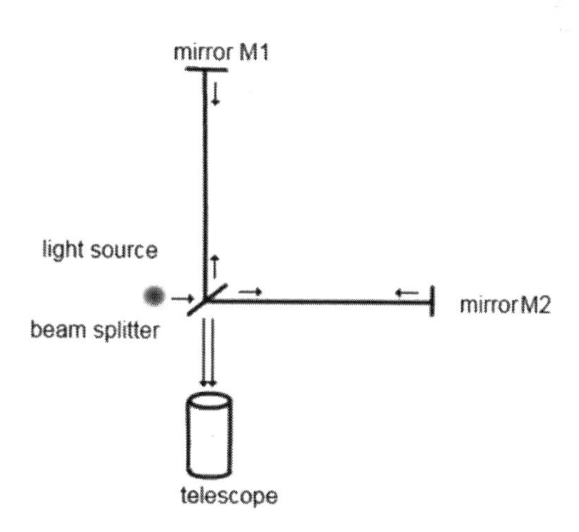

Fig. 1.4 The Michelson interferometer

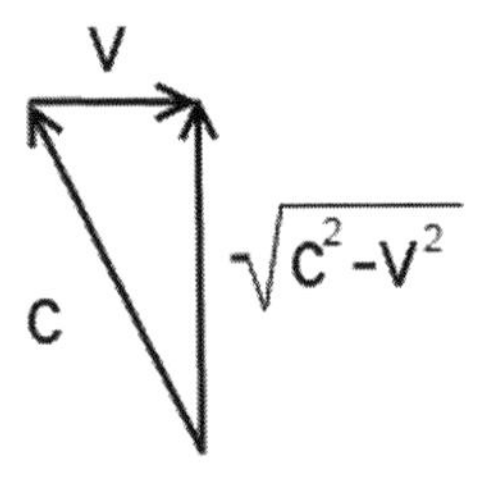

Fig. 1.5 The velocity of the light beam in the vertical arm of the interferometer with respect to the earth

$\sqrt{c^2 - v^2}$ (see Fig. 1.5).[5] The travel time for a round trip in the vertical arm is

$$t_1 = \frac{2L}{\sqrt{c^2 - v^2}} = \frac{2L}{c\sqrt{1 - \frac{v^2}{c^2}}}. \tag{1.20}$$

The time difference between the two arms is

$$\Delta t \equiv t_2 - t_1 = \frac{2L}{c}\left[\frac{1}{1 - \frac{v^2}{c^2}} - \frac{1}{\sqrt{1 - \frac{v^2}{c^2}}}\right];$$

by expanding this formula using $(1 + x)^\alpha \approx 1 + \alpha x$, one obtains[6]

$$\Delta t \simeq \frac{2L}{c}\left[1 + \frac{v^2}{c^2} - \left(1 + \frac{v^2}{2c^2}\right)\right] = \frac{Lv^2}{c^3}.$$

The two light beams start out in phase and return with a phase difference. The difference in the length of the paths is $\Delta x = c\Delta t = Lv^2/c^2$, while the resulting shift of the fringes is $\Delta\phi = \frac{\Delta x}{\lambda} = \frac{L}{\lambda}\left(\frac{v}{c}\right)^2$ to second order in v/c, as follows from the fact that the phase is $\phi = kx - \omega t = 2\pi\left(\frac{x}{\lambda} - \nu t\right)$, where λ and ν are the wavelength and frequency of the monochromatic light, respectively. If the entire apparatus is rotated by 90°, the phase difference $\Delta\phi$ is doubled. Furthermore, multiple reflection (n round trips) in the mirrors could increase the length L to nL. If $v \sim 3 \cdot 10^4$ m/s, $L = 11$ m (as in the original Michelson interferometer), $\lambda = 500$ nm, then

$$\Delta\phi = \frac{2 \cdot (11\,\text{m})}{5 \cdot 10^{-7}\,\text{m}}\left(\frac{3 \cdot 10^4\,\text{m/s}}{3 \cdot 10^8\text{m/s}}\right)^2 \approx 0.4.$$

[5] The ray traveling in the vertical direction is oblique because of reflection off the beam splitter which is moving with respect to the ether (see Refs. [7–11] for details).

[6] It appears that Michelson and Morley's original paper contained an error of order higher than $(v/c)^2$ [12]. The detailed higher order analysis of the Michelson–Morley experiment is rather complicated [13, 14], but it is not important here.

The Michelson interferometer was capable of detecting phase shifts $\Delta\phi \approx 10^{-2}$ but no shift in the fringe patterns was detected.

The experiment has now been repeated many times, at different times of the year (i.e., at different points on the earth's orbit), yielding negative results.[7] The simplest explanation of this experiment, which was probably known to Einstein although it is not mentioned explicitly in his 1905 paper, is that there is no ether and light does not require an ether medium to propagate.

To summarize the situation before 1905, Maxwell's theory predicted the existence of electromagnetic waves and de facto incorporated optics into electromagnetism. These waves were detected and generated experimentally by Hertz and constitute a major success of the theory. Since at that time physicists were used to the idea that waves need a medium to propagate, the existence of an ether in which electromagnetic waves can propagate was postulated. Then the earth should move through this ether and it should be possible to detect its absolute motion through the ether. The 1881 Michelson–Morley experiment designed to measure the speed of the earth relative to the ether provided a negative result, which leads to three possibilities:

1. The ether is attached rigidly to the earth, which is clearly an ad hoc explanation and raises more problems than it solves.
2. It was assumed that rigid bodies contract and clocks slow down when moving through the ether (an hypothesis advanced independently by Lorentz and FitzGerald in 1892 and 1889, respectively [17, 18], who could not motivate it convincingly). This ad hoc assumption does not lead to new predictions and is not falsifiable.[8]
3. There is no ether. This simple explanation, adopted by Einstein, leads to the loss of absolute time and simultaneity, as we will see in the following. The disappearance of the ether concept makes the speed of light in vacuo a fundamental quantity, instead of being merely a property of the medium in which light propagates.

1.6 The Postulates of Special Relativity

The failure of measuring the motion of the earth with respect to the ether motivates the introduction of the two basic postulates of Special Relativity, on which the entire theory is based.

1. **Principle of Relativity:** *The laws of (non-gravitational) physics assume the same form in all inertial reference frames. All inertial observers are equivalent.*

---

[7] Nowadays, the experiment is regarded as a test of the isotropy of the speed of light [15], which has been probed with laser versions of the Michelson–Morley experiment with an accuracy of 10^{-15} [16].

[8] However, the idea of motion-dependent forces is not completely incorrect: one can make a toy model of a solid with point charges bound by electrostatic forces. The latter transform in a known way under a change of inertial frame giving effectively a contraction in the direction of motion, the same which is obtained by length contraction of the inter-particle separations [19].

2. **Constancy of the speed of light:** *The speed of light in vacuo has the same value* $c \approx 3 \cdot 10^8$ m/s *in all inertial frames, regardless of the velocity of the observer or the source.*

The first postulate implies that all inertial observers are equivalent and experiments cannot distinguish between different inertial observers. This assumption is not new, in fact this postulate was already contained in Newtonian mechanics. As a consequence, there is no preferred inertial system and one cannot distinguish whether a body is at rest or in uniform motion. Absolute motion does not exist—motion is relative to a frame, position and velocity are relative concepts, and only the relative motion between two inertial frames is relevant (hence the name "relativity" given to the theory). Since it is impossible to detect absolute motion, the first postulate eliminates the ether and makes a revolution of ideas possible.

The first postulate is not only a statement about mechanics and electromagnetism but comprises all kinds of physical laws (except gravitation, which is left out of Special Relativity). In retrospect, there is no purely mechanical experiment: looking at things involves light (hence optics), and matter is made of subatomic particles, nuclei, atoms, and molecules which interact through the electromagnetic, strong, and weak forces,[9] hence the restriction of a fundamental postulate to purely mechanical phenomena would seem artificial.

The second postulate is consistent with the first one:[10] if the speed of light were not c in all inertial frames, but only in one, then a preferred frame would be selected. There is no ether and light moves with speed c with respect to *any* inertial observer, which explains the Michelson–Morley experiment in the simplest possible way (in Einstein's words, the concept of ether becomes "superfluous" [1]). The second postulate contains a feature that should not be overlooked, namely the *isotropy of space*: the speed of light has the same value in every direction in all inertial frames.

A basic feature of Special Relativity, which is ultimately unsatisfactory, is that it is based on a special class of observers, the inertial observers. It would be preferable to formulate physics in a way that is valid for *all* observers, inertial or not (in fact, strictly inertial reference frames extending to spatial infinity do not exist when gravity is included in the picture). This conceptual improvement is accomplished in the theory of General Relativity.

1.6.1 The Role of the Speed of Light

It is by now clear that the speed of light plays a major role in the theory and should rightly be regarded as a fundamental constant. The speed of light in vacuo is

---

[9] However, only the gravitational and electromagnetic interactions were known at the beginning of the twentieth century.

[10] It appears that, in the early writings on Special Relativity, the constancy of the speed of light was mostly intended as meaning that the speed of light is independent of the velocity of the source, not of the inertial frame in which it is measured [20].

independent of the motion of the sources and of the observers. The mere appearance of the constant c in a formula immediately signals that electromagnetism and/or relativity are at play.

Most relativists adopt *relativistic units* in which the speed of light assumes the value unity. In these units, the dimensions of a length are the same as those of time,

$$c = 1, \qquad [L] = [T], \qquad 1 \text{ s} = 3 \cdot 10^8 \text{ m}.$$

For clarity, we will retain the constant c in our formulas but we will set $c = 1$ in drawings ("spacetime diagrams") in which the horizontal axes represent one or two spatial dimensions x or (x, y) and the vertical axis represents the time t. Due to the large value of the constant c in ordinary units, these diagrams would be unreadable if standard units were used. For example, the straight line representing the equation of a light ray directed in the positive x-direction, $t = x/c$, would have a slope indistinguishable from zero by the eye. Choosing units in which $c = 1$ is particularly convenient for these diagrams.

Instead of using a rigid ruler to measure distances, in Special Relativity it is more useful to define distances operationally by using electromagnetic signals (light, radar, etc.).[11] The distance between an observer and an object could be measured this way: a mirror is placed on the object and the observer sends a light signal at the time $t_{emitted}$, which is reflected off the object back to the observer and received at the time $t_{received}$ (Fig. 1.6). The distance between the observer and the mirror is then

$$d = c \, \frac{(t_{received} - t_{emitted})}{2}. \tag{1.21}$$

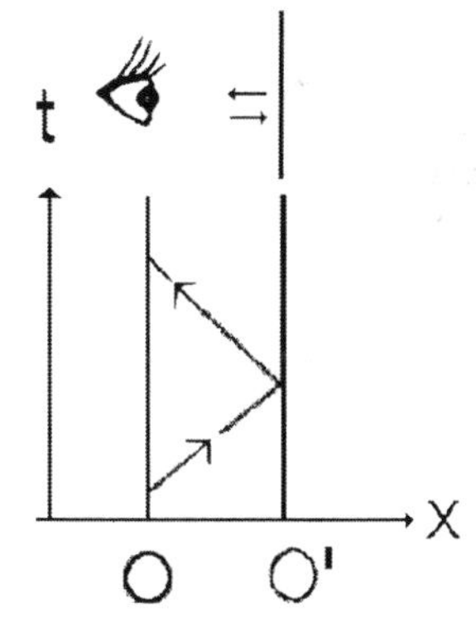

Fig. 1.6 A light signal is emitted by an observer O, bounces off a mirror O', and is reflected back to O. Units are used in which $c = 1$

[11] Indeed, interplanetary distances are measured using electromagnetic waves reflected off a target.

1.7 Consequences of the Postulates

The two postulates of Special Relativity allow one to make startling predictions concerning space, time, and simultaneity which conflict with our intuition based on everyday's experience at speeds $|v| \ll c$. These predictions originate many famous apparent paradoxes, such as the car-in-the-garage/pole-in-the-barn paradox and the twin paradox. These puzzles are not true paradoxes, they only seem paradoxical because they conflict with our intuition derived from the non-relativistic regime. Before proceeding, let us provide some definitions.

An *event* is defined as a point in space and time labeled by four coordinates, e.g., Cartesian coordinates $\{ct, x, y, z\}$. Examples of events are 2:25 pm in the centre of the Quad at Bishop's University (although it may not look very eventful in the common sense of this word); the collision of two particles at a given instant of time; or the flash of an explosion at 12 noon at a certain point in space. The same event is assigned different coordinates in different inertial frames.

A *reference frame* is a coordinate grid equipped with a set of clocks located at the grid intersections and synchronized. Synchronization can be achieved by using light signals; for example, a clock sends a laser pulse to a second clock located at a distance r. It takes a time r/c for the signal to reach the second clock. If the second clock reads the time r/c when it receives the laser pulse, then the two clocks are synchronized (note that these are two clocks *in the same inertial frame*).

1.7.1 Relativity of Simultaneity

In Newtonian physics there is an absolute time t, the same for all inertial (and non-inertial) observers. Thus, if two events are simultaneous in an inertial reference frame, they are simultaneous for all inertial observers. In Special Relativity, instead, the result of the measurement of a time interval depends on the reference frame in which the measurement is made. This fact can be illustrated by a simple thought experiment (or *gedankenexperiment*[12]): a boxcar moves with uniform velocity along a straight track. Two lightning bolts strike the ends of the boxcar simultaneously with respect to an inertial observer O who is stationary with respect to the track, leaving marks on the ground and on the boxcar (Fig. 1.7). The observer O on the ground is located midway between the marks A and B left by the lightning bolts on the ground. An inertial observer O' on the boxcar is located midway between the marks A' and B' left on the boxcar.

[12] There is a long tradition of deriving special-relativistic formulas from gedankenexperimenten [21].

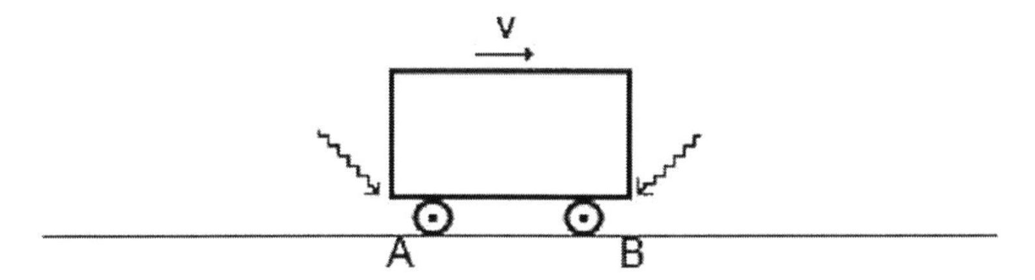

Fig. 1.7 A thought experiment demonstrating the relativity of simultaneity. Two lightning bolts strike the ends of the boxcar, leaving marks on the ground and on the boxcar

According to the inertial observer O, the two light signals reach O at the same time: the events at A and B are simultaneous for O. According to the other inertial observer O', by the time light reaches O, O' has moved.[13] Light from B' has moved past O', while light from A' has not yet reached O'. But the fact that light travels at the same speed c in both frames implies that, as seen by O', lightning struck the front of the boxcar B' *before* it struck the back A'. The two events are not simultaneous for O'. Hence,

events that are simultaneous in an inertial frame are not simultaneous in another inertial frame (relativity of simultaneity).

Simultaneity is not absolute: it depends on the state of motion of the observer. Both observers are correct: there is no preferred observer or preferred inertial frame. Together with the constancy of c, this fact leads to an apparent paradox in conflict with everyday's intuition. One can use a diagram with space x and time t on its axes (and units in which $c = 1$) to visualize the boxcar experiment, as shown in Fig. 1.8. Light signals from A and B reach O (represented by a vertical line in the diagram) simultaneously; the same light signals reach O' (represented by an oblique line) at different times according to O'.

1.7.2 Time Dilation

Now consider a vehicle (for example the boxcar of the previous thought experiment) moving to the right with constant velocity $\mathbf{v}$, and with a mirror attached to its ceiling. A passenger O' shines a flash of laser light toward the ceiling using a laser pointer. This flash is reflected and then is received by O'. Let us examine this experiment according to the inertial observer O' in the moving frame $\{t', x', y', z'\}$ (Fig. 1.9). O' carries a clock to measure the time interval $\Delta t'$ between emission and reception of the flash

---

[13] The argument holds because of the finiteness of the speed of light, in contrast with Newtonian mechanics in which the latter is infinite. It is also essential that the speed of light is the same in the two inertial frames.

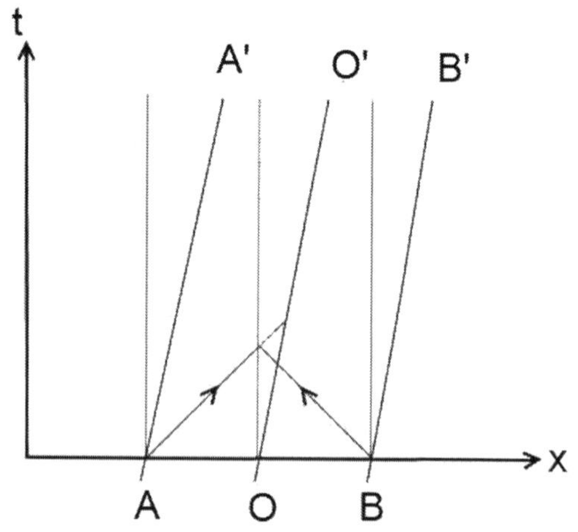

Fig. 1.8 A spacetime diagram of the boxcar gedankenexperiment. In the inertial frame S, light signals from A and B intersect the *vertical line* representing the stationary observer O at the same time t, but reach the oblique line representing the moving observer O' (which is the t'-axis) at different times

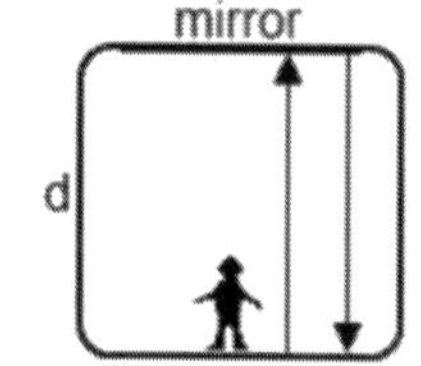

Fig. 1.9 The time dilation experiment according to the observer O'

$$\Delta t' = \frac{2d}{c}, \tag{1.22}$$

where d is the height of the boxcar travelled twice vertically by the pulse. According to the inertial observer O in the system $\{t, x, y, z\}$, for the light to reach the mirror and return to the observer, it is necessary that light is not sent vertically, but at an angle (Fig. 1.10). It takes a time Δt for light to travel to the mirror and back to O. By the time the light reaches the mirror, this has moved by $v\Delta t/2$ and, by the time the light is back to O, the laser has moved by $v\Delta t/2$ after reflection. We have

$$\left(c\frac{\Delta t}{2}\right)^2 = \left(v\frac{\Delta t}{2}\right)^2 + d^2$$

or

$$\left(\frac{\Delta t}{2}\right)^2 \left(c^2 - v^2\right) = d^2,$$

which yields

$$\Delta t = \frac{2d}{c}\frac{1}{\sqrt{1 - \frac{v^2}{c^2}}},$$

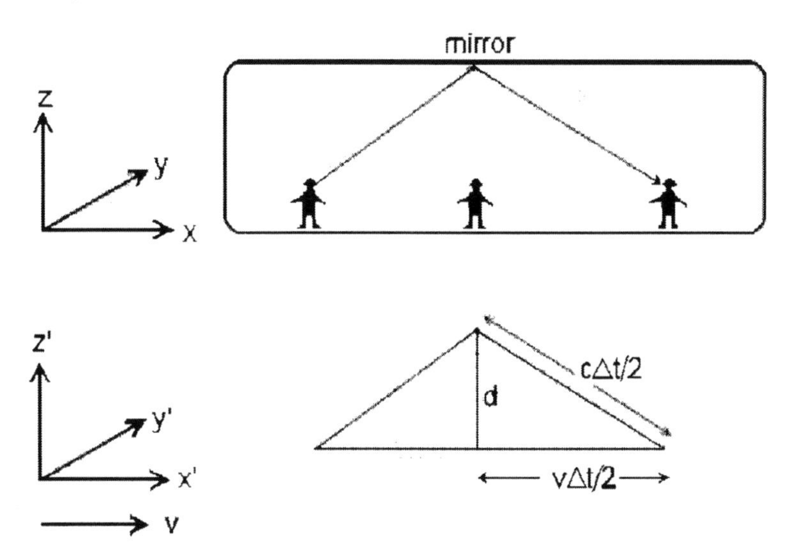

Fig. 1.10 The time dilation experiment according to the observer O

while $\Delta t'$ is given by Eq. (1.22). Then Δt and $\Delta t'$ are related by

$$\Delta t = \frac{\Delta t'}{\sqrt{1 - \frac{v^2}{c^2}}} \equiv \gamma \Delta t', \tag{1.23}$$

where

$$\gamma \equiv \frac{1}{\sqrt{1 - \frac{v^2}{c^2}}} \tag{1.24}$$

is the *Lorentz factor*, which is plotted as a function of v/c in Fig. 1.11. Clearly, if $v \neq 0$, then $\gamma > 1$ and $\Delta t > \Delta t'$. This effect is known as *time dilation*: the time interval between two events depends on the state of motion of the observer. The process in the (x, t) plane is illustrated in Fig. 1.12. We define the *proper time* interval between two events as the time interval measured by an observer who sees the two events occurring at the same point of space. For example, $\Delta t'$ in Eq. (1.23) above is the proper time. In short,

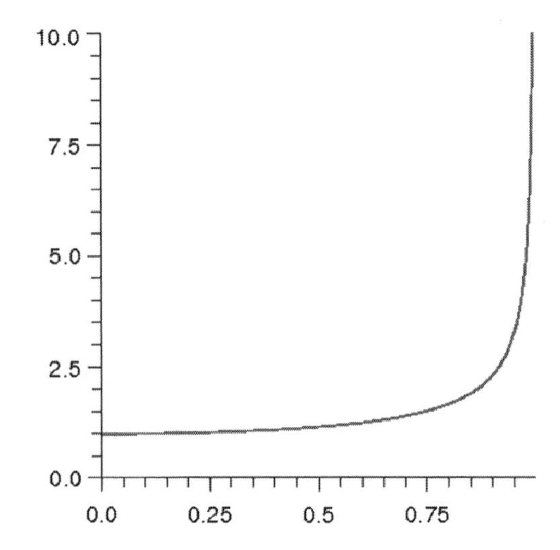

Fig. 1.11 The Lorentz factor γ as a function of $\beta \equiv v/c$

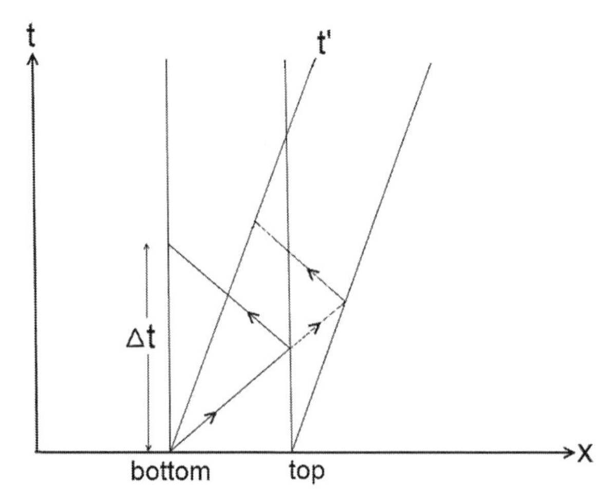

Fig. 1.12 The time dilation experiment visualized in the (x, t) plane of the observer O. This observer measures a time interval Δt on the t-axis, while the moving observer O' measures a different time interval $\Delta t'$ on the t'-axis

the proper time is the time measured by a clock in the inertial frame in which it is at rest.

(Often, the proper time is denoted by the symbol τ and Eq. (1.23) appears as $\Delta t = \gamma \Delta\tau$.) In general, a *proper* quantity relative to a body is one measured in its rest frame.

The formula for time dilation (1.23) is often interpreted by saying that time "runs slower" for a moving clock than for a clock at rest, or that a moving clock "runs slower" than a clock at rest by a factor γ. No reference is made to the nature of the clock: the derivation of Eq. (1.23) applies to any clock, whether it is mechanical, atomic, biological, or of unknown nature.

Example 1.6 An astronaut on a spacecraft moving at speed v with respect to his twin on earth will measure a different time interval between his departure and his return than the time interval measured by his twin. For the astronaut, time "runs slower" by a factor γ with respect to the time measured by his twin on earth, but the astronaut would not have any perception of this fact.

* * *

It is clear from Eq. (1.23) that time dilation is a second order effect in v/c. In fact, by expanding the Lorentz factor in powers of v/c, one obtains

$$\gamma = \left(1 - \frac{v^2}{c^2}\right)^{-1/2} \approx 1 + \frac{v^2}{2c^2} \tag{1.25}$$

for $|v|/c \ll 1$.

Example 1.7 Another example of time dilation is given by the measured lifetime of muons in cosmic rays. A muon ($\mu^{\pm}$) is an unstable particle which decays into electrons (e^-) or positrons (e^+), neutrinos (ν), and antineutrinos ($\bar{\nu}$) according to

$$\mu^- \longrightarrow e^- + \nu_\mu + \bar{\nu}_e,$$

$$\mu^+ \longrightarrow e^+ + \bar{\nu}_\mu + \nu_e$$

(here $\nu_{e,\mu}$ denote the electron and muon neutrinos, respectively) in $\tau = 2.2\ \mu s$ (proper lifetime) in a reference frame in which it is at rest. Muons are produced as secondary cosmic rays by primary cosmic rays colliding with nuclei in the upper troposphere and travel at speeds $v \approx c$. It is observed that they reach the surface of the earth before they decay, hence, in the frame in which they move at $v \approx c$, their lifetime is much > 2.2 μs. In 2.2 μs they would only cover $\left(2.2 \cdot 10^{-6} s\right) \cdot \left(3 \cdot 10^8\ m/s\right) \approx$ 660 m, while the upper troposphere reaches an elevation of approximately 10 km. It is time dilation that makes the muon lifetime longer for an observer in an inertial frame attached to the earth in which the muon moves with speed v, according to

$$t_{decay} = \gamma\, \tau = \frac{\tau}{\sqrt{1 - \frac{v^2}{c^2}}}.$$

If $v = 0.99\,c$, then $\gamma = 7.1$, $t_{decay} = \gamma\, \tau = 16\ \mu s$ in the earth's frame, and the distance travelled by the muon is $\left(16 \cdot 10^{-6}\ s\right) \cdot \left(3 \cdot 10^8\ m/s\right) \approx 5{,}000$ m. A classic experiment was performed in 1941 by Rossi and Hall by measuring muon fluxes at the top and at the bottom of Mt. Washington in New England, which are separated by an elevation difference of 2,000 m (the muon speed is $v \simeq 0.994\,c$) with results agreeing with the time dilation formula (1.23) [22]. The experiment was repeated with higher accuracy at CERN in 1971 for muons with speed $v = 0.9994\,c$ [23] and it is now feasible as a training experience for undergraduates [24]. Time dilation is tested every day in particle physics experiments; synchrotrons and modern accelerators would not work if they were based on Newtonian mechanics instead of Special Relativity.

Another classic experiment testing time dilation was performed by Hafele and Keating in 1972 [25]. Atomic clocks were flown on airplanes and, at their return to the air base, were compared with stationary clocks. The result agrees with Special Relativity (plus a correction due to General Relativity because of the different values of the gravitational field of the earth on the ground and on the airplane).

* * *

It is common in the literature to denote the ratio of a speed to the speed of light as

$$\beta \equiv \frac{v}{c} \tag{1.26}$$

and the Lorentz factor with $\gamma = \dfrac{1}{\sqrt{1-\beta^2}}$.

It is easy to see that the Lorentz factor γ satisfies the relations

$$1-\gamma^2 = -\beta^2\gamma^2, \tag{1.27}$$

$$\gamma\, v = c\sqrt{\gamma^2-1}, \tag{1.28}$$

$$c^2 d\gamma = \gamma^3 v\, dv, \tag{1.29}$$

$$d\,(\gamma\, v) = \gamma^3 dv, \tag{1.30}$$

the proof of which is left to the reader.

1.7.3 The Twin "Paradox"

A famous apparent paradox of Special Relativity is the so-called *twin paradox*.[14] Consider two identical 20 years old twins, John and Slown. John takes off in a

[14] The twin "paradox" has been discussed since the early days of Special Relativity by many people, including Paul Langevin, Albert Einstein, and Max Born. The Hafele-Keating experiment [25] can be regarded as a direct experimental verification of the twin "paradox".

spaceship travelling at $v = 0.500\,c$ relative to Slown, reaches a planet 10 light years away, and then returns. Due to time dilation, John finds that Slown is much older than himself. How is this possible?

According to Slown, the spaceship has travelled $l_0 = 20$ light years at speed $c/2$, taking $\Delta t = 40$ years; Slown is now 60 years old (this is the proper age of Slown). For John, the travel time is

$$\Delta\tau = \frac{\Delta t}{\gamma} = \sqrt{1 - \frac{v^2}{c^2}}\,\Delta t = \sqrt{1 - \frac{1}{4}}\,(40\,\text{year}) \simeq 34.6\ \text{years}.$$

John is now 54.6 years old (proper age of John). Were we to allow John to travel at speed $v = 0.900\,c$, instead, John would be 28.7 years old and Slown would be 42.2 years old.

A common objection is the following: why are the two reference frames not symmetric? Can't we invert the argument by saying that John moves at speed v with respect to Slown? The answer is negative: to invert the trip and return home, John must undergo acceleration, while Slown does not. Even though acceleration lasts only for a short period of time, it cannot be neglected. The two frames are not symmetrical, and there is no paradox.[15] Asymmetry between the two frames is the key to understanding the twin "paradox". In an (x, ct) diagram, comparing the ages of the twins amounts to comparing the proper times elapsed for two different observers, i.e., to comparing Lorentzian lengths along their representative curves in this diagram. These two curves are not symmetric.[16]

1.7.4 Length Contraction

As we have seen, time intervals are not absolute in Special Relativity. Likewise, lengths depend on the motion of the observer. We define the *proper length* l_0 of an object as the length of that object measured in a frame in which it is at rest.

The length of an object measured in a frame in which it is moving is always less than its proper length, according to

$$l = \frac{l_0}{\gamma} = l_0\sqrt{1 - \frac{v^2}{c^2}}, \tag{1.31}$$

---

[15] Locally, acceleration is equivalent to a gravitational field and is properly treated in General Relativity.

[16] Interestingly, if the (x, ct) world is curled into a cylinder so that the "rocket twin" John does not have to invert his course to come back to earth because he describes a closed curve in this closed universe, the asymmetry persists and the ages of the twins still differ [26].

a phenomenon called *Lorentz-FitzGerald contraction*. In the limit $|v|/c \ll 1$, length contraction is a second order effect in v/c. This formula was derived by Lorentz in 1892 and was interpreted by Lorentz and Poincaré as a mysterious contraction of matter related to the ether. It was only with Einstein's 1905 theory that the correct interpretation of Eq. (1.31) emerged as a consequence of the two simple postulates of Special Relativity.

To derive the formula (1.31), consider a (point-like) spaceship travelling at uniform speed v and two inertial observers, O on earth and O' on the spaceship, respectively. The observer on earth measures the distance travelled by the spaceship (e.g., the distance between two stars, fixed in the reference frame of O) as l_0 (Fig. 1.13). According to the observer O on earth, the time taken for the spaceship to go from the first to the second star is

$$\Delta t = \frac{l_0}{v}.$$

Because of time dilation, O' measures the time $\Delta t'$ given by $\Delta t = \gamma \Delta t'$ or $\Delta t' = \Delta t/\gamma$, and concludes that the second star travels toward him with speed v and, in the time $\Delta t'$, it covers the distance

$$l = v\Delta t' = v\,\frac{\Delta t}{\gamma} = \frac{l_0}{\gamma}.$$

Therefore, we obtain the formula for the Lorentz-FitzGerald contraction

$$l = \frac{l_0}{\gamma} = l_0\sqrt{1 - \frac{v^2}{c^2}} < l_0.$$

If an object has proper length l_0, its length will be l_0/γ when measured by an observer in uniform motion with velocity v. Note that the two inertial frames are not symmetric, in the sense that the object is at rest in one frame (*rest frame*), but not in the other.

Example 1.8 Let us discuss again muons generated by cosmic rays in the upper atmosphere and travelling to the surface of the earth. A muon in motion sees the length l_0 of the atmosphere Lorentz-contracted to $l = l_0\sqrt{1 - \frac{v^2}{c^2}} < l_0$ and can cover it in only 2.2 μs. In the frame of the earth there is time dilation but no length contraction; in the rest frame of the muon there is no time dilation but there is length contraction.

Fig. 1.13 The Lorentz-FitzGerald contraction of the distance between two stars seen by the inertial observer O' in a spaceship

Example 1.9 What is the length of a rod which is 2 m long in its rest frame, when seen passing by at $0.95c$?
Its length will be $l = (2\,\mathrm{m})\sqrt{1 - 0.95^2} = 0.062\,\mathrm{m}$, a significant reduction compared to the proper length.

* * *

Consider now an extended object in motion with constant speed v with respect to an inertial observer. Let V_0 be the *proper volume* of that object, i.e., the volume in its rest frame. Since, in a change of inertial frame, the transversal directions are not altered, the relation between the volume in an inertial frame moving with speed v with respect to the rest frame and the proper volume is

$$V = V_0 \sqrt{1 - \frac{v^2}{c^2}}\,. \tag{1.32}$$

1.7.4.1 The Car-in-the-Garage "Paradox"

Another of the apparent paradoxes of Special Relativity, known as *the car-in-the-garage paradox* (also known as the pole-in-the-barn paradox), arises from the phenomenon of length contraction. A man has an $l = 5$ m long garage and buys an $l_0 = 7$ m long car (proper length of the car). He reasons that, if he drives sufficiently fast, the car will fit in the garage due to length contraction (ignore the fact that he is going to ruin his new car by smashing it against the garage wall). Using the formula for length contraction (1.31), he easily finds that he must drive at speed $v = c\sqrt{1 - \left(\frac{l}{l_0}\right)^2} \simeq 0.7\,c$ (corresponding to a Lorentz factor $\gamma = 0.71$). But how can a 7 m long car fit into a 5 m long garage? This apparent paradox defies everyday's intuition, yet the conclusion is sound according to Special Relativity (which is certainly more appropriate than everyday's intuition for high speeds). The proper length of the car is still 7 m, but it will fit into the 5 m garage when the man's brother closes the door behind him.

A natural objection is: why the car, which is at rest in its own rest frame, does not see the garage approaching with velocity $-0.7\,c$ and Lorentz-contracted to 3.6 m (in which case the car won't fit in the garage)? The answer lies in the fact that the two events "front of the car at the garage wall" and "back of the car at the garage door" are not simultaneous in the rest frame of the car, while they are simultaneous in the rest frame of the garage. In the rest frame of the car, the front of the car is against the garage wall which is moving toward it. Because it takes a finite time for the shock wave to cross the length of the car and reach its back, the latter is still at rest while the approaching garage takes with it, in its motion, the front of the car for a while. During this time interval, the information that the front of the car is in contact with the garage wall has not yet reached the back of the car. Because of this time delay, the back of the car gets in the garage *after* its front. By contrast, in the frame attached to the garage, both ends of the car are in the garage simultaneously when the door is closed. The phenomenon is real as seen in any inertial frame: its *explanation*, however, is quite different in different inertial frames (in this case, the explanation is more complicated in the rest frame of the car than in the garage frame).

1.8 Conclusion

Important ingredients of Special Relativity were discovered at the end of the nineteenth century by studying the then new theory of electromagnetism. However, Einstein decided to rethink the fundamental concepts of time and space, which led him to formulating the two postulates instead of focusing on the mathematical structure of the Lorentz transformation and inventing ad hoc explanations. We have appreciated that the two postulates, which express simple ideas, have profound consequences such as the relativity of simultaneity, time dilation, and length contraction. We are now ready to express these concepts in a mathematical form by using the law relating two inertial frames, the Lorentz transformation.

Problems

1.1 Show explicitly that $l'_{(3)} = l_{(3)}$ in a Galilei transformation.

1.2 The Euclidean line element of the usual 3-dimensional space, written in Cartesian coordinates $\{x, y, z\}$ as

$$\mathrm{d}l^2{}_{(3)} = \mathrm{d}x^2 + \mathrm{d}y^2 + \mathrm{d}z^2$$

and expressing the Pythagorean theorem, can look quite different in other coordinate systems. Find the form of this line element in parabolic coordinates $\{x^1, x^2, x^3\}$ defined by

$$x = x^1 x^2 \cos(x^3),$$
$$y = x^1 x^2 \sin(x^3),$$
$$z = \frac{1}{2}\left[(x^1)^2 - (x^2)^2\right].$$

1.3 In $\mathbb{R}^3$, find the expression of the Euclidean line element $\mathrm{d}l^2_{(3)}$ in prolate spheroidal coordinates $\{\chi, \theta, \varphi\}$ related to Cartesian coordinates $\{x, y, z\}$ by

$$x = \sinh \chi \sin \theta \cos \varphi,$$
$$y = \sinh \chi \sin \theta \sin \varphi,$$
$$z = \cosh \chi \cos \theta.$$

1.4 Check that the Maxwell equations (1.12)–(1.15) are not left invariant by a Galilei transformation.

1.5 Switching from an inertial frame to another is a trick useful in many areas of (even non-relativistic) physics. For example, in environmental physics it is interesting to calculate the transport of a pollutant due to diffusion and advection. The solution of the diffusion equation for the concentration $C(t, \mathbf{x})$ of a pollutant spilled by a point-like instantaneous source at $\mathbf{x} = (0, 0, 0)$ at the time $t = 0$ is [27]

$$C(t, \mathbf{x}) = \frac{C_0}{\left(\sqrt{2\pi}\,\sigma\right)^3} \exp\left(-\frac{r^2}{2\sigma^2}\right),$$

in a frame in which the source is at rest, and where $r \equiv \sqrt{x^2 + y^2 + z^2}$, C_0 is a constant, and $\sigma = \sqrt{2Dt}$ (the diffusion coefficient D is a constant). Assume that there is a wind with constant and uniform velocity $\mathbf{v}$ in the x-direction and find the corresponding concentration.

1.6 In the inertial frame of observer A, a rod moves in standard configuration with speed v. Another inertial observer B moves with velocity v_B with respect to A, with the $x^i_{(B)}$-axes parallel to the $x^i_{(A)}$- axes of A. Observer A measures the rod to be of length l_A. What is the length of the rod as measured by observer B?

1.7 Particle physicists use a system of units in which both c and $\hbar$ (the reduced Planck constant) assume the value unity. What are the relations between length, time, and mass in these units?

1.8 Let

$$R = \begin{pmatrix} \cos\theta & \sin\theta & 0 \\ -\sin\theta & \cos\theta & 0 \\ 0 & 0 & 1 \end{pmatrix}$$

be the matrix of a rotation by an angle θ about the z-axis in the 3-dimensional Euclidean space $\mathbb{R}^3$. Find the inverse matrix R^{-1} without performing calculations. Show that R is orthogonal (i.e., $R^{-1} = R^T$) and compute Det(R).

1.9 Verify Eqs. (1.27)–(1.30).

References

1. A. Einstein, Ann. Phys. **17**, 891 (1905)
2. A. Einstein, Ann. Phys. **17**, 639 (1905)
3. A. Einstein, Nature **106**, 782 (1921)
4. R. Cerf, Am. J. Phys. **74**, 818 (2006)

5. A. Zee, *Quantum Field Theory in a Nutshell*, 2nd edn. (Princeton University Press, Princeton, 2010)
6. A.A. Michelson, E.W. Morley, Am. J. Sci. **34**, 333 (1887)
7. E.H. Kennard, D.E. Richmond, Phys. Rev. **19**, 572 (1922)
8. R.S. Shankland, Am. J. Phys. **32**, 16 (1964)
9. V.S. Soni, Am. J. Phys. **56**, 178 (1988)
10. R.A. Schumacher, Am. J. Phys. **62**, 609 (1994)
11. A. Gjurchinovski, Am. J. Phys. **72**, 1316 (2004)
12. V.S. Soni, Am. J. Phys. **57**, 1149 (1989)
13. M. Mamone Capria, F. Pambianco, Found. Phys. **24**, 885 (1994)
14. H.R. Brown, Am. J. Phys. **69**, 1044 (2001)
15. C.M. Will, in *Proceedings of the 2005 Seminaire Poincaré*, Paris, 2005, ed. by T. Damour, O. Darrigol, B. Duplantier, V. Rivasseau (Birkhauser, Basel, 2006), pp. 33–58 [preprint arXiv:gr-qc/0504085]
16. A. Brillet, J.L. Hall, Phys. Rev. Lett. **42**, 549 (1979)
17. H. Lorentz, Versl. K. Ak. Amsterdam **1**, 74 (1892)
18. G.F. FitzGerald, Science **13**, 390 (1889)
19. R.A. Sorensen, Am. J. Phys. **63**, 413 (1995)
20. R. Baierlein, Am. J. Phys. **74**, 193 (2006)
21. W.N. Mathews, Am. J. Phys. **73**, 45 (2005)
22. B. Rossi, D.B. Hall, Phys. Rev. **59**, 223 (1941)
23. J. Bailey, K. Borer, F. Combley, H. Drumm, F. Krienen, F. Lange, E. Picasso, W. von Ruden, F.J.M. Farley, J.H. Field, W. Flegel, P.M. Hattersley, Nature **268**, 301 (1977)
24. N. Easwar, D.A. MacIntire, Am. J. Phys. **59**, 589 (1991)
25. J.C. Hafele, R. Keating, Science **177**, 166 (1972)
26. T. Dray, Am. J. Phys. **58**, 822 (1990)
27. E. Boeker, R. van Grondelle, *Environmental Physics*, 3rd edn. (Wiley, Chichester, 2011)

Chapter 2
The Lorentz Transformation

Imagination is more important than knowledge.
—Albert Einstein

2.1 Introduction

Now that we have seen the main consequences of the postulates of Special Relativity, i.e., the relativity of simultaneity, time dilation, and length contraction it is clear that the Galilei transformation, with its absolute time, is incorrect. These important physical phenomena can be seen as direct consequences of the correct transformation relating inertial frames, the Lorentz transformation. This transformation is the key for the formulation of Special Relativity in an enlightening four-dimensional formalism, which we will see in the next chapter. Here we study the Lorentz transformation and its properties and derive length contraction and time dilation directly from it, in addition to the transformation property of velocities. We must emphasize that, although the Lorentz transformation was discovered by studying the Maxwell equations, its validity is more general. The Lorentz transformation relates inertial frames without reference to the kind of physics studied in them. Lorentz-invariance is a general requirement for *any* physical theory, not just for electromagnetism.

2.2 The Lorentz Transformation

The Galilei transformation is not valid for speeds which are not negligible in comparison with the speed of light. The correct transformation relating space and time coordinates in two inertial frames $\{t, x, y, z\}$ and $\{t', x', y', z'\}$ moving with relative velocity v in standard configuration was discovered by Fitzgerald in 1889 and by

V. Faraoni, *Special Relativity*, Undergraduate Lecture Notes in Physics,
DOI: 10.1007/978-3-319-01107-3_2, © Springer International Publishing Switzerland 2013

Lorentz in 1892 as the transformation which leaves the Maxwell equations invariant.[1] The *Lorentz transformation* or *Lorentz boost* is[2]

$$x' = \frac{x - vt}{\sqrt{1 - \frac{v^2}{c^2}}}, \tag{2.1}$$

$$y' = y, \tag{2.2}$$

$$z' = z, \tag{2.3}$$

$$t' = \frac{t - v\frac{x}{c^2}}{\sqrt{1 - \frac{v^2}{c^2}}}. \tag{2.4}$$

The most striking feature of this linear coordinate transformation (bear in mind that v and $\gamma \equiv \left(1 - v^2/c^2\right)^{-1/2}$ are constants) is that it mixes the space and time coordinates. As a consequence, time intervals and 3-dimensional lengths are not invariant under this transformation and, therefore, time intervals and 3-dimensional lengths are not absolute quantities. The equations of electromagnetism are invariant under this transformation (it is said that they are *Lorentz-invariant*) but the equations of Newtonian mechanics are not.

The *inverse Lorentz transformation* is obtained by the exchange $\mathbf{x} \longleftrightarrow \mathbf{x}'$ and $v \longleftrightarrow -v$ according to the Principle of Relativity[3]:

$$x = \frac{x' + vt'}{\sqrt{1 - \frac{v^2}{c^2}}}, \tag{2.5}$$

$$y = y', \tag{2.6}$$

---

[1] Lorentz was also trying to explain the null result of the Michelson-Morley experiment by a physical contraction of the apparatus in the direction of motion. His interpretation, however, is rather misleading: the Lorentz transformation relates measurements performed in two different inertial systems.

[2] A priori, the constant c appearing in the Lorentz transformation is a fundamental velocity which needs not coincide with the speed of electromagnetic waves in vacuo, and is only later identified with it. This is not the historical route, in which the Lorentz transformation was derived from the Maxwell equations. There are many facets to the quantity c in various areas of physics (see Ref. [1] for a review).

[3] This application of the Principle of Relativity is sometimes called "principle of reciprocity" and is a consequence of the isotropy of space [2, 3].

$$z = z', \tag{2.7}$$

$$t = \frac{t' + v\frac{x'}{c^2}}{\sqrt{1 - \frac{v^2}{c^2}}}. \tag{2.8}$$

Although the Lorentz transformation was obtained before the formulation of Special Relativity as the correct transformation between inertial frames which respects the Maxwell equations, its meaning was not grasped until Einstein's 1905 paper.

2.3 Derivation of the Lorentz Transformation

The Lorentz transformation can be derived on the basis of the two postulates of Special Relativity. First, due to the isotropy of space contained in the second postulate, we can orient the spatial axes of an inertial frame S' with those of another inertial frame S and limit ourselves to considering motion of the two frames in standard configuration. As a starting point for deducing the transformation relating two inertial frames $S = \{t, x, y, z\}$ and $S' = \{t', x', y', z'\}$ in relative motion with velocity v in standard configuration, it is reasonable to assume that the transformation is linear

$$x' = G(x - vt), \tag{2.9}$$

where G is a dimensionless constant that depends only on v/c. This assumption corresponds to the homogeneity of space and time since G does not depend on (x, t). Since the spatial part of the Galilei transformation $x' = x - vt$ must be recovered in the limit $v/c \to 0$, G must tend to unity in this limit.

The laws of physics must have the same form in S and S', and then one must obtain the inverse Lorentz transformation by the exchange $(t', \mathbf{x}') \longleftrightarrow (t, \mathbf{x})$ and $v \longleftrightarrow -v$ (this is the Principle of Relativity again):

$$x = G(x' + vt'). \tag{2.10}$$

There is no relative motion in the y and z directions, hence these coordinates must not be affected by the transformation,

$$y' = y, \quad z' = z. \tag{2.11}$$

Consider now a spherical pulse of electromagnetic radiation emitted at the origin of S at $t = 0$. It is received at a point on the x-axis and, during its propagation,

$$x = ct. \tag{2.12}$$

The same law must be true in S' due to the constancy of the speed of light,

$$x' = ct', \tag{2.13}$$

or $ct' = G(ct - vt)$ from Eq. (2.9), which leads to

$$t' = G\left(t - \frac{v}{c}t\right), \tag{2.14}$$

while

$$ct = G(ct' + vt') = G(c + v)t'. \tag{2.15}$$

By substituting Eq. (2.14) into Eq. (2.10), one obtains

$$ct = G(c + v)G(t - \frac{v}{c}t) = \frac{G^2}{c}(c + v)(c - v)t,$$

and

$$G^2 = \frac{c^2}{(c + v)(c - v)}.$$

Therefore, we have

$$G = \frac{1}{\sqrt{1 - \frac{v^2}{c^2}}} \equiv \gamma,$$

the Lorentz factor, and

$$x' = \gamma(x - vt). \tag{2.16}$$

Use then $x = \gamma(x' + vt')$ and $x' = \gamma(x - vt)$ and substitute x' into x to obtain

$$x = \gamma[\underbrace{\gamma(x - vt)}_{x'} + vt'],$$

$$x = \gamma^2 x - \gamma^2 vt + \gamma vt',$$

from which we obtain

$$t' = \frac{x - \gamma^2 x + \gamma^2 vt}{\gamma v}$$

and

$$t' = \frac{x}{v\gamma}(1 - \gamma^2) + \gamma t.$$

Since

$$1-\gamma^2 = 1 - \frac{1}{1-\frac{v^2}{c^2}} = \frac{1-\frac{v^2}{c^2}-1}{1-\frac{v^2}{c^2}} = \frac{-(\frac{v^2}{c^2})}{1-\frac{v^2}{c^2}},$$

we have

$$\frac{1-\gamma^2}{\gamma} = \sqrt{1-\frac{v^2}{c^2}}\,\frac{-(\frac{v^2}{c^2})}{(1-\frac{v^2}{c^2})} = \frac{-\frac{v^2}{c^2}}{\sqrt{1-\frac{v^2}{c^2}}} \equiv -\gamma\,\frac{v^2}{c^2},$$

$$t' = \frac{1-\gamma^2}{\gamma}\frac{x}{v} + \gamma t = -\gamma\,\frac{v^2}{c^2}\frac{x}{v} + \gamma t,$$

and, finally,

$$t' = \gamma\left(t - \frac{v}{c^2}x\right), \tag{2.17}$$

which completes the derivation. Equations (2.16), (2.11), and (2.17) constitute the Lorentz transformation. The fact that the Lorentz transformation can be derived from the two postulates of Special Relativity is conceptually important: it means that these two postulates constitute the physical explanation of the mathematical transformation and that this transformation should not be assumed in place of the two postulates, as Poincaré seem to have intended. While Poincaré, Lorentz, and FitzGerald stopped at the transformation (which is an important ingredient of Special Relativity and unveils the mixing of space and time of the 4-dimensional world view), they tried to explain it with an ether and with length contraction. It was Einstein's genius which reduced the physical explanation of the transformation to two simple and general postulates and led to a re-examination of the concepts of space and time, developing the full theory which was missed by other researchers.

2.4 Mathematical Properties of the Lorentz Transformation

Let us examine the properties of the Lorentz transformation.

- Qualitatively, the Lorentz transformation mixes t and x, therefore 3-dimensional lengths and time intervals cannot be left invariant. Quantitatively, length contraction and time dilation can be derived from the Lorentz transformation as a consequence, which will be done in the following sections.

The quantity

$$ds^2 = -c^2dt^2 + dx^2 + dy^2 + dz^2$$

("Minkowski line element") is invariant under Lorentz transformations.

This invariance is easy to see: by using the Lorentz transformation we have

$$\begin{aligned}
(\mathrm{d}s')^2 &\equiv -c^2(\mathrm{d}t')^2 + (\mathrm{d}x')^2 + (\mathrm{d}y')^2 + (\mathrm{d}z')^2 \\
&= -c^2\gamma^2 \left(\mathrm{d}t - \frac{v}{c^2}\,\mathrm{d}x\right)^2 + \gamma^2\,(\mathrm{d}x - v\mathrm{d}t)^2 + \mathrm{d}y^2 + \mathrm{d}z^2 \\
&= -c^2\gamma^2 \left(1 - \frac{v^2}{c^2}\right)\mathrm{d}t^2 + 2c^2\gamma^2\,\frac{v}{c^2}\,\mathrm{d}t\mathrm{d}x + \gamma^2\mathrm{d}x^2\left(1 - \frac{v^2}{c^2}\right) \\
&\qquad -2\gamma^2\,v\,\mathrm{d}t\mathrm{d}x + \mathrm{d}y^2 + \mathrm{d}z^2 \\
&= -c^2\mathrm{d}t^2 + \mathrm{d}x^2 + \mathrm{d}y^2 + \mathrm{d}z^2 \equiv \mathrm{d}s^2.
\end{aligned}$$

We will discuss extensively this aspect of Special Relativity in the following chapters.

- The Lorentz transformation is symmetric under the exchange $x \longleftrightarrow ct$:

$$\begin{aligned}
x' &= \gamma\,(x - vt)\,, \\
y' &= y, \\
z' &= z, \\
ct' &= \gamma\left(ct - \frac{vx}{c}\right),
\end{aligned}$$

becomes

$$\begin{aligned}
ct' &= \gamma\left(ct - \frac{vx}{c}\right), \\
y' &= y, \\
z' &= z, \\
x' &= \gamma\,(x - vt).
\end{aligned}$$

In standard configuration, the Lorentz transformation is also symmetric under the exchange $y \longleftrightarrow z$.

- The Galilei transformation can somehow be recovered from the Lorentz transformation in the limit of small velocities $|v|/c \ll 1$, although the derivation is a bit finicky. First, expand the Lorentz factor γ to first order in v/c:

$$\gamma \equiv \frac{1}{\sqrt{1-\frac{v^2}{c^2}}} = 1 + \frac{v^2}{2c^2} + \ldots \approx 1 \tag{2.18}$$

and

$$x' \approx x - vt,$$
$$y' = y,$$
$$z' = z.$$

Strictly speaking, the transformation of the time coordinate gives, to first order in v/c,

$$t' = t - \frac{v}{c^2}\,x,$$

not $t' = t$ as in the Galilei transformation: the relativity of simultaneity persists to first order (it is a first order effect in v/c). If the Lorentz transformation reduced to the Galilei transformation to first order, then infinitesimal Lorentz transformations and infinitesimal Galilei transformations would coincide, which is not the case.[4] However, time dilation is computed by considering time differences, recording two spatial events at the same location. Since $\Delta t' = \Delta t - \frac{v}{c^2}\,\Delta x$, by setting $\Delta x = 0$, time dilation is eliminated to first order. In practice, when speeds are small, time intervals Δt are measured over spatial distances Δx such that $c\Delta t \gg \Delta x \gg \frac{v}{c}\,\Delta x$ and the Δx term can be dropped. Although the Lorentz transformation does not quite reduce to the Galilei transformation, which is recovered only in the limit $v/c \longrightarrow 0$, Newtonian mechanics and the Galilei transformation turn out to be adequate[5] in the limit $|v| \ll c$.

- Since the Lorentz transformation is linear and homogeneous, finite coordinate differences transform in the same way as infinitesimal coordinate differences:

$$\begin{aligned} \Delta x' &= \gamma\,(\Delta x - v\Delta t), & \qquad \mathrm{d}x' &= \gamma\,(\mathrm{d}x - v\mathrm{d}t),\\ \Delta y' &= \Delta y, & \mathrm{d}y' &= \mathrm{d}y,\\ \Delta z' &= \Delta z, & \text{and} \qquad \mathrm{d}z' &= \mathrm{d}z,\\ \Delta t' &= \gamma\left(\Delta t - \tfrac{v}{c^2}\Delta x\right), & \mathrm{d}t' &= \gamma\left(\mathrm{d}t - \tfrac{v}{c^2}\mathrm{d}x\right). \end{aligned}$$

[4] This point is made clearly in Refs. [4, 5].

[5] Our derivation of the Lorentz transformation from the postulates of Special Relativity requires only that the *spatial part* of the Lorentz transformation $x' = G(v)\,(x - vt)$ reduces to the spatial part of the Galilei transformation $x' = x - vt$ in the limit $|v|/c \ll 1$, from which we deduced that $G \rightarrow 1$. We did not assume that we recover $t' = t$ in this limit, hence the proof is correct.

2.5 Absolute Speed Limit and Causality

At this point, you are certainly aware that $\gamma \equiv \frac{1}{\sqrt{1-\frac{v^2}{c^2}}}$, the ratio between coordinate and proper times, diverges as $v \to c$. The inequality $v > c$ leads to a purely imaginary γ, therefore,

the relative velocity of two inertial frames must be strictly smaller than c.

Since an inertial frame can be associated with any non-accelerated particle or object moving with subluminal (i.e., $|v| < c$) speed, this statement translates into the requirement that the speed of particles and of all physical signals be limited by c (remember that c is the sped of light in vacuo: the speed of particles traveling in a medium can be larger than the speed of light *in that medium*).[6] Never mind the fact that the Lorentz factor becomes imaginary: we can agree to define γ as the modulus $\left|\left(1-v^2/c^2\right)^{-1/2}\right|$ if $|v| > c$. What is truly important[7] is that

the restriction $|v| \leq c$ preserves the notion of cause and effect.

In fact, consider a process in which an event P causes, or affects, an event Q by sending a signal containing some information from P to Q. If a signal were sent from P to Q at superluminal speed $u > c$ in some inertial frame S, we could orient the axes of S so that both events P and Q occur on the x-axis and their time and spatial separations satisfy $\Delta t > 0$ and $\Delta x > 0$ in this frame. Then, in an inertial frame S' moving with respect to S with speed v in standard configuration, we would have

$$\Delta t' = \gamma\left(\Delta t - v\frac{\Delta x}{c^2}\right) = \gamma \Delta t\left(1 - \frac{uv}{c^2}\right), \tag{2.19}$$

where we used $\Delta x = u\Delta t$. Now, because $u > c$ it is also $-u < -c$ which, together with $0 < v < c$, implies that $-uv < -c^2$ or $-uv/c^2 < -1$, so

$$\Delta t' = \gamma \Delta t\left(1 - \frac{uv}{c^2}\right) < 0. \tag{2.20}$$

According to this result, in the frame S' the event Q precedes P: cause and effect are reversed or, the signal goes backward in time. The signal reaches Q before being emitted by P, which creates a logical problem. The fact that there is an absolute speed limit c comes to the rescue and enforces causality: if both $|u|, |v| < c$, then $\Delta t'$ in Eq. (2.19) has the same sign as Δt. The possibility of reversing cause and effect and travelling in time would lead to logical paradoxes, which have been discussed at length in the literature (see, e.g., [9] and the references therein).

---

[6] If the particle traveling faster than light in that medium is charged, Cerenkov radiation is emitted.

[7] In the past, "tachyons" traveling at speed larger than c and incapable of slowing down to speeds less than c were considered theoretically and searched for experimentally (see, e.g., [6–8]) but no tachyon has been convincingly detected.

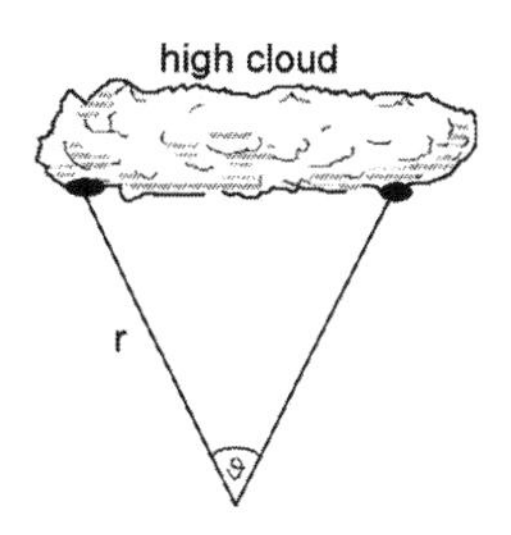

Fig. 2.1 If r becomes sufficiently large, a searchlight spot on a high cloud can attain an arbitrarily large apparent linear velocity

Another argument against superluminal travel is the following.[8] Suppose that an observer O, at rest in an inertial frame, emits a flash of laser light in a certain direction (called the "forward" direction). A second observer O' at rest in the same inertial frame sees the light, measuring that it moves with speed c. A third inertial observer O'' is moving with respect to O and O' in the "backward" direction with constant speed v. According to the second postulate, he measures light moving with speed c in the "forward" direction. He also sees O and O' moving in the "forward" direction toward him with speed v. Now, it is impossible for this speed v of the observers O and O' measured by O'' to be larger than c. For, were this possible, the flash of laser light, when emitted by O, will remain behind O. According to O'', the laser light would always remain behind O and the observer O' (who is "forward" of O) would never see this light. But then whether O' sees the light or not depends on the inertial frame, which contradicts the Principle of Relativity stating that all inertial frames are physically equivalent. Therefore, the relative speed of two inertial frames cannot be larger than c. Formally, the argument works also if the relative speed of inertial frames equals c for, in that case, light will never reach O' according to the observer O''.

If there is an absolute speed limit c then, according to the Principle of Relativity, this limit must be the same in all inertial frames, consistent with the postulate of the constancy of the speed of light.[9]

The absolute speed limit refers to *physical, propagating signals*. Apparent motions which carry no information can have arbitrarily large speeds. In these cases, certain motions *appear* to be faster than light, but they are illusions and not real motions. For example, consider a searchlight spot on high clouds (Fig. 2.1).

Let $\Delta\theta$ be the angle spanned by the light spot across a cloud in the time Δt. The linear velocity of the spot perceived by an observer on the ground is $r\,\Delta\theta/\Delta t$, where r is the distance to the cloud. If $\Delta\theta/\Delta t = 10\,\text{rad/s}$, the apparent linear velocity of the spot v_{spot} is larger than c if $r > 3 \cdot 10^7$ m. This distance is too large for a searchlight in the atmosphere but it illustrates the principle and it is not too large for astrophysical phenomena (apparent superluminal motions do occur in astrophysics and they constituted a puzzle when they were first discovered [11, 12]). In any case,

[8] This argument is due to E. F. Taylor [10].

[9] When introducing a set of axioms for a theory, one should always worry about the mutual consistency of these axioms. If two axioms are not consistent with each other, one is building an empty theory.

everyday experience shows the apparent amplification of velocity when a searchlight spot moves on clouds.[10] The point is, the perceived velocity of the spot is not related to the velocity of propagation of the light beam.

Unphysical velocities occur also when waves propagate in a dispersive medium. In general, a wave signal[11] is composed of many, or infinite, monochromatic waves of angular frequencies ω and wave vectors $\mathbf{k}$, with $k \equiv |\mathbf{k}|$. The properties of the medium are described by a *dispersion relation* $\omega = \omega(k)$, a functional relation between ω and k. The individual monochromatic waves composing the complex wave propagate at the *phase velocity*[12]

$$v_p \equiv \frac{\omega}{k}, \tag{2.21}$$

while the "envelope" composed of the individual monochromatic waves travels at the *group velocity*

$$v_g \equiv \frac{d\omega}{dk}. \tag{2.22}$$

In Eqs. (2.21) and (2.22) the right hand sides must be evaluated at a central value[13] of k.

If the dispersion relation $\omega(k)$ is linear, the medium is non-dispersive; if this relation is non-linear, it is dispersive. Then the individual component waves have phase velocities which depend on their wave vectors (or wavelengths $\lambda = \frac{2\pi}{k}$) and the waveform is altered as it propagates through the medium. Phase velocity and group velocity then differ, and the physical velocity of the wave (the velocity at which energy and information propagate) is the group velocity. It is possible that, formally, phase velocities be larger than c. This fact does not violate Special Relativity because v_p is not the true velocity of propagation of the signal. When wavepackets are not too spread out and group velocities are well defined and physically meaningful,[14] they have values which are no larger than c.

A consequence of the existence of an absolute speed limit is that the idealizations of *rigid body* and *incompressible fluid* used in Newtonian mechanics, which imply infinite sound speed, are impossible in Special Relativity. By definition, such systems

[10] This effect is the same phenomenon used advantageously in the optical lever of the torsion balance.

[11] A general wave, not necessarily electromagnetic, is discussed here. Also, quantum vacuum can behave as a medium and give rise to apparent velocities larger than c. This is not, however, the propagation velocity of a physical signal and does not threaten causality [13].

[12] The phase velocity is usually identified with the velocity of a point of constant phase, for example a point where the amplitude of the wavepacket envelope vanishes.

[13] For wavepackets which are too spread out, a situation that occurs with high absorption or near resonances, the concepts of group and phase velocity may become largely unphysical and a more detailed discussion is necessary. No less than eight "wave velocities" can be defined in this case [14, 15].

[14] Even group velocities can be larger than c if the wavepacket is too spread out in a medium with high absorption [16]. Again, we are not talking about physical velocities here.

would transmit information instantaneously by means of sound waves propagating with infinite speed.

Example 2.1 The fastest spinning pulsar known to date, PSR J1746-2446ad, spins with a frequency of 716 Hz [17]. What limit is imposed by fundamental physics on its radius R?

The answer is that the equatorial speed, which is the largest rotational speed of a particle on the pulsar, must be less than c, yielding

$$R < \frac{c}{\omega} = \frac{3 \cdot 10^8 \text{ m/s}}{2\pi \cdot 716 \text{ s}^{-1}} \simeq 67 \text{ km}.$$

Neutron stars are believed to have sizes $\sim 10 - 20$ km, which brings them fairly close to achieving the largest rotational speeds that are possible in nature.

2.6 Length Contraction from the Lorentz Transformation

Length contraction and time delay can be derived directly from the Lorentz transformation. Consider a rod at rest in the inertial frame $S' = \{t', x', y', z'\}$ and moving with speed v with respect to another inertial frame $S = \{t, x, y, z\}$ as in Fig. 2.2. The endpoints of the rod are x'_A and x'_B with $l_0 \equiv x'_B - x'_A$ the rest length of the rod. According to the Lorentz transformation, it is

$$x'_A = \frac{x_A - vt_A}{\sqrt{1 - \frac{v^2}{c^2}}}, \qquad x'_B = \frac{x_B - vt_B}{\sqrt{1 - \frac{v^2}{c^2}}}. \tag{2.23}$$

In order to measure the rod, we must find the coordinates of the endpoints *at the same time* $t_A = t_B \equiv t$ (the two endpoints are observed simultaneously). We have

$$l_0 = x'_B - x'_A = \frac{(x_B - vt) - (x_A - vt)}{\sqrt{1 - \frac{v^2}{c^2}}} = \frac{x_B - x_A}{\sqrt{1 - \frac{v^2}{c^2}}}$$

or

$$l_0 = \frac{l}{\sqrt{1 - \frac{v^2}{c^2}}},$$

and

$$l = l_0 \sqrt{1 - \frac{v^2}{c^2}}, \tag{2.24}$$

the Lorentz-FitzGerald formula for length contraction.

There is no contraction in the directions transversal to the motion. As a result of length contraction, a moving rod can be fit momentarily in a space in which it would not fit at rest (which originates the car-in-the-garage "paradox"). However, nothing

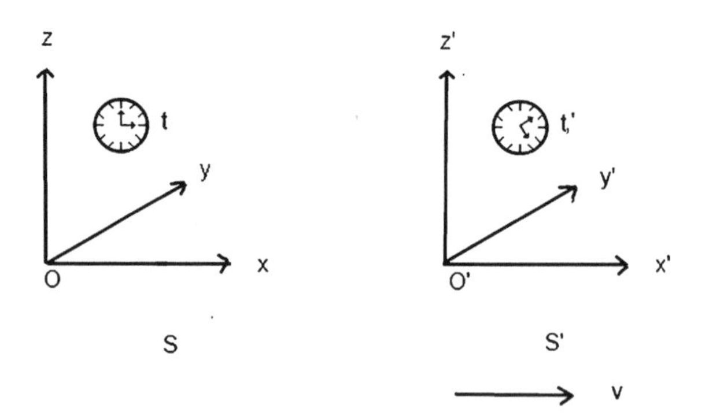

Fig. 2.2 The inertial frames S and S'

has happened to the rod. If you measure it again in its rest frame, it still has the same rest length l_0.

2.7 Time Dilation from the Lorentz Transformation

Let us now derive time dilation directly from the Lorentz transformation. Let a clock at rest at x' in the inertial frame S' record two events happening *at the same location* x' and separated by the proper time interval $\Delta\tau$. The two events have coordinates $\left(t_1', x', 0, 0\right)$ and $\left(t_2', x', 0, 0\right) \equiv \left(t_1' + \Delta\tau, x', 0, 0\right)$. What is the time interval measured by a clock in the inertial frame S which is moving with constant speed v with respect to S'? The inverse Lorentz transformation (2.5)–(2.8) gives, for these two events,

$$t_1 = \frac{t_1' + \frac{v}{c^2} x'}{\sqrt{1 - \frac{v^2}{c^2}}}, \qquad t_2 = \frac{t_2' + \frac{v}{c^2} x'}{\sqrt{1 - \frac{v^2}{c^2}}} \tag{2.25}$$

and the time interval in S is

$$\Delta t \equiv t_2 - t_1 = \frac{\left(t_2' + \frac{v}{c^2} x'\right) - \left(t_1' + \frac{v}{c^2} x'\right)}{\sqrt{1 - \frac{v^2}{c^2}}} = \frac{t_2' - t_1'}{\sqrt{1 - \frac{v^2}{c^2}}}$$

or

$$\Delta t = \frac{\Delta\tau}{\sqrt{1 - \frac{v^2}{c^2}}} \equiv \gamma \Delta\tau. \tag{2.26}$$

A moving clock "runs slower" than a static one by the Lorentz factor γ.

An *ideal clock* is defined as one that is not affected by acceleration. The finite interval of proper time recorded by an (ideal) clock between proper instants t_0 and t is

$$\Delta\tau = \int_{t_0}^{t} \mathrm{d}t \sqrt{1 - \frac{v^2}{c^2}}. \tag{2.27}$$

Example 2.2 Because they are light, electrons can be easily accelerated to become very relativistic. Consider an electron traveling at speed $v = 0.99c$ in an accelerator. The time interval $\Delta t = 1$ s in the laboratory frame corresponds, in the rest frame of this electron, to the interval $\Delta\tau = \Delta t/\gamma = \sqrt{1 - 0.99^2}\,(1\,\mathrm{s}) = 0.14$ s.

2.8 Transformation of Velocities and Accelerations in Special Relativity

Contrary to Newtonian mechanics, velocities do not simply "add up" in Special Relativity, otherwise an observer moving toward a light source would measure the speed of light to be larger than c, which contradicts the second postulate. In order to derive the correct formula for the composition of relativistic velocities,[15] suppose that a particle has velocity $u^{x'} \equiv dx'/dt'$ relative to an inertial frame $S' = \{t', x', y', z'\}$; we want to find its velocity u^x with respect to another inertial frame $S = \{t, x, y, z\}$, with respect to which $\{t', x', y', z'\}$ is moving with constant velocity v (Fig. 2.3). Remember the convention that the velocity v is positive if the inertial frame S' is moving *away* from S. Differentiate the Lorentz transformation (2.1)–(2.4) to obtain

$$\mathrm{d}x' = \gamma\,(\mathrm{d}x - v\mathrm{d}t),$$

$$\mathrm{d}y' = \mathrm{d}y,$$

$$\mathrm{d}z' = \mathrm{d}z,$$

$$\mathrm{d}t' = \gamma\left(\mathrm{d}t - \frac{v}{c^2}\mathrm{d}x\right),$$

and

$$u^{x'} \equiv \frac{dx'}{dt'} = \frac{dx - vdt}{dt - \frac{v}{c^2}dx} = \frac{\frac{dx}{dt} - v}{1 - \frac{v}{c^2}\frac{dx}{dt}} = \frac{u^x - v}{1 - \frac{vu^x}{c^2}}.$$

Analogously, $\mathrm{d}y' = \mathrm{d}y$, $\mathrm{d}z' = \mathrm{d}z$, and

[15] It is possible to derive the relativistic law of transformation of velocities without using the Lorentz transformation and relying only on the two postulates (e.g., [18–20]). Here we present only the "standard" derivation from the Lorentz transformation.

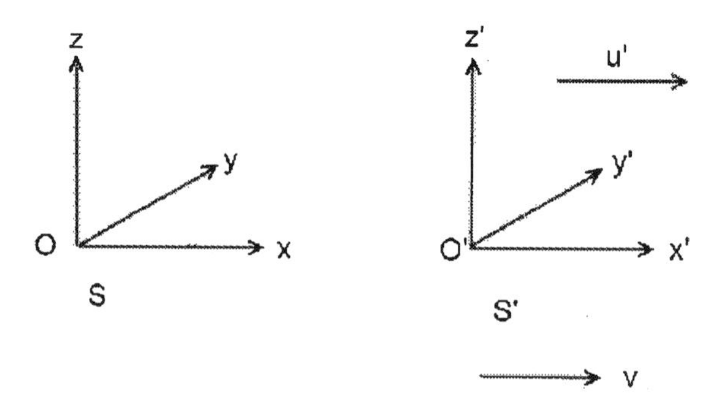

Fig. 2.3 The composition of velocities v and u' in Special Relativity

$$u^{y'} \equiv \frac{dy'}{dt'} = \frac{dy}{\gamma\left(dt - \frac{v}{c^2}dx\right)} = \frac{\frac{dy}{dt}}{\gamma\left(1 - \frac{vu^x}{c^2}\right)} = \frac{u^y}{\gamma\left(1 - \frac{vu^x}{c^2}\right)},$$

$$u^{z'} \equiv \frac{dz'}{dt'} = \frac{dz}{\gamma\left(dt - \frac{v}{c^2}dx\right)} = \frac{\frac{dz}{dt}}{\gamma\left(1 - \frac{vu^x}{c^2}\right)} = \frac{u^z}{\gamma\left(1 - \frac{vu^x}{c^2}\right)}.$$

The *relativistic velocity addition formulae* are, therefore,

$$u^{x'} = \frac{u^x - v}{1 - \frac{vu^x}{c^2}}, \tag{2.28}$$

$$u^{y'} = \frac{u^y}{\gamma\left(1 - \frac{vu^x}{c^2}\right)}, \tag{2.29}$$

$$u^{z'} = \frac{u^z}{\gamma\left(1 - \frac{vu^x}{c^2}\right)}. \tag{2.30}$$

Note that we did not assume that the particle has uniform velocity $\mathbf{u}'$ in S; the derivation is valid for *instantaneous* velocities. In addition, while $|v|$ is restricted to be less than c, u^x, u^y, and u^z can be the coordinate velocity components of a light ray.[16]

Let us consider two limiting cases. In the *Newtonian limit* $|u^x|,\ |v| \ll c$ we have, to first order,

[16] This possibility will be applied to the derivation of the laws describing the aberration of light in Chap. 7

$$u^{x'} \approx u^x - v,$$

$$u^{y'} \approx u^y,$$

$$u^{z'} \approx u^z.$$

In the *ultrarelativistic limit* $u^x \to c$ we have $u^{x'} \to \dfrac{c-v}{1-\frac{v}{c}} = c$, in agreement with the postulate that the speed of light is c in every inertial frame. This conclusion is not surprising since it is built into the Lorentz transformation used here to derive the addition law of velocities.

If the two inertial frames are in standard configuration, planar motions remain planar under the change of frame. In fact if, for example, the motion of a particle occurs in the (x, y) plane according to S, then $u^z = 0$ and, according to Eq. (2.30), also $u^{z'} = 0$. A rectilinear motion along the x-axis of O (with $u^y = u^z = 0$) appears as a rectilinear motion along the x'-axis of O' (with $u^{y'} = u^{z'} = 0$). The rectilinear motion of a particle along the y-axis of O is, of course, distorted according to O' (since $u^{x'} \neq 0$, $u^{y'} \neq 0$, and $u^{z'} = 0$), as is rectilinear motion along the z-axis (since $u^{x'} \neq 0$, $u^{y'} = 0$, and $u^{z'} \neq 0$).

According to the Principle of Relativity, the inverse velocity transformation is obtained with the exchange $\left(u^i, v\right) \longleftrightarrow \left(u^{i'}, -v\right)$ yielding

$$u^x = \frac{u^{x'} + v}{1 + \frac{vu^{x'}}{c^2}}, \tag{2.31}$$

$$u^y = \frac{u^{y'}}{\gamma\left(1 + \frac{vu^{x'}}{c^2}\right)}, \tag{2.32}$$

$$u^z = \frac{u^{z'}}{\gamma\left(1 + \frac{vu^{x'}}{c^2}\right)}. \tag{2.33}$$

2.8.1 *Relative Velocity of Two Particles*

Consider two particles moving instantaneously along the same line with *speeds* v_1 and v_2 in an inertial frame $S = \{t, x, y, z\}$. Their relative velocity is computed by using an inertial frame $S' = \left\{t', \mathbf{x}'\right\}$ in which particle 1 is at rest. In this frame, which has a velocity (and speed) v_1 with respect to S, particle 2 has the velocity

$$u^{x'}_{(2)} = \frac{u^x - v_1}{1 - \frac{u^x v_1}{c^2}} \tag{2.34}$$

where $u^x = -v_2$ is the velocity of particle 2 in the frame $\{t', x', y', z'\}$. This is the *relativistic law of composition of velocities*. Then, it is

$$u^{x'}_{(2)} = -\frac{(v_1 + v_2)}{1 + \frac{v_1 v_2}{c^2}}. \tag{2.35}$$

The *relative speed* of the two particles is given by

$$\beta = \frac{\beta_1 + \beta_2}{1 + \beta_1 \beta_2}. \tag{2.36}$$

Example 2.3 In a science fiction movie two spaceships are moving head-on toward each other with speeds $0.65c$ and $0.90c$ with respect to an observer on earth. What is the relative speed measured by the astronauts on each ship?
The relative speed is

$$\frac{0.65c - (-0.90c)}{1 - \frac{0.65c(-0.90c)}{c^2}} = 0.98c,$$

which is obviously larger than the speed of each spaceship with respect to earth but still less than c.

* * *

Let us study now the function of two variables

$$f(x, y) \equiv \frac{x + y}{1 + xy} \tag{2.37}$$

appearing in the law of composition of relativistic speeds, in the relevant range $(x, y) \in [0, 1] \times [0, 1]$. Here $x \equiv v_1/c$ and $y \equiv v_2/c$. Physics tells us that the value of this function should never exceed unity, which is confirmed by the following mathematical considerations. Note that $f(0, 0) = 0$, f is continuous with all its derivatives of any order in $[0, 1] \times [0, 1]$, $f(y, x) = f(x, y)$,

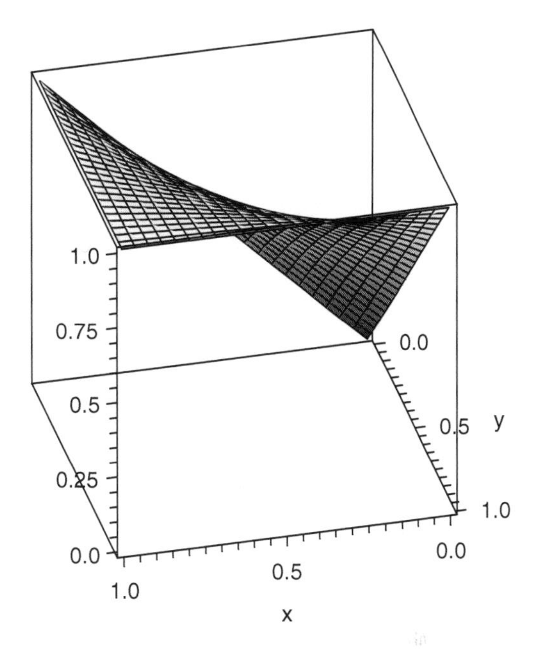

Fig. 2.4 The function $f(x,y) = \dfrac{x+y}{1+xy}$ of $x = v_1/c$ and $y = v_2/c$

$$\frac{\partial f}{\partial x} = \frac{1-y^2}{(1+xy)^2} \geq 0 \quad \text{if} \quad y \leq 1,$$
$$\frac{\partial f}{\partial y} = \frac{1-x^2}{(1+xy)^2} \geq 0 \quad \text{if} \quad x \leq 1,$$

and $\nabla f = (0,0)$ at $(x,y) = (1,1)$. The differential of f is

$$\mathrm{d}f = \nabla f \cdot \mathrm{d}\mathbf{x} = \frac{\left(1-y^2\right)\mathrm{d}x + \left(1-x^2\right)\mathrm{d}y}{(1+xy)^2}.$$

The maximum of the function f is attained at $(x,y) = (1,1)$ and

$$f\,(1,1) = 1$$

hence $0 \leq f\,(x,y) < 1 \;\; \forall \;\; (x,y) \in [0,1) \times [0,1)$. The function $f(x,y)$ is plotted in Fig. 2.4.

2.8.2 Relativistic Transformation Law of Accelerations

In a way similar to how the relativistic transformation law of velocities is derived, one can obtain the relativistic law of transformation of accelerations under a change of inertial frames (found by Tolman in 1912 [21])

$$a^{x'} = \frac{a^x}{\gamma^3 \left(1 - \frac{vu^x}{c^2}\right)^3}, \tag{2.38}$$

$$a^{y'} = \frac{1}{\gamma^2 \left(1 - \frac{vu^x}{c^2}\right)^3} \left[\left(1 - \frac{vu^x}{c^2}\right) a^y + \frac{v}{c^2} u^y a^x \right], \tag{2.39}$$

$$a^{z'} = \frac{1}{\gamma^2 \left(1 - \frac{vu^x}{c^2}\right)^3} \left[\left(1 - \frac{vu^x}{c^2}\right) a^z + \frac{v}{c^2} u^z a^x \right], \tag{2.40}$$

where $\gamma = \gamma(v)$ (the detailed derivation is left as an exercise). From these transformation properties it follows that all inertial observers agree on whether a particle is accelerated or not. Moreover, if a particle has 3-dimensional acceleration $\mathbf{a}$ =constant in one inertial frame, its acceleration is necessarily non-constant in another inertial frame. Finally, we note that in the Newtonian limit $v/c \longrightarrow 0$ the acceleration is Galilei-invariant, $\mathbf{a}' = \mathbf{a}$, and Newton's second law is invariant under Galilei transformations, as already discussed.

2.9 Matrix Representation of the Lorentz Transformation

The Lorentz transformation

$$\hat{L}_v : \begin{pmatrix} ct \\ x \\ y \\ z \end{pmatrix} \rightarrow \begin{pmatrix} ct' \\ x' \\ y' \\ z' \end{pmatrix}$$

is a linear homogeneous coordinate transformation in the space (ct, x, y, z) and can be represented by a symmetric 4×4 matrix $\hat{L}_v$ with components given by

$$\hat{L}_v = \left(L_{(v)}{}^{\alpha}{}_{\beta}\right) \equiv \begin{pmatrix} \gamma & -\gamma\beta & 0 & 0 \\ -\gamma\beta & \gamma & 0 & 0 \\ 0 & 0 & 1 & 0 \\ 0 & 0 & 0 & 1 \end{pmatrix}, \tag{2.41}$$

where $\beta \equiv v/c$ and $\gamma = \frac{1}{\sqrt{1-\beta^2}}$. This is a real symmetric 4×4 matrix parametrized by the parameter v. To check that this representation is correct, take the product

$$\begin{pmatrix} \gamma & -\gamma\beta & 0 & 0 \\ -\gamma\beta & \gamma & 0 & 0 \\ 0 & 0 & 1 & 0 \\ 0 & 0 & 0 & 1 \end{pmatrix} \begin{pmatrix} ct \\ x \\ y \\ z \end{pmatrix} = \begin{pmatrix} \gamma ct - \gamma\beta x \\ -\gamma\beta ct + \gamma x \\ y \\ z \end{pmatrix} = \begin{pmatrix} \gamma\left(ct - \frac{v}{c}x\right) \\ \gamma\,(x - vt) \\ y \\ z \end{pmatrix},$$

therefore,

$$ct' = \gamma\left(ct - \frac{v}{c}x\right),$$

$$x' = \gamma\,(x - vt),$$

$$y' = y,$$

$$z' = z,$$

which is the Lorentz transformation.

The matrix $\hat{L}_v$ of the Lorentz transformation has unit determinant:

$$\begin{aligned} \mathrm{Det}\left(\hat{L}_v\right) &= \mathrm{Det}\begin{pmatrix} \gamma & -\gamma\beta & 0 & 0 \\ -\gamma\beta & \gamma & 0 & 0 \\ 0 & 0 & 1 & 0 \\ 0 & 0 & 0 & 1 \end{pmatrix} \\ &= \gamma \begin{vmatrix} \gamma & 0 & 0 \\ 0 & 1 & 0 \\ 0 & 0 & 1 \end{vmatrix} - (-\gamma\beta) \begin{vmatrix} -\gamma\beta & 0 & 0 \\ 0 & 1 & 0 \\ 0 & 0 & 1 \end{vmatrix} \\ &= \gamma \cdot \gamma + \gamma\beta\,(-\gamma\beta) = \gamma^2\left(1 - \beta^2\right) = \left(\frac{1}{\sqrt{1-\beta^2}}\right)^2 \left(1 - \beta^2\right) \\ &= 1. \end{aligned}$$

The inverse of the matrix $\hat{L}_v$ is the matrix $\hat{L}_{(-v)}$ corresponding to the inverse Lorentz transformation,

$$\hat{L}_{(-v)} = \begin{pmatrix} \gamma & \gamma\beta & 0 & 0 \\ \gamma\beta & \gamma & 0 & 0 \\ 0 & 0 & 1 & 0 \\ 0 & 0 & 0 & 1 \end{pmatrix}. \tag{2.42}$$

In fact, we have

$$\hat{L}_v \hat{L}_{(-v)} = \begin{pmatrix} \gamma & -\gamma\beta & 0 & 0 \\ -\gamma\beta & \gamma & 0 & 0 \\ 0 & 0 & 1 & 0 \\ 0 & 0 & 0 & 1 \end{pmatrix} \begin{pmatrix} \gamma & \gamma\beta & 0 & 0 \\ \gamma\beta & \gamma & 0 & 0 \\ 0 & 0 & 1 & 0 \\ 0 & 0 & 0 & 1 \end{pmatrix}$$

$$= \begin{pmatrix} 1 & 0 & 0 & 0 \\ 0 & 1 & 0 & 0 \\ 0 & 0 & 1 & 0 \\ 0 & 0 & 0 & 1 \end{pmatrix}$$

and

$$\hat{L}_{(-v)} \hat{L}_v = \begin{pmatrix} \gamma & \gamma\beta & 0 & 0 \\ \gamma\beta & \gamma & 0 & 0 \\ 0 & 0 & 1 & 0 \\ 0 & 0 & 0 & 1 \end{pmatrix} \begin{pmatrix} \gamma & -\gamma\beta & 0 & 0 \\ -\gamma\beta & \gamma & 0 & 0 \\ 0 & 0 & 1 & 0 \\ 0 & 0 & 0 & 1 \end{pmatrix}$$

$$= \begin{pmatrix} 1 & 0 & 0 & 0 \\ 0 & 1 & 0 & 0 \\ 0 & 0 & 1 & 0 \\ 0 & 0 & 0 & 1 \end{pmatrix}.$$

The matrix corresponding to $v = 0$ and $\gamma = 1$ is obviously the identity matrix.

2.10 *The Lorentz Group

As Poincaré realized, the Lorentz transformations $\hat{L}_v$ form a group with respect to the composition of transformations $\circ$. In fact,

- if $\hat{L}_u$, $\hat{L}_v$ are Lorentz transformations, then $\hat{L}_u \circ \hat{L}_v$ is a Lorentz transformation $\hat{L}_w$ with parameter

$$w = -\frac{(u + v)}{1 + \frac{uv}{c^2}} \tag{2.43}$$

  given by the relativistic law of composition of velocities. Of course, one can obtain this result directly using the matrix representation of $\hat{L}_u$ and $\hat{L}_v$, which has the advantage of providing an alternative derivation of the law of composition of velocities. Let $\hat{L}_v$ and $\hat{L}_u$ be Lorentz transformations represented by

$$\begin{pmatrix} \gamma_v & -\gamma_v \frac{v}{c} & 0 & 0 \\ -\gamma_v \frac{v}{c} & \gamma_v & 0 & 0 \\ 0 & 0 & 1 & 0 \\ 0 & 0 & 0 & 1 \end{pmatrix}$$

and

$$\begin{pmatrix} \gamma_u & -\gamma_u \frac{u}{c} & 0 & 0 \\ -\gamma_u \frac{u}{c} & \gamma_u & 0 & 0 \\ 0 & 0 & 1 & 0 \\ 0 & 0 & 0 & 1 \end{pmatrix},$$

respectively. Then

$$\hat{L}_v \hat{L}_u = \begin{pmatrix} \gamma_v \gamma_u \left(1 + \frac{uv}{c^2}\right) & -\gamma_v \gamma_u \frac{(v+u)}{c} & 0 & 0 \\ -\gamma_v \gamma_u \frac{(v+u)}{c} & \gamma_v \gamma_u \left(\frac{uv}{c^2} + 1\right) & 0 & 0 \\ 0 & 0 & 1 & 0 \\ 0 & 0 & 0 & 1 \end{pmatrix}$$

and letting $\gamma_u \gamma_v \left(1 + \dfrac{uv}{c^2}\right) \equiv \gamma_w$, we have

$$\gamma_w \equiv \frac{1}{\sqrt{1 - \frac{w^2}{c^2}}} = \frac{1 + \frac{uv}{c^2}}{\sqrt{1 - \frac{u^2}{c^2}}\sqrt{1 - \frac{v^2}{c^2}}},$$

$$1 - \frac{w^2}{c^2} = \frac{\left(1 - \frac{u^2}{c^2}\right)\left(1 - \frac{v^2}{c^2}\right)}{\left(1 + \frac{uv}{c^2}\right)^2},$$

$$1 - \frac{w^2}{c^2} = \frac{1 - \frac{u^2}{c^2} - \frac{v^2}{c^2} + \frac{u^2 v^2}{c^4}}{\left(1 + \frac{uv}{c^2}\right)^2}.$$

Then

$$\left(1 + \frac{uv}{c^2}\right)^2 - \frac{w^2}{c^2}\left(1 + \frac{uv}{c^2}\right)^2 = 1 - \frac{u^2}{c^2} - \frac{v^2}{c^2} + \frac{u^2 v^2}{c^4},$$

$$1 + \frac{u^2 v^2}{c^4} + 2\frac{uv}{c^2} - \frac{w^2}{c^2}\left(1 + \frac{uv}{c^2}\right)^2 = 1 - \frac{u^2}{c^2} - \frac{v^2}{c^2} + \frac{u^2 v^2}{c^4},$$

$$\frac{w^2}{c^2}\left(1 + \frac{uv}{c^2}\right)^2 = \left(\frac{u+v}{c}\right)^2,$$

and finally

$$w = -\frac{(u+v)}{1 + \frac{uv}{c^2}}, \tag{2.44}$$

which is the law of composition of velocities. Two consecutive Lorentz transformations *with parallel velocity vectors* $\mathbf{v}_1$ and $\mathbf{v}_2$ commute, i.e., the result of the combined transformations does not depend on the order in which they are performed. This is no longer true if $\mathbf{v}_1$ and $\mathbf{v}_2$ are not parallel, contrary to the case of Galilei transformations. To conclude, the composition of two Lorentz transformations is a Lorentz transformation.

- The operation $\circ$ is *associative*.
- The transformation $\hat{I}_d = (\delta^{\alpha}_{\beta}) = \hat{L}_0$ corresponding to $v = 0$ is the *neutral element* of the group.
- For any Lorentz transformation $\hat{L}_v$ there is an *inverse* Lorentz transformation

$$\left(\hat{L}_v\right)^{-1} = \hat{L}_{(-v)}. \tag{2.45}$$

The fact that $\hat{L}_v$ is invertible follows from the fact that its determinant is unity.

To conclude, the Lorentz transformations $\left\{\hat{L}_v\right\}$ form a group. Since they depend on a continuous parameter v, this is called a continuous *1-parameter group*.

Pure Lorentz transformations in standard configuration form a group but other linear coordinate transformations which leave the interval ds^2 invariant can be added, including continuous transformations such as purely spatial rotations (which themselves form a 3-parameter group called *special orthogonal group* SO(3)), spatial translations $\mathbf{x} \longrightarrow \mathbf{x} + \mathbf{x}_{(0)}$, and time translations $t \longrightarrow t + t_{(0)}$; and discrete transformations such as reflections of the spatial axes and time reflection. A *proper* transformation is one with determinant equal to unity and an *orthochronous* transformation is one which preserves the time orientation, i.e., $L^0{}_0 \geq 0$. The *proper orthochronous Lorentz group* is a 6-parameter continuous group consisting of one Lorentz boost in standard configuration (parametrized by one continuous parameter v), two spatial rotations needed to align the x-axis of the inertial frame S with the velocity v of the inertial observer O' (which needs two angles as continuous parameters), and three spatial rotations to rotate the frame S of the inertial observer O in the same orientation of the frame S' of the inertial observer O', accounting for the remaining three continuous parameters, which are rotation angles about the three spatial axes).

By adding the translations in space and time $x^{\mu} \longrightarrow x^{\mu'} = x^{\mu} + x^{\mu}_{(0)}$, where the $x^{\mu}_{(0)}$ are constants, one obtains the ten-parameter *Poincaré group* consisting of linear inhomogeneous transformations $x^{\mu} \longrightarrow x^{\mu'} = L^{\mu}{}_{\alpha} x^{\alpha} + x^{\mu}_{(0)}$ which leave the interval ds^2 invariant.

2.11 The Lorentz Transformation as a Rotation by an Imaginary Angle with Imaginary Time

An interesting mathematical representation of the Lorentz transformation is the following. Considering again two inertial frames in relative motion with speed v in standard configuration, it is straightforward to check that the quantity $-c^2t^2 + x^2$ is invariant under Lorentz transformations. Define the imaginary "times" $T \equiv ict$ and $T' \equiv ict'$ in the two inertial frames. Then $T^2 + x^2$ is a Lorentz invariant, i.e.,

$$\underbrace{T^2 + x^2}_{\substack{\text{distance from}\\ \text{the origin in}\\ \text{the (x, T) plane}}} = \underbrace{\left(T'\right)^2 + \left(x'\right)^2}_{\substack{\text{distance from}\\ \text{the origin in}\\ \text{the } (x', T') \text{ plane}}}.$$

The distance from the origin is invariant under a rotation in the (x, T) plane described by

$$x' = x\cos\theta + T\sin\theta,$$
$$T' = -x\sin\theta + T\cos\theta, \tag{2.46}$$

or, in matrix form,

$$\begin{pmatrix} x' \\ T' \end{pmatrix} = \begin{pmatrix} \cos\theta & \sin\theta \\ -\sin\theta & \cos\theta \end{pmatrix} \begin{pmatrix} x \\ T \end{pmatrix}. \tag{2.47}$$

We see that the T'-axis has equation $x' = 0$ equivalent to $x = vt = v\dfrac{T}{ic}$ and that a rotation producing this axis satisfies

$$\overbrace{0}^{x'} = \overbrace{\frac{vT}{ic}\cos\theta}^{x\cos\theta} + T\sin\theta.$$

Therefore,

$$\tan\theta = -\frac{v}{ic} = \frac{iv}{c} \tag{2.48}$$

corresponds to an imaginary rotation angle θ. Since

$$\underbrace{1 + \tan^2\theta}_{1-\frac{v^2}{c^2}} = \frac{1}{\cos^2\theta},$$

we have $\cos\theta = \left(1 - \dfrac{v^2}{c^2}\right)^{-1/2}$ and

$$x' = x\cos\theta + T\sin\theta = (x + T\tan\theta)\cos\theta$$

$$= \gamma\left(x + \tfrac{\mathrm{i}v}{c}\,T\right) = \gamma\,(x - vt),$$

while

$$T' = -x\sin\theta + T\cos\theta = (-x\tan\theta + T)\cos\theta$$

$$= \gamma\left(-x\,\tfrac{\mathrm{i}v}{c} + \mathrm{i}ct\right),$$

or

$$t' = \gamma\left(t - \frac{v}{c^2}\,x\right)$$

so that

$$ct' = \gamma\left(ct - \tfrac{v}{c}\,x\right),$$

$$x' = \gamma\,(x - vt).$$

Then $\tan\theta = \mathrm{i}\,v/c$, $\cos\theta = \gamma$, and $\sin\theta = \mathrm{i}\gamma\,v/c$. It is customary to define the *rapidity* ϕ by[17]

$$\tanh\phi \equiv \beta \quad \text{or} \quad \phi \equiv \tanh^{-1}\left(\frac{\mathrm{v}}{\mathrm{c}}\right); \tag{2.49}$$

then the relation $\tan\theta = \mathrm{i}\,\dfrac{v}{c}$ gives $\tanh\phi \equiv \dfrac{v}{c} = -\mathrm{i}\,\tan\theta$ and, using the identity $\tanh(\mathrm{i}\theta) = \mathrm{i}\tan\theta$, one obtains $-\tanh\phi = \tanh(-\phi) = \mathrm{i}\tan\theta = \tanh(\mathrm{i}\theta)$, or

$$\phi = -\mathrm{i}\theta \ \in \mathbb{R}$$

and

$$\theta = \mathrm{i}\phi. \tag{2.50}$$

We can revisit the fact that Lorentz transformations form a group by viewing Lorentz boosts as rotations by imaginary angles in a space with imaginary time.

The composition of $\hat{L}_{v_1}$ and $\hat{L}_{v_2}$ is a Lorentz boost $\hat{L}_w$ with speed $w = -\frac{v_1+v_2}{1+\frac{v_1 v_2}{c^2}}$. To prove this statement, note that

[17] The name "rapidity" arises from the one-to-one correspondence of ϕ with the velocity v and the fact that $\phi \approx \beta$ for $|v| \ll c$.

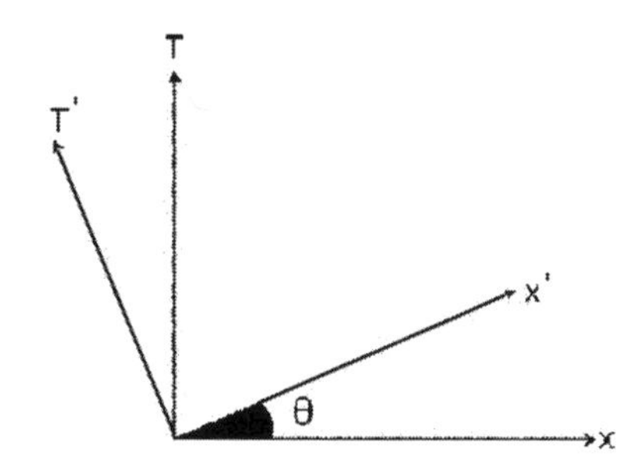

Fig. 2.5 A rotation by an imaginary angle θ in the $(x, T = \mathrm{i}ct)$ plane

$$\tan\theta = \tan(\theta_1 + \theta_2) = \frac{\tan\theta_1 + \tan\theta_2}{1 - \tan\theta_1 \tan\theta_2}, \tag{2.51}$$

or

$$\theta = \tan^{-1}\left(\mathrm{i}\frac{w}{c}\right) = -\tan^{-1}\left(\frac{\mathrm{i}\frac{v_1}{c} + \mathrm{i}\frac{v_2}{c}}{1 - \mathrm{i}^2\frac{v_1}{c}\frac{v_2}{c}}\right) = -\tan^{-1}\left[\frac{\mathrm{i}(v_1 + v_2)}{c + \frac{v_1 v_2}{c}}\right] = \theta_1 + \theta_2, \tag{2.52}$$

in other words, the rapidity $\phi = -\mathrm{i}\theta$ is additive, like all angles (Fig 2.5).

The trivial transformation $\hat{L}_0 = \hat{I}_d$ is a rotation by an angle $\theta = 0$. Moreover, $\left(\hat{L}_v\right)^{-1} = \hat{L}_{-v}$ because $\hat{L}_v$ corresponds to $\tan\theta = \mathrm{i}\,v/c$ and $\hat{L}_{-v}$ corresponds to $-\tan\theta = -\mathrm{i}\,v/c$.

2.12 *The GPS System

The Global Positioning System (GPS) nowadays used for navigating aircrafts, shipping, in private and commercial vehicles, and for urban navigation and wilderness recreation, originated in the 1970s for military navigation purposes following a few predecessors and early ideas dating back to the 1940s [22, 23]. More exotic applications include the monitoring of shifts in plate tectonics and more mundane applications include the precise time-stamping of financial transactions. GPS receivers available in outdoor stores have a typical position accuracy of 15 m, while differential techniques using multiple receivers next to each other can potentially achieve an accuracy of centimeters ("survey grade GPS").

The GPS system consists of a constellation of twenty-four satellites (plus spares) in six orbital planes, in each of which reside four satellites, in high ($\sim$20000 km radius) orbits all with a period of twelve hours. Each satellite carries a stable atomic clock which keeps track of time with fractional stability better than one part in 10^{13}. This network of satellites is designed so that, from any point on the earth with unobstructed line of sight, at least four satellites are visible above the horizon at any time. GPS receivers on the ground, in flight, or on a ship detect signals emitted by these satellites and determine their own position by means of triangulation.

When a signal is emitted from a satellite, it carries encoded with it information about the precise time and position at which it was emitted. The distance between the satellite and a GPS receiver detecting it is the time difference between emission and detection of this signal multiplied by c. The receiver communicates with several satellites and it is a straightforward triangulation problem to compute the receiver's location using four or more satellites. However, for this task to be performed, the time must be kept to high accuracy, and this is where relativistic effects come into play. In order to achieve an accuracy of 15 m, times must be known with an error not less than $15\,\mathrm{m/c} = 5 \cdot 10^{-8}\,\mathrm{s}$ (50 ns). Since the satellites are moving with respect to an observer on the ground, a ship, or an aircraft, the relativistic time dilation effect is present. The linear speed of a satellite in a circular orbit of radius r and angular velocity ω with respect to the ground is

$$v = \omega r \simeq \frac{2\pi}{12 \cdot (3600\,\mathrm{s})} \cdot (2 \cdot 10^7\,\mathrm{m}) \simeq 3 \cdot 10^3\,\frac{\mathrm{m}}{\mathrm{s}} \simeq 10^{-5}\mathrm{c}.$$

The proper time τ of the satellite and the time t of the observer on the ground are related by $\Delta t = \gamma \Delta \tau$; the ratio $\Delta\tau/\Delta t = \sqrt{1 - v^2/c^2} \simeq 1 - \frac{v^2}{2c^2}$ is the percent frequency shift of a signal $\delta\nu/\nu$ and $\dfrac{v^2}{2c^2} \simeq 5 \cdot 10^{-11}$. Over 24 h, the time error is $5 \cdot 10^{-11} \cdot (24 \cdot 3600\,\mathrm{s}) = 4.3 \cdot 10^{-6}\,\mathrm{s}$. A more precise calculation takes into account the fact that the earth rotates and the position is referred to a rotating reference frame and not an inertial frame. Moreover, the satellites' orbits are not exactly circular but elliptical and are perturbed by the moon and the sun, while the earth is not perfectly spherical and has local overdensities and underdensities affecting these orbits. When all these effects are taken into account, the error arising from neglecting Special Relativity would amount to 7 ms per day. Even more important is a general-relativistic effect which consists in the slowing down of clocks in a gravitational potential well with respect to clocks far away. If Φ is the Newtonian gravitational potential and $\Delta\Phi$ is the difference in the values of this quantity at the emission and detection points, the frequency shift of a signal is $\Delta\Phi/c^2$. This second effect, if not accounted for, would be responsible for an error of $45\,\mu\mathrm{s}$ per day. The two effects subtract from each other, since a moving click ticks slower than a stationary one while a clock far away from a mass ticks faster than one closer to it. As a result, there would be a net error of $38\,\mu\mathrm{s}$ per day in neglecting these effects. Since errors larger than $5 \cdot 10^{-8}$ s ruin the required 15 m precision, at the rate of $3.8 \cdot 10^{-5}\,\mathrm{s/day}$ it would take $114\,\mathrm{s} \simeq 2$ minutes to build up the necessary error for the GPS to fail. In a full day, the accumulated time error of $3.8 \cdot 10^{-5}$ s would correspond to a position error of $(3.8 \cdot 10^{-5}\,\mathrm{s})\mathrm{c} = 11.4\,\mathrm{km}$, certainly not what you want when landing a commercial aircraft in poor visibility, piloting a large ship in narrow straits, or trying to find a precise spot on an arctic expedition or crossing a desert. The GPS system automatically corrects for the general- and special- relativistic effects. Without these corrections the GPS system would become completely useless in a matter of minutes or hours. This example shows how the seemingly abstract theory of Special Relativity and the seemingly even more abstract theory of General Relativity (whose effects are actually about six times more pronounced than Special Relativity in the GPS system), have become essential to the functioning of modern life.

2.13 Conclusion

The theoretical discovery of the Lorentz transformation was an important step of the learning process leading to Special Relativity, but its deep meaning was not understood before Einstein. In our presentation we have made it clear that the Lorentz transformation can be derived from the two postulates of Special Relativity, which are physically more transparent than what, at first sight, appears "only" as a mathematical transformation. From the physical point of view it is more satisfactory to construct the theory beginning from two very clear ideas rather than from a telling, but less transparent, mathematical symmetry. However, the Lorentz group constitutes the symmetry group of Special Relativity and suggests a unified view of space and time, a new way of looking at nature which we present in the next chapter.

Problems

2.1. Write the mathematical expression of a Lorentz boost with the $\{ct', x', y', z'\}$ inertial frame sliding along the z- (or z'-) axis, and with the x'- and the y'- axes parallel to the x-axis and the y-axis, respectively.

2.2. Find eigenvalues and eigenspaces of the matrix describing the Lorentz transformation and interpret them physically.

2.3. In an inertial frame S, two laser pulses are emitted by points on the x-axis 10 km apart and separated by 3 μs. They reach an inertial observer O' travelling in standard configuration with velocity v away from S. O' receives the two laser pulses simultaneously. Find v.

2.4. Show that the rapidity ϕ satisfies the relations

$$\begin{aligned} e^{\phi} &= \gamma\,(1+\beta)\,, \\ e^{-\phi} &= \gamma\,(1-\beta)\,, \\ \gamma &= \cosh\phi, \\ \beta\gamma &= \sinh\phi, \end{aligned}$$

so that the Lorentz transformation can be written as

$$ct' = ct \cosh\phi - x \sinh\phi,$$
$$x' = -ct \sinh\phi + x \cosh\phi.$$

2.5. Show that, given the two events $x^{\mu}_{(1,2)} = \left(ct_{(1,2)}, \mathbf{x}_{(1,2)}\right)$, the quantity

$$\mathcal{I}\left(x^{\mu}_{(1)}, x^{\mu}_{(2)}\right) \equiv \frac{(x_1 - ct_1)(x_2 + ct_2)}{(x_1 + ct_1)(x_2 - ct_2)}$$

is an invariant of the Lorentz transformation (2.1)–(2.4) [24, 25].

2.6. Derive the inverse law of addition of velocities (2.31)–(2.33) without invoking the Principle of Relativity, i.e., without the exchange $\left(u^i, v\right) \longleftrightarrow \left(u^{i'}, -v\right)$.

2.7. Derive the relativistic law of transformation of accelerations (2.38)–(2.40) under a change of inertial frames, and its inverse. Argue that all inertial observers agree on whether a particle is accelerated or not, however, if a particle has uniform acceleration in one inertial frame, its acceleration is necessarily non-uniform in another inertial frame.

2.8. A laser beam is shone from the surface of the earth onto the moon and the laser spot sweeps the surface of the full moon in the time $\Delta t = 0.010$ s. The radius of the moon is $R_m = 1.737 \cdot 10^6$ m and the earth-moon distance is $d = 3.844 \cdot 10^8$ m. What is the linear velocity of the laser spot? Comment.

2.9. The dispersion relation of electromagnetic waves propagating in a dilute plasma is

$$\omega(k) = \sqrt{c^2k^2 + \omega_p^2},$$

where the constant ω_p (*plasma frequency*) is given by $\sqrt{\frac{4\pi e^2 n_e}{m_e}}$ for non-relativistic electrons and by $\frac{1}{\gamma}\sqrt{\frac{4\pi e^2 n_e}{m_e}}$ for relativistic electrons, where n_e is the number density of electrons (with charge e and mass m_e). Compute the phase velocity and group velocity as functions of k and sketch their graphs. Discuss their magnitudes with respect to c and compute their geometric average $\sqrt{v_p v_g}$. Discuss the propagation of a plane monochromatic electromagnetic wave with electric field $\mathbf{E} = \mathbf{E}_0\, \mathrm{e}^{\mathrm{i}(\mathrm{kx} - \omega \mathrm{t})}$ as the ratio ω/ω_p varies (here $\mathbf{E}_0$ is a constant amplitude).

References

1. G.F.R. Ellis, J.-P. Uzan, Am. J. Phys. **73**, 240 (2005)
2. V. Berzi, V. Gorini, J. Math. Phys. **13**, 665 (1969)
3. J.-M. Lévy-Leblond, Am. J. Phys. **44**, 271 (1977)
4. W.H. Furry, Am. J. Phys. **23**, 517 (1955)

5. R. Baierlein, Am. J. Phys. **74**, 193 (2006)
6. M.P. Bilaniuk, V.K. Deshpande, E.C.G. Sudarshan, Am. J. Phys. **30**, 718 (1962)
7. G. Feinberg, Phys. Rev. **159**, 1089 (1967)
8. R. Ehrlich, Am. J. Phys. **71**, 1109 (2003)
9. M. Lockwood, *The Labyrinth of Time* (Oxford University Press, Oxford, 2005)
10. E.F. Taylor, Am. J. Phys. **58**, 889 (1990)
11. A.R. Whitney, I.I. Shapiro, A.E.E. Rogers, D.S. Robertson, C.A. Knight, T.A. Clark, R.M. Goldstein, G.E. Marandino, N.R. Vandenberg, Science **173**, 225 (1971)
12. M.H. Cohen, W. Cannon, G.H. Purcell, D.B. Schaffer, J.J. Broderick, K.I. Kellermann, D.L. Jauncey, Astrophys. J. **170**, 207 (1971)
13. S. Liberati, S. Sonego, M. Visser, Ann. Phys. (NY) **298**, 167 (2002)
14. R.L. Smith, Am. J. Phys. **38**, 978 (1970)
15. S.C. Bloch, Am. J. Phys. **45**, 538 (1977)
16. P.C. Peters, Am. J. Phys. **56**, 129 (1988)
17. J.W.T. Hessels, S.M. Ransom, I.H. Stairs, P.C.C. Freire, V.M. Kaspi, F. Camilo, Science **311**, 1901 (2006)
18. M. Stautberg Greenwood, Am. J. Phys. **50**, 1156 (1982)
19. N.D. Mermin, Am. J. Phys. **51**, 1130 (1983)
20. L. Sartori, Am. J. Phys. **63**, 81 (1995)
21. R.C. Tolman, Phil. Mag. **25**(125), 150 (1912)
22. http://www.spaceandtech.com/spacedata/constellations/navstar-gps_consum.shtml
23. N. Ashby, Physics Today, May 2002, p. 41
24. G.B. Gurevich, *Foundations of the Theory of Algebraic Invariants* (Noordhoff, Groningen, 1964)
25. G. Treviño, Am. J. Phys. **56**, 185 (1988)

Chapter 3
The 4-Dimensional World View

Now he has departed from this strange world a little ahead of me. That means nothing. People like us, who believe in physics, know that the distinction between past, present, and future is only a stubbornly persistent illusion.

—Albert Einstein

3.1 Introduction

We have seen the basic physical consequences of the two postulates of Special Relativity and we know how to derive them from the mathematical transformation relating two inertial frames, the Lorentz transformation. The mathematics gives us an insight into how space and time are inextricably mixed and the most natural way to see this is in a representation of the world with four dimensions, three spatial and one temporal. This is not just mathematics: the physics just makes so much more sense when viewed in four dimensions than in three spatial dimensions with time as a parameter (because that's all time is in Newtonian physics). In this chapter we begin to study the geometry of the 4-dimensional continuum. The description of physics in this continuum requires the formalism of tensors, which is presented in the next chapter. After learning this formalism, we will study the kinematics and dynamics of point particles and systems of particles. However, it is not until Chap. 9 (when continuous distributions of matter and fields are introduced into the spacetime arena), that the power and elegance of the 4-dimensional world view are fully revealed.

V. Faraoni, *Special Relativity*, Undergraduate Lecture Notes in Physics,
DOI: 10.1007/978-3-319-01107-3_3,

3.2 The 4-Dimensional World

Let us compare the Galilei and the Lorentz transformations:

$$x' = x - vt, \qquad x' = \gamma\,(x - vt)\,,$$

$$y' = y, \qquad y' = y,$$

$$z' = z, \qquad z' = z,$$

$$t' = t, \qquad ct' = \gamma\left(ct - \tfrac{v}{c}\,x\right).$$

Galilei transformation	**Lorentz transformation**
leaves Newtonian mechanics invariant	leaves Maxwell's theory invariant
intervals of absolute time are invariant	time intervals are not invariant
3-D lengths are invariant	3-D lengths are not invariant

Since the Lorentz transformation mixes the time and space coordinates, it implicitly suggests to treat these quantities on the same footing and to contemplate a 4-dimensional space (ct, x, y, z). The 4-dimensional world view was developed by Hermann Minkowski after the publication of Einstein's theory. In Minkowski's words,[1] "Henceforth space by itself and time by itself are doomed to fade away into mere shadows, and only a kind of union of the two will preserve an independent reality".

In the Newtonian picture of the world, space and time are separate entities with time playing the role of a parameter in the Newtonian equations of motion. An *event* is now simply a point in the 4-dimensional Minkowski spacetime. If we consider two simultaneous events (t, x_1, y_1, z_1) and (t, x_2, y_2, z_2), the 3-dimensional *Euclidean distance* (squared)

$$l_{(3)}^2 = (x_2 - x_1)^2 + (y_2 - y_1)^2 + (z_2 - z_1)^2 \geq 0 \tag{3.1}$$

between the two spatial points is invariant under Galilei transformations. In the 4-dimensional view of the universe of Special Relativity, time and space merge into a continuum called *spacetime*. Given any two events (ct_1, x_1, y_1, z_1) and (ct_2, x_2, y_2, z_2), the quantity that is invariant under Lorentz transformations is not the 3-dimensional length $l_{(3)}$ nor the time separation Δt between these events, but rather the 4-dimensional *spacetime interval* (squared)

$$\Delta s^2 \equiv -c^2\,(t_2 - t_1)^2 + (x_2 - x_1)^2 + (y_2 - y_1)^2 + (z_2 - z_1)^2\,, \tag{3.2}$$

[1] Reported from a famous 1908 lecture given by Minkowski at the Polytechnic of Zurich [1].

which is *not positive-definite* (in spite of the use of the symbol Δs^2). For infinitesimally close events $(ct, \mathbf{x})$ and $(ct + c\mathrm{d}t, \mathbf{x} + \mathrm{d}\mathbf{x})$, the *infinitesimal interval* or *line element* in Cartesian coordinates is

$$\mathrm{d}s^2 \equiv -c^2\mathrm{d}t^2 + \mathrm{d}x^2 + \mathrm{d}y^2 + \mathrm{d}z^2. \tag{3.3}$$

3-dimensional lengths and time intervals are *relative* to the inertial observer, while the 4-dimensional $\mathrm{d}s^2$ is *absolute* in Special Relativity; it is the same for all inertial observers related by a Lorentz transformation,

$$-c^2(\mathrm{d}t')^2 + (\mathrm{d}x')^2 + (\mathrm{d}y')^2 + (\mathrm{d}z')^2 = -c^2\mathrm{d}t^2 + \mathrm{d}x^2 + \mathrm{d}y^2 + \mathrm{d}z^2 \tag{3.4}$$

or

$$\mathrm{d}s'^2 = \mathrm{d}s^2, \tag{3.5}$$

as we already checked in Sect. 2.4. A 4-dimensional spacetime continuum equipped with the line element (3.3) which is invariant is called *Minkowski spacetime*. The background geometry for Special Relativity is the space $\mathbb{R}^4$ but not with the usual Euclidean notion of distance between points. The notion of distance needs to be generalized by the line element $\mathrm{d}s^2$ given by Eq. (3.3) because Lorentz transformations leave invariant $\mathrm{d}s^2$ but not the quantities $c^2\mathrm{d}t^2 + \mathrm{d}x^2 + \mathrm{d}y^2 + \mathrm{d}z^2$, $\mathrm{d}l^2_{(3)}$, or $\mathrm{d}t^2$. It is $\mathrm{d}s^2$ which is fundamental—in fact, the geometry of Minkowski space is even more fundamental than the theory of electromagnetism which led us to Special Relativity. Electromagnetic waves in vacuo, or zero-mass particles, and any signal which propagates at light speed satisfy the equation $c^2\mathrm{d}t^2 = \mathrm{d}\mathbf{x}^2$ or

$$\mathrm{d}s^2 = 0 \qquad \text{(massless particles)}. \tag{3.6}$$

Example 3.1 We are now ready to revisit our example of Chap. 1 in which a flash of light is emitted at the origin of an inertial system at $t = 0$ and, at time $t > 0$, is found on the surface of a sphere of radius ct. We have seen that the Galilei transformation to another inertial frame does not preserve this sphere, nor the equation $-c^2t^2 + r^2 = 0$. However, Lorentz transformations leave this equation invariant because it corresponds to $\Delta s^2 = 0$. Since the Maxwell equations are left invariant by Lorentz transformations, so is the propagation of light in this example.

* * *

We have already seen that the speed of light is a boundary that can never be crossed, or reached, by adding velocities smaller than c. Geometrically, this means that signals emitted at a spacetime point do not go out of the *light cone* emanating from that point, which is a surface generated by light rays passing through that point. In a two-dimensional (x, t) diagram using units such that $c = 1$, the light cone through a point is the locus of spacetime points forming two straight lines inclined at 45° and passing through that point (Fig. 3.1).

Massive particles are *subluminal*, i.e., they travel at speeds $v < c$; *luminal* particles travel at the speed of light c and comprise photons, gravitons, and possibly unknown particles with zero rest mass. *Superluminal* particles, also called *tachyons*, travelling at speeds $v > c$ have never been observed. Either they don't exist, or they don't interact with ordinary matter; they could travel backward in time. The trajectory of an hypothetical particle beconing tachyonic and exiting the light cone is depicted in Fig. 3.1. We will regard tachyon trajectories as physically forbidden. The fact that an event can only influence events lying inside or along its future light cone gives rise to the notion of *causality*. If particles which can become tachyonic and travel outside their light cone could exist, the notion of cause and effect would be in jeopardy.

Let O be a spacetime point; the set of points *causally connected* to O (i.e., the set of spacetime points which can affect O or be affected by O via signals propagating at speed c form the *light cone* or *null cone* through O. This surface comprises

- the *future light cone*, which is the set of all events that can be reached by O with light signals; and
- the *past light cone*, which is the set of all events that can influence O from the past by sending light signals.

We also identify

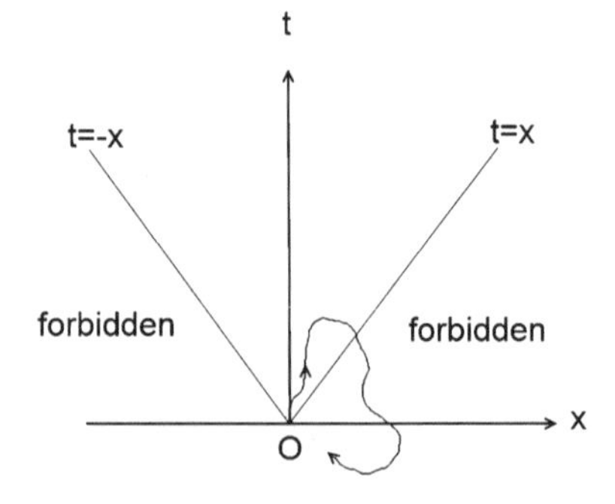

Fig. 3.1 The future light cone of the spacetime point O and a tachyonic trajectory. The region $|x| > |t|$ marked as "forbidden" cannot be reached from O using particles traveling at speeds $|v| \leq c$

- the *causal future* $J^+(O)$ of O, which is the set of all events that can be influenced by O through signals travelling[2] at speeds $v \leq c$; and the
- the *causal past* $J^-(O)$ of O, the set of all events which can influence O from the past by sending signals travelling[3] at speeds $v \leq c$.

Let S be a set of points O in Minkowski spacetime; then $J^{\pm}(S) \equiv \bigcup_{O \in S} J^{\pm}(O)$ is the union of the causal futures [pasts] of the events O as O varies in S.

Although events that are simultaneous in an inertial frame are not simultaneous in other inertial frames, the notion of event A preceding event B cannot be altered by Lorentz boosts with $v < c$: if the time separation between two events is $\Delta t > 0$, it will be $\Delta t' > 0$ in any other inertial frame S'. The concept of causality is then Lorentz-invariant. All inertial observers agree that a certain event is in the absolute future or past of O because the time interval $\Delta t'$ between O and this event has the same sign of Δt if $|v| < c$.

Formally, one can also consider the *elsewhere*, the complement of the causal past and causal future $J(O)$ of O. This set consists of all the events which cannot be connected with O by signals travelling at speed $v \leq c$. Different observers disagree that an event in the elsewhere of O is in the past or the future of O. Events in the elsewhere of O have no causal connection with O.

3.3 Spacetime Diagrams

Consider an (x, t) diagram obtained by suppressing the y and z directions in Minkowski spacetime in Cartesian coordinates and setting $c = 1$ (*spacetime* or *Minkowski diagram*).[4] The space x at a constant time is represented by a straight line parallel to the x-axis (a "moment of time"), see Fig. 3.2. A point of space is represented by a vertical line of constant x (with the convention that one can only move forward in time, or upward along this line).

Consider another inertial frame S', which has axes x' and t': since $t' = \gamma\,(t - vx)$, the "moment of time" $t' = D$ in S' corresponds to

$$\gamma t - \gamma v x = D \quad \text{or} \quad t = vx + \frac{D}{\gamma}$$

in S, which is represented by a straight line with slope v and intercept at the origin D/γ. The x'-axis of equation $t' = 0$ corresponds to

---

[2] In the language of Chap. 5, this is the set of all events which can be connected with O by curves starting from O and have as tangent a future-directed causal vector.

[3] This is the set of all events which can be connected with O by curves ending at O which have as tangent a future-directed causal vector.

[4] See Ref. [2] for a detailed pedagogical introduction to spacetime diagrams.

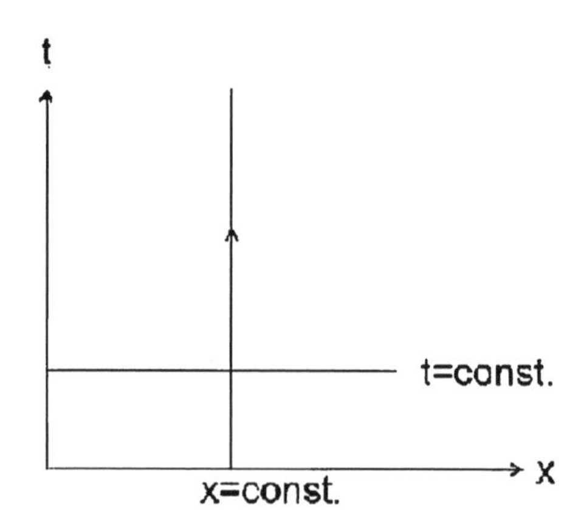

Fig. 3.2 The worldline of a particle which is stationary at $x = $ const. in the inertial frame $\{t, x\}$, and a moment of time $t = $ const

$$t = vx \qquad (t'\text{-axis}). \tag{3.7}$$

The line $x' = A$ in the S' frame corresponds to (using $x' = \gamma\,(x - vt)$)

$$\gamma x - \gamma vt = A, \qquad t = \frac{x}{v} - \frac{A}{\gamma v}.$$

In particular, the t'-axis of equation $x' = 0$ corresponds to

$$t = \frac{x}{v} \qquad (x'\text{-axis}). \tag{3.8}$$

The t'-axis and the x'-axis given by Eqs. (3.7) and (3.8), represented by lines with slopes v and $1/v$ which are the inverse of each other, are symmetric with respect to the diagonal $t = x$ (Fig. 3.3). The x'- and t'-axes can make any angle between 0° and 180°.

Since in these spacetime diagrams we use units in which $c = 1$, a light ray travelling at speed c is represented by a line at 45° with the coordinate axes. For example, two photons travelling along the positive or negative x-axis, respectively,

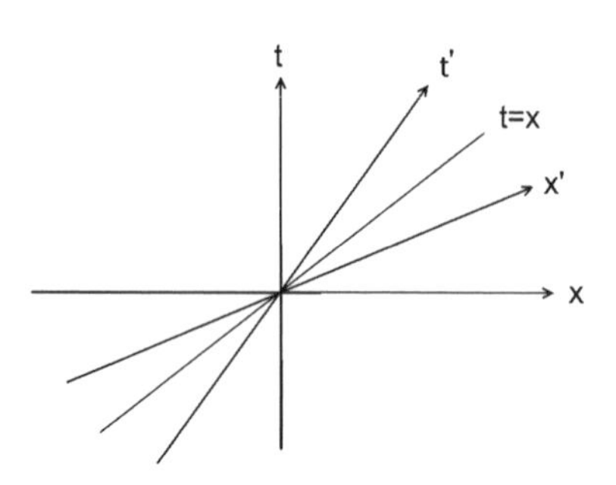

Fig. 3.3 Two inertial frames $\{t, x\}$ and $\{t', x'\}$. The x'- and t'-axes are symmetric with respect to the line $t = x$

and going through $x = 0$ at the time $t = 0$ are represented by the lines $t = \pm x$ (since objects can only travel forward in time, in the following we will often omit the arrows denoting the time direction).

A point of coordinates (x, t) in a spacetime diagram is an event; the history of a point-like particle is described by its spacetime trajectory, called a *worldline*. The history of an extended object is described by a *worldtube*, the collection of all the worldlines of the constituent particles. Worldlines in the (x, t) plane can be given by an equation of the form $t(x)$, or $f(t, x) = 0$, or by a parametric representation $\left(x(\lambda), t(\lambda)\right)$ in terms of a parameter λ (for massive particles, this is usually taken to be the proper time τ along the worldline). Since c is an absolute barrier in Special Relativity and no particle or physical signal travels faster than light, their worldlines have tangents with slopes larger than, or equal to, unity in the (x, t) plane.

Spacetime diagrams are useful to visualize simple processes in Special Relativity, for example, the emission of two consecutive light pulses from an inertial observer A and its reception and reflection back to A from a mirror located at B are described by Fig. 3.4, in which B is moving with constant velocity away from A in standard configuration.

The worldline of a massive particle which is accelerated in coordinates (x, t) will be a curve which is not straight (Fig. 3.5). Since $|v| < c$ at all times for a massive particle, the tangent to its worldline will always have slope larger than unity.

Since the squared interval is invariant under Lorentz transformations,

$$-c^2t'^2 + x'^2 = -c^2t^2 + x^2, \tag{3.9}$$

it is of interest to draw the hyperbolae (in units $c = 1$)

$$-t^2 + x^2 = \pm 1; \tag{3.10}$$

these curves coincide with the hyperbolae $-t'^2 + x'^2 = \pm 1$ and always intersect the axes at unit values of t, x, t', or x' (Fig. 3.6).

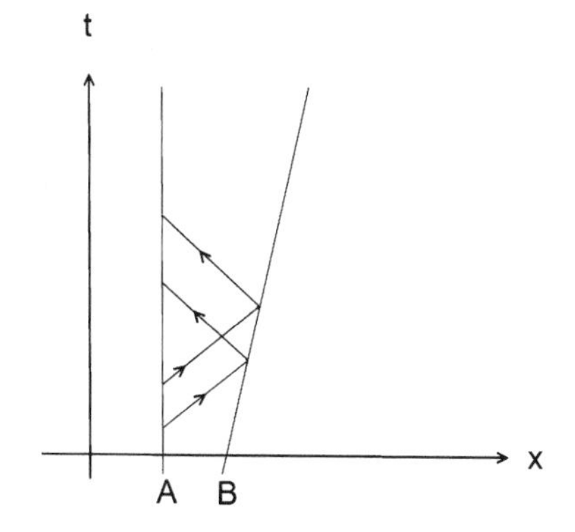

Fig. 3.4 Spacetime diagram of two light signals emitted by A, arriving at B, and reflected back to A. A is at rest in the (x, t) frame, while B moves with constant velocity

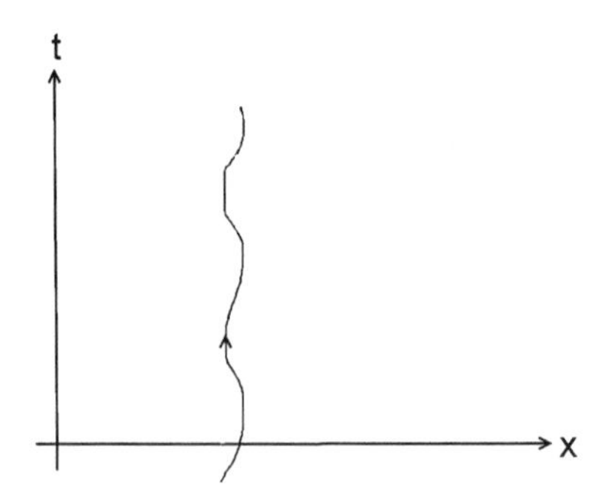

Fig. 3.5 The worldline of an accelerated particle

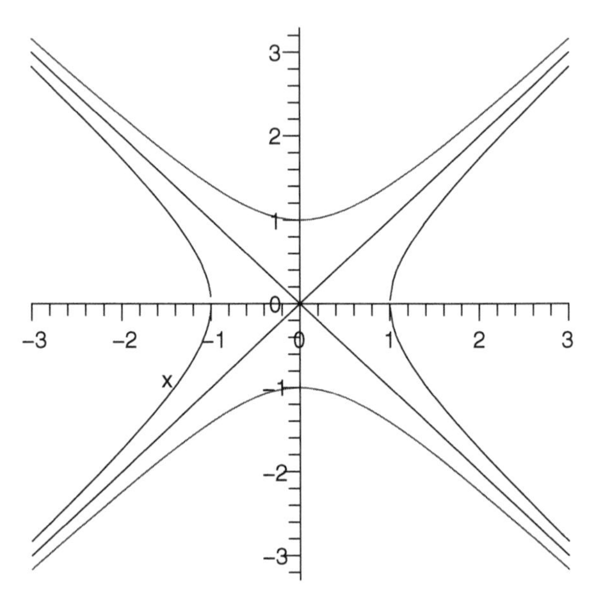

Fig. 3.6 The invariant hyperbolae $-t^2 + x^2 = -t'^2 + x'^2 = \pm 1$ asymptotic to the lines $t = \pm x$

The Lorentz transformation (2.1)–(2.4) allows us to conclude that:

- the x'-axis of equation $ct' = 0$ is the straight line $ct = \frac{v}{c}x$ of slope < 1 in units in which $c = 1$.
- The t'-axis of equation $x' = 0$ is the straight line of equation $ct = \frac{c}{v}x$ with slope larger than unity (when $c = 1$). This is the line symmetric to the x'-axis with respect to the diagonal $t = x$ in the (x, t) plane. It is also the worldline of an observer at rest in the $\{t', x'\}$ frame, which has zero 3-dimensional velocity in this frame but for which time keeps going on.
- The lines parallel to the t'-axis are the worldlines of particles at rest with fixed x' coordinate in S'. The lines parallel to the x'-axis are lines of constant t' (*lines of simultaneity* of S').

- The x'- and t'-axes make equal angles, but measured in opposite directions, with the line $t = x/c$ (in units $c = 1$, of course). As v gets closer and closer to c, these axes get closer and closer to the symmetry line $t = x/c$ in the spacetime diagram, and they merge with it in the limit $v \to c$. Formally this line is an invariant of the Lorentz transformation: $x' = ct' \quad \Leftrightarrow \quad x = ct$.

Note that the x'- and t'-axes do not appear perpendicular in the (x, t) plane in which the observer S' is not at rest. Also, the length scales along the axes are not the same. To relate a length on the x'-axis to that on the x-axis we use the invariant hyperbolae (3.10) (see Fig. 3.7). Since these hyperbolae intersect the axes at unit length, the intersection of this curve with the x-axis mapped into its intersection with the x'-axis gives the unit of length in S', which can be used to calibrate lengths. The same procedure applies to the time axes t and t'.

The coordinates $\left(x_0', t_0'\right)$ in S' of an event of coordinates (x_0, t_0) in S can be obtained by projecting the event parallel to the coordinate axes x' and t' (Fig. 3.8).

It is now easy to understand graphically why two events that are simultaneous in the inertial frame S are not simultaneous in another inertial frame S' (Fig. 3.9). Simultaneous events according to S all lie on a horizontal line $t =$ const. in the (x, t) plane. Projecting two events P_1 and P_2 lying on this line parallel to the x'-axis yields two distinct intersections with the t'-axis, i.e., two events which are not simultaneous according to S'.

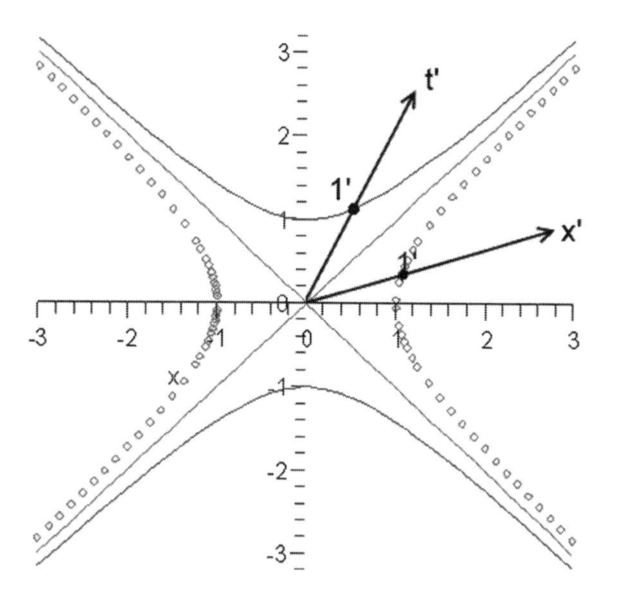

Fig. 3.7 A segment of unit length on the x-axis corresponds to a segment of different length (1') on the x'-axis, given by its intersection with the invariant hyperbola

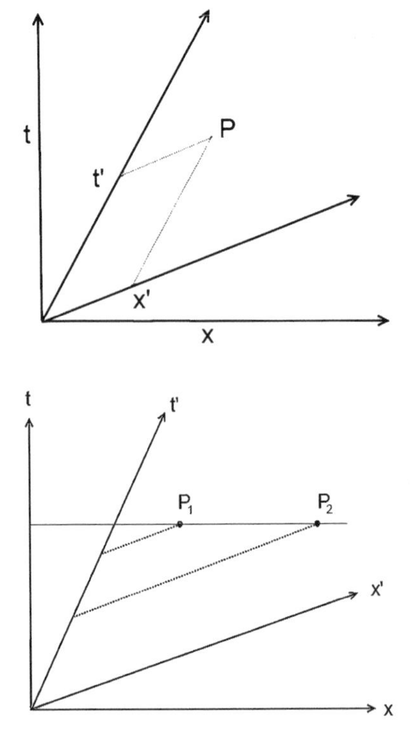

Fig. 3.8 The coordinates (x', t') of an event P in the inertial frame S' are obtained geometrically by projecting P parallel to the x'- and t'-axes

Fig. 3.9 The relativity of simultaneity in a spacetime diagram

3.4 Conclusion

The 4-dimensional world view provides much insight into the essence of Special Relativity, the intimate relations between space and time, the momentum and energy of a particle or of a physical system, the wave vector and the frequency of a wave, and the electric and magnetic fields. Before we uncover these relations, which we have just begun to see, we need to become acquainted with the necessary mathematics: the formalism of tensors.

Problems

3.1. Verify Eq. (3.5).

3.2. A particle moving with constant velocity with respect to a certain inertial observer O decays into two particles. Draw a spacetime diagram of the process in the reference frame of O.

3.3. Draw the causal future $J^+(S)$ and the causal past $J^-(S)$ of the set of events

$$S = \left\{ (x, t) : \quad t = 0, \quad x \in [0, 1] \cup [2, 3] \right\}$$

in 2-dimensional Minkowski spacetime. Write expressions for the boundaries $\partial J^+(S)$ of $J^+(S)$, $\partial J^-(S)$ of $J^-(S)$, and $\partial J(S)$ of $J^-(S) \cup J^+(S)$.

3.4. Draw the wordline of a particle moving with speed $c/2$ along the negative x-axis, and a series of light cones emanating from this particle at proper times $\tau = 1, 3$, and 6 seconds.

3.5. Consider two particles at rest along the x-axis at $x = x_1$ and $x = x_2$. Draw the past light cones of the events $(ct, x_1, 0, 0)$ and $(ct, x_2, 0, 0)$ with $t > 0$ in an (x, t) spacetime diagram. Under what conditions can the particles have interacted[5] at times $t \geq 0$?

3.6. Newtonian mechanics corresponds to the limit of Special Relativity when the speed of light becomes infinite (this is a formal limit, of course: c is in fact a constant). Discuss the causal structure of Minkowski spacetime in this limit: for example, what does the light cone through the origin of the (x, t) Minkowski spacetime become in this limit?

3.7. A particle oscillates along the x-axis with simple harmonic motion

$$x(t) = x_0 \cos(\omega t), \qquad y = z = 0$$

(where x_0 and ω are positive constants). Draw the particle worldline in an (x, t) spacetime diagram. Given the amplitude x_0 of the motion, what is the upper bound on its frequency?

3.8. Show that the line element

$$ds^2 = -\frac{a^2 x^2}{c^4} c^2 dt^2 + dx^2,$$

where a is a constant with the dimensions of an acceleration and $x > 0$ and $t \in (-\infty, +\infty)$, is nothing but the line element of the 2-dimensional Minkowski spacetime in accelerated coordinates $\{cT, X\}$ given by

$$cT = x \sinh\left(\frac{at}{c}\right),$$

$$X = x \cosh\left(\frac{at}{c}\right).$$

Which portion of the 2-dimensional Minkowski spacetime is covered by the coordinates $\{cT, X\}$?

[5] This exercise is related to the *horizon problem* of Big Bang cosmology [3, 4].

References

1. H. Minkowski, translated by W. Perrett and G.B, Jeffery in *The Principle of Relativity* (Dover, New York, 1952)
2. T. Takeuchi, *An Illustrated Guide to Relativity* (Cambridge University Press, Cambridge, 2010)
3. C.B. Collins, S.W. Hawking, Astrophys. J. **180**, 317 (1973)
4. A. Liddle, *An Introduction to Cosmology* (Wiley, Chichester, 2003)

Chapter 4
The Formalism of Tensors

Do not worry about your difficulties in mathematics.
I can assure you mine are still greater.
—Albert Einstein

4.1 Introduction

At this point it is of great advantage to use the language of tensors applied to the 4-dimensional Minkowski spacetime. Tensors are mathematical objects, described by their components in a given coordinate system, which separate true geometric and physical properties from properties that are specific to certain coordinate systems and only valid there. The formalism of tensors is constructed by studying how tensorial quantities transform under coordinate changes and, since Special Relativity is based on changes of inertial frame (the Lorentz transformations) which are coordinate changes, it is natural to use 4-tensors to formulate Special Relativity. The formalism of tensors, however, is not specific to Special Relativity; it is used in many other areas of physics. For example, tensors occur in the study of the Newtonian mechanics of rigid bodies, fluids, elasticity, electromagnetism, and diffusion. This chapter is quite general and does not refer specifically to tensors in Special Relativity. The adaptation of the formalism to Minkowski spacetime and to spacetime quantities will be discussed in the next chapter.

4.2 Vectors and Tensors

In various areas of non-relativistic physics one encounters 3-dimensional tensors. They generalize the notion of vector and, at a first glance, they look to the student as "symbols with two or more indices" as opposed to vectors which have only one

V. Faraoni, *Special Relativity*, Undergraduate Lecture Notes in Physics,
DOI: 10.1007/978-3-319-01107-3_4, © Springer International Publishing Switzerland 2013

index, or scalars which have no indices. In order to understand tensors, we first need to revisit the notion of vector. A 3-dimensional vector is usually defined as an element of the 3-dimensional real vector space $\mathbb{R}^3$ (a quantity characterized by its magnitude and direction) and can be represented on a basis of this vector space by giving all its components. For example, by choosing the orthonormal basis of unit vectors $\left\{\hat{i}, \hat{j}, \hat{k}\right\} \equiv \left\{\hat{e}_x, \hat{e}_y, \hat{e}_z\right\}$ and projecting along the x-, y-, and z-axes, respectively, a vector $\mathbf{A}$ can be represented by means of its components in Cartesian coordinates $\{x, y, z\}$,

$$\mathbf{A} = A^x\, \hat{i} + A^y\, \hat{j} + A^z\, \hat{k} = \left(A^x, A^y, A^z\right) = \sum_{i=1}^{3} A^i\, \mathbf{e}_i, \tag{4.1}$$

i.e., it is equivalent to use the symbol $\mathbf{A}$ or the coordinate representation A^i (where $i = 1, 2, 3$). We are familiar with operations on a vector or between vectors, such as the multiplication $\lambda\mathbf{A}$ of a vector $\mathbf{A}$ by a scalar λ, the addition and subtraction $\mathbf{A} \pm \mathbf{B}$ of two vectors $\mathbf{A}$ and $\mathbf{B}$, the scalar product $\mathbf{A} \cdot \mathbf{B}$, and the vector product $\mathbf{A} \times \mathbf{B}$.

A simple example of a 2-index tensor can be constructed by taking the product of the components A^i of a vector $\mathbf{A}$ and B^j of a vector $\mathbf{B}$: the symbol $C^{ij} \equiv A^i B^j$ denotes the components of this "bivector", which can be represented by the 3×3 matrix

$$\left(C^{ij}\right) = \begin{pmatrix} A^x B^x & A^x B^y & A^x B^z \\ A^y B^x & A^y B^y & A^y B^z \\ A^z B^x & A^z B^y & A^z B^z \end{pmatrix}. \tag{4.2}$$

Similarly, one can construct the tensor $D^{ijk} = A^i B^j C^k$ out of the vectors $\mathbf{A}$, $\mathbf{B}$, and $\mathbf{C}$, and so on. These objects, however, are very special tensors and the most general tensor cannot be obtained as a product of vector components.

Another example of a tensor is obtained by differentiating a vector field $\mathbf{A}(\mathbf{x})$, i.e., a vector that is a function of the position $\mathbf{x}$ in space (both its magnitude and direction can change from point to point, together with its components $A^i(\mathbf{x})$, which are functions of the position). One can consider the 2-index object

$$A^i{}_j \equiv \frac{\partial A^i}{\partial x^j} \qquad (i, j = 1, 2, 3) \tag{4.3}$$

and indeed such objects are used, for example, in fluid dynamics and field theory. If $\mathbf{v} = \mathbf{v}(t, \mathbf{x})$ is the velocity field in a fluid, the velocity gradient tensor

$$v_{ij} \equiv \frac{\partial v_i}{\partial x_j} \tag{4.4}$$

can be considered at any point, along with the *shear tensor*

$$\sigma_{ij} \equiv \frac{\partial v_i}{\partial x_j} + \frac{\partial v_j}{\partial x_i} - \frac{1}{3} (\nabla \cdot \vec{v}) \, \delta_{ij} , \tag{4.5}$$

and the *vorticity tensor*

$$\omega_{ij} \equiv \frac{\partial v_i}{\partial x_j} - \frac{\partial v_j}{\partial x_i} . \tag{4.6}$$

Another example of a 3-dimensional tensor encountered in electromagnetism is the *Maxwell stress tensor*, defined by

$$T_{ij} = \frac{1}{4\pi} \left[E_i E_j + B_i B_j - \frac{1}{2} \left(E^2 + B^2 \right) \delta_{ij} \right] , \tag{4.7}$$

where $\mathbf{E}$ and $\mathbf{B}$ are the electric and magnetic field, respectively. This 3-dimensional tensor is analogous to the tensor describing stresses in an elastic medium or pressures in a (possibly viscous) fluid. Yet another example occurring in a continuous medium or in a fluid is encountered when a 2-dimensional surface in the medium is identified by its unit normal $\mathbf{n}$ and a force $\mathbf{f}$ acts on it, in general not parallel to $\mathbf{n}$. Let $\mathrm{d}\mathbf{S} = \mathbf{n} \, \mathrm{d}S$ be a vector with magnitude equal to an element $\mathrm{d}S$ of area with unit normal $\mathbf{n}$ and with the direction of $\mathbf{n}$. Then the i-th component of the force acting on $\mathrm{d}S$ is $\sum_{j=1}^{3} \Sigma_{ij} \, \mathrm{d}S^j$, where Σ_{ij} is the *stress tensor*.

One can change the coordinates $\left\{x^i\right\}_{i=1,2,3}$ used in the space $\mathbb{R}^3$ and correspondingly change the basis of unit vectors $\{\mathbf{e}_i\}_{i=1,2,3}$ associated with the coordinates: $x^i \longrightarrow x^{i'}$ and $\mathbf{e}_i \longrightarrow \mathbf{e}_{i'}$ (in fact, one could also consider a new basis of vectors not associated with a coordinate system). By doing this, the components of a vector $\mathbf{A}$ will change, $A^i \longrightarrow A^{i'}$. The *vector* itself is a geometric object which exists independent of the coordinates used in the space in which it lives, whereas a *coordinate representation* of $\mathbf{A}$ (for example $\left(A^x, A^y, A^z\right)$ in Cartesian coordinates $\{x, y, z\}$) depends on the coordinates used. Under a coordinate change the components of $\mathbf{A}$ transform in a well-specified way which will be used below as a *definition* of vector and later generalized to define a tensor. It is not obvious why an object (vector, tensor, etc.) should be defined according to the way in which it transforms under coordinate changes, even though it seems reasonable that its properties under coordinate changes may characterize it. There is, in fact, a more direct and geometric way of characterizing vectors and tensors, which can be proven to be equivalent to the one based on coordinate transformations which we will use. However, the price to pay for this alternative definition is a higher degree of abstraction which, although appealing to the mathematician, is less attractive for the beginning physics student and we will not base our discussion on it. Before proceeding to define vectors and tensors in the space $\mathbb{R}^n$ (where $n \geq 2$), let us introduce some notations and conventions universally adopted in relativity.

4.2.1 Coordinate Transformations

Let $\{x^{\mu}\}_{\mu=1,2,\ldots,n}$ be a coordinate system on $\mathbb{R}^n$, or on an open region of $\mathbb{R}^n$ and let $\left\{x^{\mu'}\right\}_{\mu'=1,2,\ldots,n}$ be another coordinate system on $\mathbb{R}^n$, or on the same open region, and consider the coordinate change

$$x^{\alpha} \longrightarrow x^{\alpha'} = x^{\alpha'}\left(x^{\beta}\right), \qquad \alpha, \alpha', \beta = 1, 2, \ldots, n. \tag{4.8}$$

The *transformation matrix* is the $n \times n$ matrix

$$\left(\frac{\partial x^{\alpha'}}{\partial x^{\beta}}\right) = \begin{pmatrix} \frac{\partial x^{1'}}{\partial x^1} & \frac{\partial x^{1'}}{\partial x^2} & \cdots & \frac{\partial x^{1'}}{\partial x^n} \\ \frac{\partial x^{2'}}{\partial x^1} & \frac{\partial x^{2'}}{\partial x^2} & \cdots & \frac{\partial x^{2'}}{\partial x^n} \\ \vdots & \vdots & \vdots & \vdots \\ \frac{\partial x^{n'}}{\partial x^1} & \frac{\partial x^{n'}}{\partial x^2} & \cdots & \frac{\partial x^{n'}}{\partial x^n} \end{pmatrix} \tag{4.9}$$

and its determinant (called *Jacobian* of the transformation) is

$$J \equiv \mathrm{Det}\left(\frac{\partial x^{\alpha'}}{\partial x^{\beta}}\right). \tag{4.10}$$

It must be $J \neq 0$ in order for the transformation to be one-to-one and invertible (i.e., to be an admissible coordinate transformation), which guarantees the existence of the *inverse transformation*

$$x^{\alpha'} \longrightarrow x^{\alpha} = x^{\alpha}\left(x^{\mu'}\right), \qquad \alpha, \alpha', \mu' = 1, 2, \ldots, n \tag{4.11}$$

with Jacobian determinant

$$\mathrm{Det}\left(\frac{\partial x^{\alpha}}{\partial x^{\beta'}}\right) = J^{-1}. \tag{4.12}$$

Now differentiate the equation $x^{\alpha'} = x^{\alpha'}\left(x^{\beta}\right)$:

$$\mathrm{d}x^{\alpha'} = \frac{\partial x^{\alpha'}}{\partial x^1}\,\mathrm{d}x^1 + \frac{\partial x^{\alpha'}}{\partial x^2}\,\mathrm{d}x^2 + \ldots + \frac{\partial x^{\alpha'}}{\partial x^n}\,\mathrm{d}x^n = \sum_{\mu=1}^{n} \frac{\partial x^{\alpha'}}{\partial x^{\mu}}\,\mathrm{d}x^{\mu}. \tag{4.13}$$

Vice-versa, by differentiating the inverse relation $x^{\alpha} = x^{\alpha}(x^{\beta'})$, one obtains

$$\mathrm{d}x^{\alpha} = \sum_{\nu'=1}^{n} \frac{\partial x^{\alpha}}{\partial x^{\nu'}} \mathrm{d}x^{\nu'}. \tag{4.14}$$

4.2.2 *Einstein Convention*

We adopt the *Einstein summation convention* according to which the summation symbol $\sum$ is dropped every time that an index is repeated, appearing once in a lowered and once in a raised position. For example, $A_{\alpha} B^{\alpha}$ denotes $\sum_{\alpha=1}^{n} A_{\alpha} B^{\alpha}$ and

$$\begin{aligned} T_{\alpha\beta\gamma}\, Q^{\alpha\beta} &= \sum_{\alpha=1}^{n} \sum_{\beta=1}^{n} T_{\alpha\beta\gamma}\, Q^{\alpha\beta} \\ &= T_{11\gamma}\, Q^{11} + T_{12\gamma}\, Q^{12} + \ldots + T_{1n\gamma}\, Q^{1n} \\ &\quad + T_{21\gamma}\, Q^{21} + T_{22\gamma}\, Q^{22} + \ldots + T_{2n\gamma}\, Q^{2n} \\ &\quad + \ldots \\ &\quad + T_{n1\gamma}\, Q^{n1} + T_{n2\gamma}\, Q^{n2} + \ldots + T_{nn\gamma}\, Q^{nn}. \end{aligned}$$

Example 4.1 Let $A_{\mu} = (1, 2, 1/2)$ and $B^{\mu} = (0, 3, 6)$ be two vectors in $\mathbb{R}^3$. Then

$$A_{\mu} B^{\mu} = A_1 B^1 + A_2 B^2 + A_3 B^3 = 1 \cdot 0 + 2 \cdot 3 + \frac{1}{2} \cdot 6 = 9.$$

* * *

The repeated index is a *dummy index* and its name is unimportant: it can be changed at will, for example

$$A_{\alpha} B^{\alpha} = A_{\mu} B^{\mu} = A_{\rho} B^{\rho}$$

and

$$A_{\mu\nu} B^{\nu} = A_{\mu\sigma} B^{\sigma}.$$

Often, one needs to change the name of a dummy index in order to avoid repetition and ambiguities, for example the expression $A_{\alpha\alpha} B^{\alpha}$ is unclear and is to be avoided because one does not know whether the first or the second index of the tensor $A_{\mu\nu}$ is contracted with that of the vector B^{σ}. If it is the second index of $A_{\mu\nu}$ which is contracted, one should write instead $A_{\alpha\rho} B^{\rho}$. An index that is not repeated (e.g., α in A_{α} or μ in $A_{\mu\nu} B^{\nu}$) is called a *free index*. We will also use the *Kronecker delta*

$$\delta_\alpha^\beta \equiv \begin{cases} 1 & \text{if } \alpha = \beta, \\ 0 & \text{if } \alpha \neq \beta. \end{cases} \tag{4.15}$$

For example, we have

$$\frac{\partial x^\alpha}{\partial x^\beta} = \delta_\beta^\alpha$$

and

$$\frac{\partial x^{\mu'}}{\partial x^{\nu'}} = \delta_{\nu'}^{\mu'}.$$

The chain rule of differentiation is

$$\frac{\partial x^\alpha}{\partial x^{\beta'}} \frac{\partial x^{\beta'}}{\partial x^{\gamma''}} = \frac{\partial x^\alpha}{\partial x^{\gamma''}} \tag{4.16}$$

and

$$\frac{\partial x^\alpha}{\partial x^{\beta'}} \frac{\partial x^{\beta'}}{\partial x^\gamma} = \frac{\partial x^\alpha}{\partial x^\gamma} = \delta_\gamma^\alpha.$$

A useful relation in a space of dimension n is

$$\delta_\alpha^\alpha = n. \tag{4.17}$$

4.3 Contravariant and Covariant Vectors

Definition 4.1 A *contravariant vector* (or *contravariant tensor of rank 1*) is a set of n components A^α in a specified coordinate system $\{x^\mu\}_{\mu=1,2,\ldots,n}$ that transform according to

$$A^{\alpha'} = \frac{\partial x^{\alpha'}}{\partial x^\beta} A^\beta \tag{4.18}$$

under the coordinate change $x^\mu \longrightarrow x^{\mu'}$.

The vector is a geometric object defined independently of the coordinate system, while its components $A^{\alpha'}$ change with the coordinate system adopted and merely constitute a *representation* of that vector on a certain basis. A vector can be represented by an $n \times 1$ column $\begin{pmatrix} A^1 \\ A^2 \\ \vdots \\ A^n \end{pmatrix}$.

* * *

Example 4.2 The differentials of the coordinates $\mathrm{d}x^{\mu}$ transform as contravariant vectors:

$$\mathrm{d}x^{\mu'} = \frac{\partial x^{\mu'}}{\partial x^{\nu}} \mathrm{d}x^{\nu}, \tag{4.19}$$

which is the reason why we conventionally write the coordinates x^{μ} using superscripts and not subscripts.

Example 4.3 The position vector $\mathbf{x}$ of a particle in three dimensions has coordinate components $x^{\mu} = (x, y, z)$ in a Cartesian coordinate system. The velocity vector $\mathbf{v}$ of the same particle has components $v^{\mu} = (v^x, v^y, v^z)$. Under the change from Cartesian to cylindrical coordinates $x^{\mu} = \{x, y, z\} \longrightarrow x^{\mu'} = \{r, \varphi, z\}$ with

$$\begin{aligned} r &= \sqrt{x^2 + y^2}, \\ \varphi &= \tan^{-1}(y/x), \\ z &= z, \end{aligned} \tag{4.20}$$

the vector components transform according to

$$v^{\mu'} = \frac{\partial x^{\mu'}}{\partial x^{\nu}} v^{\nu}. \tag{4.21}$$

Let us compute the transformation matrix of this coordinate change:

$$\left(\frac{\partial x^{\mu'}}{\partial x^{\nu}}\right) = \begin{pmatrix} \frac{\partial r}{\partial x} & \frac{\partial r}{\partial y} & \frac{\partial r}{\partial z} \\ \frac{\partial \varphi}{\partial x} & \frac{\partial \varphi}{\partial y} & \frac{\partial \varphi}{\partial z} \\ \frac{\partial z}{\partial x} & \frac{\partial z}{\partial y} & \frac{\partial z}{\partial z} \end{pmatrix} = \begin{pmatrix} \frac{x}{\sqrt{x^2+y^2}} & \frac{y}{\sqrt{x^2+y^2}} & 0 \\ \frac{-y}{x^2+y^2} & \frac{x}{x^2+y^2} & 0 \\ 0 & 0 & 1 \end{pmatrix}$$

$$= \begin{pmatrix} \cos\varphi & \sin\varphi & 0 \\ -\frac{\sin\varphi}{r} & \frac{\cos\varphi}{r} & 0 \\ 0 & 0 & 1 \end{pmatrix},$$

therefore,

$$v^r \equiv v^{1'} = \frac{\partial x^{1'}}{\partial x^{\nu}} v^{\nu} = \frac{\partial r}{\partial x} v^x + \frac{\partial r}{\partial y} v^y + \frac{\partial r}{\partial z} v^z = v^x \cos\varphi + v^y \sin\varphi, \tag{4.22}$$

$$v^{\varphi} \equiv v^{2'} = \frac{\partial x^{2'}}{\partial x^{\nu}} v^{\nu} = \frac{\partial \varphi}{\partial x} v^x + \frac{\partial \varphi}{\partial y} v^y + \frac{\partial \varphi}{\partial z} v^z = -v^x \frac{\sin\varphi}{r} + v^y \frac{\cos\varphi}{r}, \tag{4.23}$$

$$v^z \equiv v^{3'} = \frac{\partial x^{3'}}{\partial x^{\nu}} v^{\nu} = \delta^3_{\nu} v^{\nu} = v^z. \tag{4.24}$$

Definition 4.2 A *covariant vector* (or *dual vector, 1-form*, or *covariant tensor of rank 1*) is a set of n components A_α (note the position of the index) that transform according to

$$A_{\alpha'} = \frac{\partial x^{\beta}}{\partial x^{\alpha'}} A_{\beta} \tag{4.25}$$

under the coordinate change $x^\mu \longrightarrow x^{\mu'}$.

Example 4.4 If a contravariant vector in $\mathbb{R}^n$ is represented in Cartesian coordinates by an $n \times 1$ column $\begin{pmatrix} A^1 \\ A^2 \\ \vdots \\ A^n \end{pmatrix}$, the dual vector can be represented by an $1 \times n$ row $(A^1, A^2, \ldots, A^n)$.

Example 4.5 The gradient of a scalar function $\nabla_\alpha f = \partial f / \partial x^\alpha$ is a covariant or dual vector. In fact,

$$\nabla_{\alpha'} f \equiv \frac{\partial f}{\partial x^{\alpha'}} = \frac{\partial f}{\partial x^{\beta}} \frac{\partial x^{\beta}}{\partial x^{\alpha'}} \equiv \frac{\partial x^{\beta}}{\partial x^{\alpha'}} \nabla_{\beta} f. \tag{4.26}$$

Now we can consider quantities with two or more indices.

4.4 Contravariant and Covariant Tensors

Definition 4.3 A *contravariant tensor of rank 2* is a set of n^2 components $T^{\alpha\beta}$ which transform according to

$$T^{\alpha'\beta'} = \frac{\partial x^{\alpha'}}{\partial x^{\rho}} \frac{\partial x^{\beta'}}{\partial x^{\sigma}} T^{\rho\sigma} \tag{4.27}$$

under the coordinate change $x^{\mu} \longrightarrow x^{\mu'}$.

A 2-tensor can be represented by a $n \times n$ matrix

$$\left(T^{ij}\right) = \begin{pmatrix} T^{11} & T^{12} & \dots & T^{1n} \\ T^{21} & T^{22} & \dots & T^{2n} \\ \vdots & \vdots & \vdots & \vdots \\ T^{n1} & T^{n2} & \dots & T^{nn} \end{pmatrix}. \tag{4.28}$$

Example 4.6 The vectors A^i and B^j can be used to form the bivector $V^{ij} \equiv A^i B^j$, which can be represented by the matrix

$$\left(V^{ij}\right) = \begin{pmatrix} A^1 B^1 & A^1 B^2 & \dots & A^1 B^n \\ A^2 B^1 & A^2 B^2 & \dots & A^2 B^n \\ \vdots & \vdots & \vdots & \vdots \\ A^n B^1 & A^n B^2 & \dots & A^n B^n \end{pmatrix}. \tag{4.29}$$

Definition 4.4 A *covariant tensor of rank 2* is a set of n^2 components $T_{\alpha\beta}$ which transform according to

$$T_{\alpha'\beta'} = \frac{\partial x^{\rho}}{\partial x^{\alpha'}} \frac{\partial x^{\sigma}}{\partial x^{\beta'}} T_{\rho\sigma} \tag{4.30}$$

under the transformation $x^{\mu} \longrightarrow x^{\mu'}$.

Example 4.7 The second derivative of a scalar function $\frac{\partial^2 f}{\partial x^\alpha \partial x^\beta}$ is not a covariant 2-index tensor. In fact, it is

$$\begin{aligned}\frac{\partial^2 f}{\partial x^{\alpha'}\partial x^{\beta'}} &= \frac{\partial}{\partial x^{\beta'}}\left(\frac{\partial f}{\partial x^{\alpha'}}\right) = \frac{\partial}{\partial x^{\beta'}}\left(\frac{\partial f}{\partial x^{\mu}}\frac{\partial x^{\mu}}{\partial x^{\alpha'}}\right)\\ &= \frac{\partial^2 f}{\partial x^{\beta'}\partial x^{\mu}}\frac{\partial x^{\mu}}{\partial x^{\alpha'}} + \frac{\partial f}{\partial x^{\mu}}\frac{\partial^2 x^{\mu}}{\partial x^{\alpha'}\partial x^{\beta'}}\\ &= \frac{\partial^2 f}{\partial x^{\mu}\partial x^{\nu}}\frac{\partial x^{\nu}}{\partial x^{\beta'}}\frac{\partial x^{\mu}}{\partial x^{\alpha'}} + \frac{\partial f}{\partial x^{\mu}}\frac{\partial^2 x^{\mu}}{\partial x^{\alpha'}\partial x^{\beta'}}\,.\end{aligned} \tag{4.31}$$

Unless $x^\mu \longrightarrow x^{\mu'}$ is a linear transformation or $f =$ constant, which are degenerate cases, we have $\frac{\partial^2 x^\mu}{\partial x^{\alpha'}\partial x^{\beta'}} \neq 0$ and

$$\frac{\partial^2 f}{\partial x^{\alpha'}\partial x^{\beta'}} \neq \frac{\partial x^{\mu}}{\partial x^{\alpha'}}\frac{\partial x^{\nu}}{\partial x^{\beta'}}\frac{\partial^2 f}{\partial x^{\mu}\partial x^{\nu}}\,. \tag{4.32}$$

This example shows that not all objects with indices are tensors, i.e., they do not transform as tensors under coordinate changes.

Definition 4.5 A *contravariant tensor of rank k* is a set of n^k components $T^{\alpha_1\alpha_2\,\ldots\,\alpha_k}$ which transform according to

$$T^{\alpha_1'\alpha_2'\,\ldots\,\alpha_k'} = \frac{\partial x^{\alpha_{1'}}}{\partial x^{\beta_1}}\frac{\partial x^{\alpha_{2'}}}{\partial x^{\beta_2}}\cdots\frac{\partial x^{\alpha_{k'}}}{\partial x^{\beta_k}}\,T^{\beta_1\beta_2\,\ldots\,\beta_k} \tag{4.33}$$

under the coordinate change $x^\mu \longrightarrow x^{\mu'}$.

Definition 4.6 A *covariant tensor of rank k* is a set of n^k components $T_{\alpha_1\alpha_2\,\ldots\,\alpha_k}$ which transform according to

$$T_{\alpha_1'\alpha_2'\,\ldots\,\alpha_k'} = \frac{\partial x^{\beta_1}}{\partial x^{\alpha_{1'}}}\frac{\partial x^{\beta_2}}{\partial x^{\alpha_{2'}}}\cdots\frac{\partial x^{\beta_k}}{\partial x^{\alpha_{k'}}}\,T_{\beta_1\beta_2\,\ldots\,\beta_k} \tag{4.34}$$

under the coordinate change $x^\mu \longrightarrow x^{\mu'}$.

Example 4.8 A *scalar* ϕ (a 0-index tensor) is a quantity that does not change under the coordinate transformation $x^\mu \longrightarrow x^{\mu'}$:

$$\phi' = \phi. \tag{4.35}$$

For this reason, a scalar is also called an *invariant*.

Example 4.9 A vector is a contravariant tensor of rank 1:

$$v^{\alpha'} = \frac{\partial x^{\alpha'}}{\partial x^{\beta}} v^{\beta}, \tag{4.36}$$

while a dual vector is a covariant tensor of rank 1:

$$a_{\alpha'} = \frac{\partial x^{\beta}}{\partial x^{\alpha'}} a_{\beta}. \tag{4.37}$$

To summarize:

CO−	means	LOWERINDEX	(subscript)
CONTRA−	means	UPPERINDEX	(superscript)
RANK	=	number of indices	

The coordinates themselves are usually written as x^{μ} (and not as x_{μ}) because the position vector $\mathbf{x} = \left(x^1, \ldots, x^n\right)$ with the coordinates as components is a contravariant vector. Scalars are tensors of rank 0, while vectors are tensors of rank 1. Finally,

Definition 4.7 a *mixed tensor of contravariant rank l and covariant rank m* is a set of n^{l+m} components $T^{\alpha_1\alpha_2\ldots\alpha_l}{}_{\beta_1\beta_2\ldots\beta_m}$ that transform according to

$$T^{\alpha'_1\alpha'_2\ldots\alpha'_l}{}_{\beta'_1\beta'_2\ldots\beta'_m} = \frac{\partial x^{\alpha'_1}}{\partial x^{\rho_1}}\frac{\partial x^{\alpha'_2}}{\partial x^{\rho_2}}\cdots\frac{\partial x^{\alpha'_l}}{\partial x^{\rho_l}}\frac{\partial x^{\sigma_1}}{\partial x^{\beta'_1}}\frac{\partial x^{\sigma_2}}{\partial x^{\beta'_2}}\cdots\frac{\partial x^{\sigma_m}}{\partial x^{\beta'_m}} T^{\rho_1\rho_2\ldots\rho_l}{}_{\sigma_1\sigma_2\ldots\sigma_m} \tag{4.38}$$

under the coordinate change $x^{\mu} \longrightarrow x^{\mu'}$.

Example 4.10 $T^{\alpha\beta}{}_{\gamma}$ transforms according to

$$T^{\alpha'\beta'}{}_{\gamma'} = \frac{\partial x^{\alpha'}}{\partial x^{\mu}} \frac{\partial x^{\beta'}}{\partial x^{\nu}} \frac{\partial x^{\tau}}{\partial x^{\gamma'}} T^{\mu\nu}{}_{\tau} . \tag{4.39}$$

* * *

A *tensor equation*

$$A^{\alpha_1\alpha_2\ldots\alpha_l}{}_{\beta_1\beta_2\ldots\beta_k} = B^{\alpha_1\alpha_2\ldots\alpha_l}{}_{\beta_1\beta_2\ldots\beta_k} \tag{4.40}$$

holds in every coordinate system even though the components of the tensor appearing on both sides change (according to the rules that you now know). If the tensor equation is satisfied in one coordinate system, it is satisfied in *any* coordinate system. This statement can be rephrased by saying that, if all the components of a tensor (the left hand side minus the right hand side of a tensor equation) vanish in a certain coordinate system, then they vanish in *any* coordinate system. The proof of this statement is trivial for, if $T^{\alpha_1\alpha_2\ldots\alpha_l}{}_{\beta_1\beta_2\ldots\beta_m} = 0$ in a certain coordinate system $\{x^{\mu}\}$, then in another coordinate system $\left\{x^{\mu'}\right\}$ it will be

$$\begin{aligned} & T^{\alpha_1'\alpha_2'\ldots\alpha_l'}{}_{\beta_1'\beta_2'\ldots\beta_m'} \\ & = \frac{\partial x^{\alpha_{1'}}}{\partial x^{\rho_1}} \frac{\partial x^{\alpha_{2'}}}{\partial x^{\rho_2}} \cdots \frac{\partial x^{\alpha_{l'}}}{\partial x^{\rho_l}} \frac{\partial x^{\sigma_1}}{\partial x^{\beta_{1'}}} \frac{\partial x^{\sigma_2}}{\partial x^{\beta_{2'}}} \cdots \frac{\partial x^{\sigma_m}}{\partial x^{\beta_{m'}}} T^{\rho_1\rho_2\ldots\rho_l}{}_{\sigma_1\sigma_2\ldots\sigma_m} \\ & = 0 . \end{aligned}$$

If a free index appears on the left hand side of a tensor equation, a free index with the same name and in the same position must appear on the right hand side. For example, $T_{\alpha\beta} A^{\gamma} B^{\rho} = X_{\alpha\beta}{}^{\gamma\rho}$ is a correct equation, while $A^{\rho} = B^{\alpha}$ and $A^{\rho} B_{\mu} = A^{\rho} B^{\mu}$ are wrong.

Physics must be expressed in terms of tensor equations.

Any coordinate transformation between inertial frames is in principle admissible and physics does not know about the coordinates used by humans (for us, a coordinate system can be more economical to perform calculations, e.g., spherical coordinates for a system with spherical symmetry such as the sun-earth system or the hydrogen atom, but that is another story which physics does not know). One wants the *form* of a physical law not to be tied to a particular coordinate system but, instead, that physical laws assume the same form in all coordinate systems. This desirable property is obtained if the laws of physics are expressed as tensor equations.

Example 4.11 The vector equation in three dimensions

$$\mathbf{a} = \mathbf{b}$$

is true in Cartesian coordinates $\{x, y, z\}$ and it means

$$a^x = b^x,$$

$$a^y = b^y,$$

$$a^z = b^z;$$

changing coordinates as in $x^\mu \longrightarrow x^{\mu'}$, it is still true that

$$a^{x'} = b^{x'},$$

$$a^{y'} = b^{y'},$$

$$a^{z'} = b^{z'},$$

because the components of both **a** and **b** transform in the same way.

4.4.1 Tensor Symmetries

In general, the order of the indices in a tensor is important: $P_{\alpha\beta} \neq P_{\beta\alpha}$ and $T^{\alpha\beta}{}_{\gamma} \neq T^{\beta\alpha}{}_{\gamma} \neq T_{\gamma}{}^{\alpha\beta}$. However, certain multi-index tensors exhibit symmetry properties under permutation of their indices.

Definition 4.8 A tensor of rank 2 (covariant or contravariant) $T_{\alpha\beta}$ or $T^{\alpha\beta}$ is *symmetric* if

$$T_{\beta\alpha} = T_{\alpha\beta} \quad \text{or} \quad T^{\beta\alpha} = T^{\alpha\beta} \tag{4.41}$$

for all values of the indices α and β, and it is *antisymmetric* or *skew-symmetric* if

$$T_{\beta\alpha} = -T_{\alpha\beta} \quad \text{or} \quad T^{\beta\alpha} = -T^{\alpha\beta}. \tag{4.42}$$

A rank 2 symmetric tensor in a space of dimension n has at most $\frac{n(n+1)}{2}$ independent components. An antisymmetric rank 2 tensor has at most $\frac{n(n-1)}{2}$ independent components and all diagonal components necessarily vanish in order to satisfy $T_{\alpha\alpha} = -T_{\alpha\alpha}$.

A 2-index contravariant (or covariant) tensor can always be decomposed into a *symmetric part* $T_{(\alpha\beta)}$ and an *antisymmetric part* $T_{[\alpha\beta]}$ as follows:

$$T_{\alpha\beta} = \frac{T_{\alpha\beta} + T_{\beta\alpha}}{2} + \frac{T_{\alpha\beta} - T_{\beta\alpha}}{2} \equiv T_{(\alpha\beta)} + T_{[\alpha\beta]}, \tag{4.43}$$

where

$$T_{(\alpha\beta)} \equiv \frac{T_{\alpha\beta} + T_{\beta\alpha}}{2} \tag{4.44}$$

and

$$T_{[\alpha\beta]} \equiv \frac{T_{\alpha\beta} - T_{\beta\alpha}}{2} \tag{4.45}$$

are the *symmetric part* and the *antisymmetric part* of $T_{\alpha\beta}$, respectively. Eq. (4.43) is a trivial identity, but

the decomposition $T_{\alpha\beta} = T_{(\alpha\beta)} + T_{[\alpha\beta)]}$ *is unique*

(the proof is left as an exercise).

In general, the symmetric and antisymmetric parts of a tensor $T_{\mu_1\mu_2 \dots \mu_n}$ are defined by

$$T_{(\mu_1\mu_2 \dots \mu_k)} \equiv \frac{1}{k!}(\text{sum over all permutations of } \mu_1 \dots \mu_{\mathrm{k}}) \tag{4.46}$$

$$T_{[\mu_1\mu_2 \dots \mu_k]} \equiv \frac{1}{k!}(\text{sum over all permutations with alternating sign}) \tag{4.47}$$

(remember that the number of permutations of k objects is $k!$). For example, a 3-index tensor $T_{\mu\nu\rho}$ has symmetric part

$$T_{(\alpha\beta\gamma)} = \frac{1}{3!}\left(T_{\alpha\beta\gamma} + T_{\alpha\gamma\beta} + T_{\gamma\alpha\beta} + T_{\gamma\beta\alpha} + T_{\beta\gamma\alpha} + T_{\beta\alpha\gamma}\right) \tag{4.48}$$

and antisymmetric part

$$T_{[\alpha\beta\gamma]} = \frac{1}{3!}\left(T_{\alpha\beta\gamma} - T_{\alpha\gamma\beta} + T_{\gamma\alpha\beta} - T_{\gamma\beta\alpha} + T_{\beta\gamma\alpha} - T_{\beta\alpha\gamma}\right). \tag{4.49}$$

A *totally symmetric* [*antisymmetric*] *tensor* is one equal to its symmetric [antisymmetric] part. Tensors of rank higher than 2 can posses symmetries with respect to pairs of indices, for example $T_{ijk} = T_{jik}$ or $T_{ijk} = -T_{ikj}$. The symmetry of a tensor is preserved by coordinate transformations or, in relativity language, it is an *invariant property*.

Example 4.12 The shear tensor (4.5) is, by construction, symmetric while the vorticity tensor (4.6) is antisymmetric. The Kronecker delta δ_{ij} is symmetric.

* * *

The notation denoting symmetrization or antisymmetrization of indices is used also for indices belonging to different tensors, for example:

$$A_{(\mu}B_{\nu)} \equiv \frac{A_\mu B_\nu + A_\nu B_\mu}{2},$$

$$X^{[\alpha}Y^{\beta]} \equiv \frac{X^\alpha Y^\beta - X^\beta Y^\alpha}{2},$$

$$A_{\mu(\nu}B_{\rho)\sigma} = \frac{A_{\mu\nu}B_{\rho\sigma} + A_{\mu\rho}B_{\nu\sigma}}{2},$$

or

$$\begin{aligned} X_\mu{}^{(\nu}Y^{\rho)}{}_{\sigma(\alpha}Z_{\beta)}{}^\gamma &= \frac{1}{2}\left(X_\mu{}^\nu Y^\rho{}_{\sigma(\alpha}Z_{\beta)}{}^\gamma + X_\mu{}^\rho Y^\nu{}_{\sigma(\alpha}Z_{\beta)}{}^\gamma\right) \\ &= \frac{1}{4}\left(X_\mu{}^\nu Y^\rho{}_{\sigma\alpha}Z_\beta{}^\gamma + X_\mu{}^\nu Y^\rho{}_{\sigma\beta}Z_\alpha{}^\gamma + X_\mu{}^\rho Y^\nu{}_{\sigma\alpha}Z_\beta{}^\gamma + X_\mu{}^\rho Y^\nu{}_{\sigma\beta}Z_\alpha{}^\gamma\right). \end{aligned}$$

4.5 Tensor Algebra

Definition 4.9 The *multiplication of a* $\begin{pmatrix} p_1 \\ q_1 \end{pmatrix}$ *tensor by a* $\begin{pmatrix} p_2 \\ q_2 \end{pmatrix}$ *tensor* (called *direct product*) produces a tensor of type $\begin{pmatrix} p_1 + p_2 \\ q_1 + q_2 \end{pmatrix}$ with components equal to the product of the components

$$A^{\alpha_1 \dots \alpha_{p_1}}{}_{\beta_1 \dots \beta_{q_1}} B^{\gamma_1 \dots \gamma_{p_2}}{}_{\delta_1 \dots \delta_{q_2}} \equiv P^{\alpha_1 \dots \alpha_{p_1} \gamma_1 \dots \gamma_{p_2}}{}_{\beta_1 \dots \beta_{q_1} \delta_1 \dots \delta_{q_2}} . \tag{4.50}$$

Example 4.13 Using the tensors $Y^\alpha{}_\beta$ and $Z_{\gamma\delta}$, one constructs the direct product

$$P^\alpha{}_{\beta\gamma\delta} \equiv Y^\alpha{}_\beta Z_{\gamma\delta} .$$

Definition 4.10 The *contraction of two indices* of a tensor $T^{\alpha_1 \dots \alpha_l}{}_{\beta_1 \dots \beta_m}$ reduces its rank by 2 and is defined by

$$T^{\alpha_1 \dots \alpha_j \dots \alpha_l}{}_{\beta_1 \dots \alpha_j \dots \beta_m}, \tag{4.51}$$

a tensor of type $\begin{pmatrix} l-1 \\ m-1 \end{pmatrix}$ in which a pair of indices is reduced to two dummy indices.

Example 4.14 Consider the 4-index tensor $R^{\alpha}{}_{\beta\gamma\delta}$ and contract the first and third index:

$$R^{\alpha}{}_{\beta\alpha\delta} \equiv R_{\beta\delta};$$

this operation reduces the rank by two, from a $\binom{1}{3}$ tensor to a $\binom{0}{2}$ tensor.

* * *

The contraction is equivalent to multiplication by a Kronecker delta:

$$X^{\alpha}{}_{\alpha\gamma\delta} = \delta^{\alpha}_{\beta}\, X^{\beta}{}_{\alpha\gamma\delta} = \delta^{\beta}_{\alpha}\, X^{\alpha}{}_{\beta\gamma\delta}\,.$$

Another operation that one can do on a multi-index tensor is taking the trace.

Definition 4.11 The *trace* of a mixed tensor $T^{\alpha}{}_{\beta}$ of type $\binom{1}{1}$ is

$$T \equiv T^{\alpha}{}_{\alpha} \equiv \delta^{\beta}_{\alpha}\, T^{\alpha}{}_{\beta} \tag{4.52}$$

and is a scalar.

Any antisymmetric 2-index tensor is *trace-free*, i.e., its trace vanishes.

Example 4.15 The trace of the velocity gradient tensor $v_{ij} \equiv \partial v_i/\partial x^j$ is the divergence $\nabla \cdot \mathbf{v} = \sum_{\mathbf{i}=\mathbf{1}}^{\mathbf{3}} \partial \mathbf{v_i}/\partial \mathbf{x^i}$ of the vector $\mathbf{v}$. The shear tensor (4.5) is trace-free by construction, $\sigma^i{}_i = 0$, while the vorticity tensor (4.6) is trace-free because it is antisymmetric.

Example 4.16 The trace of the Kronecker delta δ^{μ}_{ν} is equal to the dimension n of the space:

$$\delta^{\mu}_{\mu} = \sum_{\mu=1}^{n} \delta^{\mu}_{\mu} = \sum_{\mu=1}^{n} 1 = n. \tag{4.53}$$

Definition 4.12 The *sum of two tensors of the same type* $A^{\mu_1 \ldots \mu_k}{}_{\nu_1 \ldots \nu_l}$ and $B^{\mu_1 \ldots \mu_k}{}_{\nu_1 \ldots \nu_l}$ is the tensor of the same type

$$S^{\mu_1 \ldots \mu_k}{}_{\nu_1 \ldots \nu_l} \equiv A^{\mu_1 \ldots \mu_k}{}_{\nu_1 \ldots \nu_l} + B^{\mu_1 \ldots \mu_k}{}_{\nu_1 \ldots \nu_l} \tag{4.54}$$

(i.e., the sum of tensors is defined by adding the components).

The sum of two tensors transforms in the same way as the two tensors A and B:

$$S^{\mu_{1'} \dots \mu_{k'}}{}_{\nu_{1'} \dots \nu_{l'}} = \frac{\partial x^{\mu_{1'}}}{\partial x^{\rho_1}} \dots \frac{\partial x^{\mu_{k'}}}{\partial x^{\rho_k}} \frac{\partial x^{\sigma_1}}{\partial x^{\nu_{1'}}} \dots \frac{\partial x^{\sigma_l}}{\partial x^{\nu_{l'}}} A^{\rho_1 \dots \rho_k}{}_{\sigma_1 \dots \sigma_l}$$

$$+ \frac{\partial x^{\mu_{1'}}}{\partial x^{\rho_1}} \dots \frac{\partial x^{\mu_{k'}}}{\partial x^{\rho_k}} \frac{\partial x^{\sigma_1}}{\partial x^{\nu_{1'}}} \dots \frac{\partial x^{\sigma_l}}{\partial x^{\nu_{l'}}} B^{\rho_1 \dots \rho_k}{}_{\sigma_1 \dots \sigma_l}$$

$$= \frac{\partial x^{\mu_{1'}}}{\partial x^{\rho_1}} \dots \frac{\partial x^{\mu_{k'}}}{\partial x^{\rho_k}} \frac{\partial x^{\sigma_1}}{\partial x^{\nu_{1'}}} \dots \frac{\partial x^{\sigma_l}}{\partial x^{\nu_{l'}}} \left(A^{\rho_1 \dots \rho_k}{}_{\sigma_1 \dots \sigma_l} + B^{\rho_1 \dots \rho_k}{}_{\sigma_1 \dots \sigma_l} \right)$$

$$\equiv \frac{\partial x^{\mu_{1'}}}{\partial x^{\rho_1}} \dots \frac{\partial x^{\mu_{K'}}}{\partial x^{\rho_k}} \frac{\partial x^{\sigma_1}}{\partial x^{\nu_{1'}}} \dots \frac{\partial x^{\sigma_l}}{\partial x^{\nu_{l'}}} S^{\rho_1 \dots \rho_k}{}_{\sigma_1 \dots \sigma_l} .$$

In the same way, one defines the *subtraction of two tensors of the same type* as

$$A^{\mu_1 \dots \mu_k}{}_{\nu_1 \dots \nu_l} - B^{\mu_1 \dots \mu_k}{}_{\nu_1 \dots \nu_l} \tag{4.55}$$

and the *multiplication of a tensor by a scalar λ*

$$\lambda\, C^{\mu_1 \dots \mu_k}{}_{\nu_1 \dots \nu_l} . \tag{4.56}$$

Taking the inverse of the components of a tensor does not create a tensor, in the same way that the writing $1/\mathbf{v}$ is not defined for a vector $\mathbf{v}$ in the Euclidean space $\mathbb{R}^3$ and $1/v^i$ are not the components of a vector.

4.6 Tensor Fields

A vector or a tensor are constant quantities: they are defined at a point $P \in \mathbb{R}^n$ and are analogous to constant functions. However, one can let the point P vary spanning a region of $\mathbb{R}^n$, thus obtaining *vector fields* or *tensor fields*, in the same way that by varying the position $\mathbf{x}$ of the point P one obtains a function $f(\mathbf{x})$. Tensor fields are tensors whose values depend on the position, $T^{\alpha_1 \dots \alpha_l}{}_{\beta_1 \dots \beta_m} = T^{\alpha_1 \dots \alpha_l}{}_{\beta_1 \dots \beta_m}(\mathbf{x})$.

Example 4.17 The electric and magnetic fields of Maxwell's theory $\mathbf{E}(t, \mathbf{x})$ and $\mathbf{B}(t, \mathbf{x})$ depend on both space and time, as does the velocity field of a fluid $\mathbf{v}(\mathbf{t}, \mathbf{x})$.

Definition 4.13 A *continuous tensor field* on a region $\mathscr{D} \subseteq \mathbb{R}^n$ is one such that all its components $T^{\alpha_1 \dots \alpha_k}{}_{\beta_1 \dots \beta_l}(x^\mu)$ are continuous functions of the coordinates x^μ at all points of $\mathscr{D}$.

Definition 4.14 Let $\mathscr{D}$ be an open region; then a *differentiable*, $\mathscr{C}^1, \mathscr{C}^2, \dots\dots, \mathscr{C}^k$ *tensor field* is one whose components are all differentiable, $\mathscr{C}^1, \mathscr{C}^2, \dots, \mathscr{C}^k$ on $\mathscr{D}$, respectively.

Definition 4.15 A *smooth tensor field* on $\mathscr{D}$ is one such that its components are $\mathscr{C}^\infty$ (i.e., it is differentiable with all its derivatives of all orders on $\mathscr{D}$).

One can introduce *tensor calculus* by taking derivatives, gradients, etc. of sufficiently regular tensor fields. Common notations for tensor fields include

$$\frac{\partial A_{\alpha\beta}}{\partial x^{\gamma}} \equiv A_{\alpha\beta,\gamma}, \tag{4.57}$$

$$\frac{\partial T_{\alpha\beta}{}^{\gamma}}{\partial x^{\delta}} \equiv T_{\alpha\beta}{}^{\gamma}{}_{,\delta}. \tag{4.58}$$

In general, however,

the partial derivative of a tensor is not a tensor.

In fact, consider how the partial derivative $\partial/\partial x^{\gamma}$ of a tensor $T^{\alpha_1 \ldots \alpha_l}{}_{\beta_1 \ldots \beta_m}$ transforms:

$$\begin{aligned}
\frac{\partial T^{\alpha_{1'} \ldots \alpha_{l'}}{}_{\beta_{1'} \ldots \beta_{m'}}}{\partial x^{\gamma'}} &= \frac{\partial x^{\delta}}{\partial x^{\gamma'}} \frac{\partial}{\partial x^{\delta}} \left[\frac{\partial x^{\alpha_{1'}}}{\partial x^{\mu_1}} \cdots \frac{\partial x^{\alpha_{l'}}}{\partial x^{\mu_l}} \frac{\partial x^{\nu_1}}{\partial x^{\beta_1'}} \cdots \frac{\partial x^{\nu_m}}{\partial x^{\beta_m'}} T^{\mu_1 \ldots \mu_l}{}_{\nu_1 \ldots \nu_m} \right] \\
&= \frac{\partial x^{\delta}}{\partial x^{\gamma'}} \left[\frac{\partial}{\partial x^{\delta}} \left(\frac{\partial x^{\alpha_{1'}}}{\partial x^{\mu_1}} \cdots \frac{\partial x^{\alpha_l'}}{\partial x^{\mu_l}} \frac{\partial x^{\nu_1}}{\partial x^{\beta_1'}} \cdots \frac{\partial x^{\nu_m}}{\partial x^{\beta_m'}} \right) \right] T^{\mu_1 \ldots \mu_l}{}_{\nu_1 \ldots \nu_m} \\
&\quad + \frac{\partial x^{\delta}}{\partial x^{\gamma'}} \frac{\partial x^{\alpha_{1'}}}{\partial x^{\mu_1}} \cdots \frac{\partial x^{\alpha_l'}}{\partial x^{\mu_l}} \frac{\partial x^{\nu_1}}{\partial x^{\beta_{1'}}} \cdots \frac{\partial x^{\nu_m}}{\partial x^{\beta_{m'}}} \frac{\partial T^{\mu_1 \ldots \mu_l}{}_{\nu_1 \ldots \nu_m}}{\partial x^{\delta}};
\end{aligned} \tag{4.59}$$

because of the second last line, the tensor transformation property is spoiled. It is preserved only by *linear* transformations for which $\frac{\partial^2 x^{\alpha'}}{\partial x^{\beta} \partial x^{\delta}}$ and $\frac{\partial^2 x^{\alpha}}{\partial x^{\beta'} \partial x^{\delta'}}$ vanish.

Higher order partial derivatives of a tensor also do not transform as tensors (except under linear coordinate transformations). Therefore, we will need to define a new notion of derivative ("covariant derivative" ∇_{μ}) such that, in this new sense, "derivatives" of tensors are tensors. This definition is given in Chap. 10.

Consider a differentiable curve in $\mathbb{R}^n$ (for example the worldline of a particle in Minkowski spacetime) given by a parametric representation

$$x^{\mu} = x^{\mu}(\lambda)$$

with parameter λ (a scalar, invariant under coordinate transformations). Then the tangent $T^{\mu} \equiv dx^{\mu}/d\lambda$ to the curve is a contravariant vector under arbitrary coordinate transformations. In fact, under the change $x^{\alpha} \longrightarrow x^{\alpha'}$, it is

$$T^{\mu'} \equiv \frac{dx^{\mu'}}{d\lambda} = \frac{\partial x^{\mu'}}{\partial x^{\alpha}} \frac{dx^{\alpha}}{d\lambda} \equiv \frac{\partial x^{\mu'}}{\partial x^{\alpha}} T^{\alpha}. \tag{4.60}$$

The basic vectors of special-relativistic mechanics that we will use in the next few chapters, 4-velocity, 4-acceleration, 4-momentum and 4-force, are constructed using only differentiation with respect to a parameter and, possibly, multiplication by the particle mass, which is another scalar:

$$u^\mu \equiv \frac{dx^\mu}{d\lambda}, \quad a^\mu \equiv \frac{du^\mu}{d\lambda} = \frac{d^2x^\mu}{d\lambda^2}, \quad p^\mu = mu^\mu, \quad f^\mu = ma^\mu$$

are true vectors. However, the total derivative

$$\frac{dT^{\mu_1 \dots \mu_l}{}_{\nu_1 \dots \nu_m}}{d\lambda} = \frac{\partial T^{\mu_1 \dots \mu_l}{}_{\nu_1 \dots \nu_m}}{\partial x^\alpha} \frac{dx^\alpha}{d\lambda} \tag{4.61}$$

of a tensor $T^{\mu_1 \dots \mu_l}{}_{\nu_1 \dots \nu_m}$ with respect to the scalar parameter λ is a tensor with respect to *linear* transformations only.

Example 4.18 In fluid mechanics in $\mathbb{R}^3$ the velocity of a fluid particle $\mathbf{v}\,(\mathrm{t}, \mathbf{x})$ is a vector field. The tensor with components

$$v_{ij} \equiv \frac{\partial v_i}{\partial x^j} \tag{4.62}$$

in Cartesian coordinates $\left\{x^i\right\}$ is a 2-tensor and is decomposed into a symmetric part θ_{ij} and an antisymmetric part ω_{ij} as

$$v_{ij} = \frac{v_{ij} + v_{ji}}{2} + \frac{v_{ij} - v_{ji}}{2} \equiv \theta_{ij} + \omega_{ij} \tag{4.63}$$

with symmetric part

$$\theta_{ij} \equiv \frac{v_{ij} + v_{ji}}{2}$$

(*expansion tensor*) and antisymmetric part

$$\omega_{ij} \equiv \frac{v_{ij} - v_{ji}}{2}$$

(*vorticity tensor*). The symmetric part of θ_{ij} is further decomposed into a trace-free part and a pure trace part as

$$\theta_{ij} \equiv \left(\theta_{ij} - \frac{\theta}{3}\delta_{ij}\right) + \frac{\theta}{3}\delta_{ij} \equiv \sigma_{ij} + \frac{\theta}{3}\delta_{ij}, \tag{4.64}$$

where $\theta \equiv \theta^i{}_i = \partial v_i/\partial x^i = \nabla \cdot \mathbf{v}$ is the trace of θ_{ij} and

$$\sigma_{ij} \equiv \theta_{ij} - \frac{\theta}{3}\delta_{ij} = \frac{\partial v_i}{\partial x^j} - \frac{1}{3}(\nabla \cdot \mathbf{v})\,\delta_{ij}.$$

The shear tensor σ_{ij} is symmetric and trace-free,

$$\sigma^i{}_i = \theta^i{}_i - \frac{1}{3}\delta^i_i\,\theta = \theta - \frac{3}{3}\theta = 0.$$

The action of the tensors σ_{ij}, $\theta\,\delta_{ij}/3$, and ω_{ij} on a sphere of particles in the fluid describes the deformations occurring during the motion in an infinitesimal time interval δt. Consider a sphere of fluid particles at time t (Fig. 4.1); then at time $t+\delta t$ the expansion $\frac{\theta}{3}\,\delta_{ij}$ will have increased its volume without deformation or rotation. If only the shear σ_{ij} is non-zero, it will have deformed the sphere into an ellipsoid without changing its volume or rotating it; if only the vorticity ω_{ij} is different from zero it will have rotated the sphere of particles without changing its shape or its volume.

4.7 *Index-Free Description of Tensors

For the reader who is not satisfied with the previous definition of tensor based on coordinate transformations, we provide in this section a more abstract characterization of covariant vector fields, which can be generalized to define tensors of any type. However, we will not make use of this new definition and of this geometric approach in this book.[1]

One can think of vectors and tensors as geometric objects defined in a way independent of their components and without referring to these. Think of a vector field

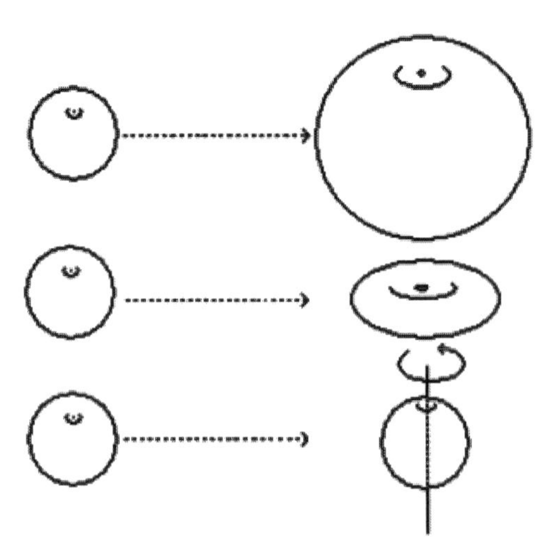

Fig. 4.1 The action of the expansion, shear, and vorticity tensors on a sphere of fluid particles

[1] For a more comprehensive discussion see Refs. [1–4].

V on a region $\mathscr{D} \subseteq \mathbb{R}^n$ as an *operator acting on functions* defined on that region. For each such function f, the operator produces another function

$$\mathsf{V} : f \longrightarrow \mathsf{V}(f) = g \quad \text{(a function on } \mathscr{D}\text{)}$$

with V *linear*,

$$\mathsf{V}(\alpha f_1 + \beta f_2) = \alpha \mathsf{V}(f_1) + \beta \mathsf{V}(f_2) \quad \forall\, \alpha, \beta \in \mathbb{R} \;\; \forall f_1, f_2 \text{ on } \mathscr{D}.$$

Clearly V is a *geometric map* defined independently of the coordinates. Let us relate it now to coordinates. Let $\{x^\mu\}$ be a coordinate system on $\mathscr{D}$ and consider the operators (vectors)

$$\partial_\alpha \equiv \frac{\partial}{\partial \mathsf{x}^\alpha} \tag{4.65}$$

acting on functions regular on $\mathscr{D}$; then define the *components* V^α of V by

$$\mathsf{V} \equiv V^\alpha \partial_\alpha .$$

With this definition, for any smooth function f on $\mathscr{D}$, the vector V is

$$\mathsf{V}(f) = V^\alpha \partial_\alpha f = V^\alpha \frac{\partial f}{\partial x^\alpha},$$

where V^α are the components of the vector V with respect to the basis (4.65) (*coordinate basis*). In another coordinate system $\left\{x^{\mu'}\right\}$ it is

$$\mathsf{V} = V^{\alpha'} \partial_{\alpha'} = V^{\alpha'} \frac{\partial}{\partial \mathsf{x}^{\alpha'}}$$

Now, for the vector V to remain an object defined independently of the coordinates, the components V^α must transform according to $V^{\alpha'} = \dfrac{\partial x^{\alpha'}}{\partial x^\gamma} V^\gamma$; then it is

$$\begin{aligned} V^{\alpha'} \partial_{\alpha'} &= V^{\alpha'} \left(\frac{\partial}{\partial \mathsf{x}^\beta} \right) \frac{\partial x^\beta}{\partial x^{\alpha'}} \\ &= \frac{\partial x^{\alpha'}}{\partial x^\gamma} V^\gamma \frac{\partial x^\beta}{\partial x^{\alpha'}} \frac{\partial}{\partial \mathsf{x}^\beta} = \frac{\partial x^{\alpha'}}{\partial x^\gamma} \frac{\partial x^\beta}{\partial x^{\alpha'}} V^\gamma \partial_\beta \\ &= \frac{\partial x^\beta}{\partial x^\gamma} V^\gamma \partial_\beta = \delta^\beta_\gamma V^\gamma \partial_\beta = V^\beta \partial_\beta , \end{aligned}$$

so

$$V^{\alpha'} \partial_{\alpha'} = V^\alpha \partial_\alpha ,$$

which makes it clear that the vector V is defined in a way independent of the coordinate system although its components with respect to a coordinate basis are not. By considering as operations on vectors the usual multiplication by a scalar and addition of two vectors, the vectors V form a vector space of operators with dimension n (*tangent space*) denoted by T or $\tau\begin{pmatrix}1\\0\end{pmatrix}$, a basis of which (*coordinate basis*) is given by the vectors

$$\left\{\frac{\partial}{\partial \mathsf{x}^1}, \frac{\partial}{\partial \mathsf{x}^2}, \dots, \frac{\partial}{\partial \mathsf{x}^n}\right\} \equiv \left\{\partial_\alpha\right\}_{\alpha=1,\dots,n}.$$

Let us consider now the set of linear maps $\omega : T \longrightarrow \mathbb{R}$ which associate real numbers to vectors of the tangent space. These maps can be added and multiplied by a scalar and, with these two operations, the set of all such linear maps forms a vector space, called *dual space* of T or *cotangent space*, denoted by T^* or $\tau\begin{pmatrix}0\\1\end{pmatrix}$. The operations of sum and multiplication by a scalar are defined, as usual, by

$$\begin{aligned}(\omega_1 + \omega_2)(\mathsf{V}) &\equiv \omega_1(\mathsf{V}) + \omega_2(\mathsf{V}),\\ (\lambda\omega)(\mathsf{V}) &\equiv \lambda\,\omega(\mathsf{V}).\end{aligned}$$

The elements of the cotangent space are called *dual vectors* or *1-forms*.[2] The basis of the tangent space $\left\{\frac{\partial}{\partial \mathsf{x}^1}, \frac{\partial}{\partial \mathsf{x}^2}, \dots, \frac{\partial}{\partial \mathsf{x}^n}\right\}$ induces a basis

$$\left\{\omega^{(1)}, \omega^{(2)}, \dots, \omega^{(n)}\right\}$$

of T^* as follows. For any vector $\mathsf{V} \in T$ and for any dual vector $\omega \in T^*$, it is

$$\omega(\mathsf{V}) = \omega\left(V^\alpha \frac{\partial}{\partial \mathsf{x}^\alpha}\right) = V^\alpha\,\omega\left(\frac{\partial}{\partial \mathsf{x}^\alpha}\right).$$

By denoting

$$\omega\left(\frac{\partial}{\partial \mathsf{x}^\alpha}\right) \equiv \omega_\alpha,$$

we have

$$\omega(\mathsf{V}) = V^\alpha \omega_\alpha$$

and

$$\omega\left(\frac{\partial}{\partial \mathsf{x}^\mu}\right) = \delta^\alpha_\mu\,\omega_\alpha = \omega_\mu.$$

[2] If vectors are visualized graphically as arrows, then dual vectors can be visualized as hyperplanes (in $\mathbb{R}^2$, straight lines) perpendicular to these vectors [1, 5].

By defining the 1-forms $\omega^{(\rho)}$ so that

$$\omega^{(\rho)}\left(\frac{\partial}{\partial \mathsf{x}^{\sigma}}\right) \equiv \delta^{\rho}_{\sigma}, \tag{4.66}$$

we have, for any 1-form $\omega = \omega_{\alpha}\, \omega^{(\alpha)}$,

$$\omega\left(\frac{\partial}{\partial \mathsf{x}^{\sigma}}\right) = \omega_{\sigma}, \tag{4.67}$$

so ω_{α} are the components of the 1-form ω on the basis $\{\omega^{(1)}, \omega^{(2)}, \ldots, \omega^{(n)}\}$. A widely used notation for the dual basis of the cotangent space is $\omega^{(\mu)} = \mathsf{dx}^{\mu}$, which emphasizes the duality between vectors and 1-forms.

It would be equivalent to study three-dimensional physics using vectors $\mathsf{V} \in T$ or dual vectors $\omega \in T^*$.

Definition 4.16 A *tensor* of type $\binom{l}{m}$ is an operator T acting on the l tensor products of T with itself times m tensor products of T^* with itself and producing a real number,

$$\mathsf{T} : \underbrace{T \times \ldots \times T}_{l \text{ times}} \times \underbrace{T^* \times \ldots \times T^*}_{m \text{ times}} \longrightarrow \mathbb{R},$$

which is linear in all its arguments.

The tensors of type $\binom{l}{m}$ form a vector space denoted by $\tau\binom{l}{m}$ with the usual operations of addition and multiplication by a scalar. A basis of this space is obtained by taking l outer products of the basis vectors of T and m outer products of the dual vectors forming the induced basis of T^*. Such a tensor T has components $T^{\mu_1 \ldots \mu_l}{}_{\nu_1 \ldots \nu_m}$ which satisfy

$$\mathsf{T} = T^{\mu_1 \ldots \mu_l}{}_{\nu_1 \ldots \nu_m} \frac{\partial}{\partial \mathsf{x}^{\mu_1}} \cdots \frac{\partial}{\partial \mathsf{x}^{\mu_l}}\, \omega^{(\nu_1)} \ldots \omega^{(\nu_m)}. \tag{4.68}$$

A coordinate change $x^{\mu} \longrightarrow x^{\mu'}$ induces the change of bases in T and T^*

$$\frac{\partial}{\partial \mathsf{x}^{\alpha'}} \longrightarrow \frac{\partial x^{\beta}}{\partial x^{\alpha'}} \frac{\partial}{\partial \mathsf{x}^{\beta}},$$

$$\omega^{(\mu)} \longrightarrow \omega^{(\mu')} = \frac{\partial x^{\mu'}}{\partial x^{\mu}}\, \omega^{(\mu)},$$

or

$$\mathsf{dx}^{\mu} \longrightarrow \mathsf{dx}^{\mu'} = \frac{\partial x^{\mu'}}{\partial x^{\mu}}\, \mathsf{dx}^{\mu},$$

while the respective tensor components must transform in the opposite way to keep the geometric object T invariant. Since

$$\mathsf{T} = T^{\mu'_1 \dots \mu'_l}{}_{\nu'_1 \dots \nu'_m} \frac{\partial}{\partial \mathsf{x}^{\mu'_1}} \cdots \frac{\partial}{\partial \mathsf{x}^{\mu'_l}} \omega^{(\nu'_1)} \dots \omega^{(\nu'_m)},$$

the equality of T' and T is guaranteed if the components transform according to Eq. (4.38); then

$$\mathsf{T} = T^{\mu_1 \dots \mu_l}{}_{\nu_1 \dots \nu_m} \frac{\partial}{\partial \mathsf{x}^{\mu_1}} \cdots \frac{\partial}{\partial \mathsf{x}^{\mu_l}} \omega^{(\nu_1)} \dots \omega^{(\nu_m)}.$$

It is clear that the definition of vector, 1-form, and tensor are given in a coordinate-independent way and that one needs to use coordinates only to specify the components with respect to assigned bases of T and T^*. In practice, however, many calculations are performed more conveniently in coordinates than in abstract coordinate-free notation.

One can define the tangent space at any point of the n-dimensional space and, by varying the point, a vector in the tangent space becomes a *vector field* and tensors depending on the position become *tensor fields*.

Let X and Y be vector fields. The *commutator* (or *Lie bracket*) of X and Y is the vector field

$$[\mathsf{X}, \mathsf{Y}] \equiv \mathsf{XY} - \mathsf{YX}, \tag{4.69}$$

which is defined as follows. For any smooth test function f,

$$[\mathsf{X}, \mathsf{Y}] f \equiv \mathsf{X}(\mathsf{Y}(f)) - \mathsf{Y}(\mathsf{X}(f))$$

$$= \mathsf{X}\left(Y^\alpha \partial_\alpha f\right) - \mathsf{Y}\left(X^\alpha \partial_\alpha f\right)$$

$$= X^\beta \partial_\beta \left(Y^\alpha \partial_\alpha f\right) - Y^\beta \partial_\beta \left(X^\alpha \partial_\alpha f\right).$$

Now consider, e.g., the term $X^\beta Y^\alpha \partial_\beta \partial_\alpha f$: since f is a smooth function,

$$\partial_\alpha \partial_\beta f = \partial_\beta \partial_\alpha f = \partial_{(\alpha} \partial_{\beta)} f$$

and

$$X^\beta Y^\alpha \partial_\beta \partial_\alpha f = \left(X^{(\beta} Y^{\alpha)} + X^{[\beta} Y^{\alpha]}\right) \partial_{(\beta} \partial_{\alpha)} f$$

$$= X^{(\beta} Y^{\alpha)} \partial_{(\beta} \partial_{\alpha)} f$$

because $X^{[\beta} Y^{\alpha]} \partial_{(\beta} \partial_{\alpha)} f = 0$ (the contraction of an antisymmetric tensor with a symmetric one is always zero) and

$$[\mathsf{X}, \mathsf{Y}] f = \left(X^\beta \partial_\beta Y^\alpha\right)(\partial_\alpha f) + X^\beta Y^\alpha \left(\partial_\beta \partial_\alpha f\right)$$

$$- \left(Y^\beta \partial_\beta X^\alpha\right)(\partial_\alpha f) - Y^\beta X^\alpha \left(\partial_\beta \partial_\alpha f\right)$$

Now, it is $X^\beta Y^\alpha \partial_\beta \partial_\alpha f = X^\alpha Y^\beta \partial_\beta \partial_\alpha f$ and we are left with

$$[\mathsf{X}, \mathsf{Y}] f = \left(X^\beta \partial_\beta Y^\alpha - Y^\beta \partial_\beta X^\alpha\right)(\partial_\alpha f)$$

and

$$[\mathsf{X}, \mathsf{Y}] = X^\beta \partial_\beta Y^\alpha - Y^\beta \partial_\beta X^\alpha .$$

The commutator satisfies the properties (the proof of which is left as an exercise)

$$[\mathsf{X}, \mathsf{X}] = 0, \tag{4.70}$$

$$[\mathsf{Y}, \mathsf{X}] = -[\mathsf{X}, \mathsf{Y}], \tag{4.71}$$

$$\Big[\mathsf{X}, [\mathsf{Y}, \mathsf{Z}]\Big] + \Big[\mathsf{Y}, [\mathsf{Z}, \mathsf{X}]\Big] + \Big[\mathsf{Z}, [\mathsf{X}, \mathsf{Y}]\Big] = 0 \quad \text{(Jacobi identity)}. \tag{4.72}$$

If X and Y are two vector fields of a coordinate basis $\left\{\frac{\partial}{\partial \mathsf{x}^1}, \ldots, \frac{\partial}{\partial \mathsf{x}^n}\right\}$, they always commute:

$$\left[\frac{\partial}{\partial \mathsf{x}^i}, \frac{\partial}{\partial \mathsf{x}^j}\right] = 0.$$

This property follows from the fact that for any smooth test function f it is $\frac{\partial^2 f}{\partial x^i \partial x^j} = \frac{\partial^2 f}{\partial x^j \partial x^i}$. Conversely, given a collection $\mathsf{X}_1, \ldots, \mathsf{X}_n$ of non-vanishing vector fields which commute with each other and are linearly independent, one can always find a coordinate system $\{x^\mu\}$ for which they are coordinate basis vector fields. However, *not all bases* $\{\mathbf{e}_{(i)}\}_{i=1,\ldots,n}$ *of* $\mathbb{R}^n$ *are coordinate bases*, i.e., not always there is a coordinate system such that

$$e^{(1)}_\mu = \frac{\partial}{\partial \mathsf{x}^\mu_{(1)}}, \quad e^{(2)}_\mu = \frac{\partial}{\partial \mathsf{x}^\mu_{(2)}}, \ldots, \quad e^{(n)}_\mu = \frac{\partial}{\partial \mathsf{x}^\mu_{(n)}}. \tag{4.73}$$

In fact, for a basis to be a coordinate basis it must be $\left[e^{(i)}_\mu, e^{(j)}_\nu\right] = 0$ for all vectors of the basis because, if $e^{(i)}_\mu = \partial_\mu$, then for any regular function it is $\frac{\partial^2 f}{\partial x^\mu \partial x^\nu} = \frac{\partial^2 f}{\partial x^\nu \partial x^\mu}$.

Example 4.19 The gradient of a scalar ϕ is the covariant vector

$$\frac{\partial \phi}{\partial x^{\mu}} \equiv \partial_{\mu}\phi$$

and the differential $\mathrm{d}\phi = \left(\partial_{\mu}\phi\right)\mathrm{d}\mathbf{x}^{\mu}$ is the contraction of the covariant vector of components $\partial_{\mu}\phi$ and of $\mathrm{d}\mathbf{x}^{\mu}$. In 3-dimensional notation we have $\mathrm{d}\phi = \boldsymbol{\nabla}\phi \cdot \mathbf{dx}$.

4.8 The Metric Tensor

A special tensor of type $\begin{pmatrix} 0 \\ 2 \end{pmatrix}$, the *metric tensor* or simply "the metric", provides us with the notions of distance between points and of norm of a vector and introduces a significant amount of structure on what is otherwise a set of points with only tensor fields defined on it.[3]

Definition 4.17 A *metric* is a tensor field $g_{\alpha\beta}$ of type $\begin{pmatrix} 0 \\ 2 \end{pmatrix}$ which is

- symmetric, $g_{\alpha\beta} = g_{\beta\alpha} \quad \forall\ \alpha, \beta$;
- non-degenerate, i.e., $g_{\alpha\beta}X^{\alpha}Y^{\beta} = 0 \quad \forall\, X^{\alpha} \Rightarrow Y^{\alpha} = 0$.

The infinitesimal *distance squared between two points* of coordinates x^{μ} and $x^{\mu} + \mathrm{d}x^{\mu}$, or *line element* (squared), is

$$\mathrm{d}s^2 = g_{\mu\nu}\mathrm{d}x^{\mu}\mathrm{d}x^{\nu}. \tag{4.74}$$

Definition 4.18 The *scalar product* between two vectors A^{μ} and B^{μ} is[4]

$$g_{\mu\nu}A^{\mu}B^{\nu}.$$

Definition 4.19 Two vectors are *orthogonal* if their scalar product is zero.

In particular, if $B^{\mu} = A^{\mu}$, one has the notion of length squared of a vector

$$g_{\mu\nu}A^{\mu}A^{\nu}.$$

Example 4.20 In $\mathbb{R}^n$ with Cartesian coordinates $\left\{x^1, \ldots, x^n\right\}$, the Euclidean metric is given by

[3] The definitions of metric and of distance between points used in relativity are not the same used in topology and in the theory of metric spaces. There, the distance between two points is positive-definite while here, in view of the application to Special Relativity it is, in general, indefinite. In index-free notation, the scalar product of two vectors X and Y is denoted with $\mathsf{g}(\mathsf{X}, \mathsf{Y})$.

[4] Again, it is not required that the scalar product of a vector with itself be non-negative.

$$e_{\mu\nu} = \delta_{\mu\nu}.$$

The distance squared between two infinitesimally close points x^{μ} and $x^{\mu} + \mathrm{d}x^{\mu}$ is

$$\mathrm{d}l_{(n)}^{2} = \delta_{\mu\nu}\,\mathrm{d}x^{\mu}\mathrm{d}x^{\nu} = \left(\mathrm{d}x^{1}\right)^{2} + \left(\mathrm{d}x^{2}\right)^{2} + \ldots + \left(\mathrm{d}x^{n}\right)^{2}, \tag{4.75}$$

the Pythagorean theorem in n dimensions. The Euclidean metric is positive-definite since $\mathrm{d}l_{(3)}^{2} \geq 0$ and $\mathrm{d}l_{(3)} = 0$ if and only if $\left(\mathrm{d}x^{1}, \ldots, \mathrm{d}x^{n}\right) = (0, \ldots, 0)$. The scalar product between two vectors defined by $e_{\mu\nu}$ is the usual dot product. In Cartesian coordinates, we have

$$e_{\mu\nu}A^{\mu}B^{\nu} = \delta_{\mu\nu}A^{\mu}B^{\nu} = A^{1}B^{1} + A^{2}B^{2} + \ldots + A^{n}B^{n} = \mathbf{A}\cdot\mathbf{B}$$

and the length squared of a vector $\mathbf{A} = \left(A^{1}, \ldots, A^{n}\right)$ is

$$\|A\|^{2} = e_{\mu\nu}A^{\mu}A^{\nu} = \sum_{\mu=1}^{n}\left(A^{\mu}\right)^{2} = \mathbf{A}\cdot\mathbf{A} \tag{4.76}$$

and is non-negative.

4.8.1 Inverse Metric

Let the metric tensor $g_{\mu\nu}$ be represented by a matrix of elements $g_{\mu\nu}$ in some coordinate system. Since the metric is non-degenerate, the inverse matrix exists. Denote its elements by $g^{\mu\nu}$; since $g_{\mu\nu}$ is symmetric, so is $g^{\mu\nu}$. The tensor represented by the matrix $(g^{\mu\nu})$ is called the *inverse metric* of $g_{\mu\nu}$ and by definition satisfies the relation

$$g_{\mu\nu}\,g^{\nu\alpha} = \delta_{\mu}^{\alpha}. \tag{4.77}$$

This tensor equation is valid in any coordinate system: in fact, if $x^{\mu} \longrightarrow x^{\mu'}$ is a coordinate transformation, we have

$$
\begin{aligned}
g_{\mu'\nu'} g^{\nu'\alpha'} &= \frac{\partial x^{\rho}}{\partial x^{\mu'}} \frac{\partial x^{\sigma}}{\partial x^{\nu'}} g_{\rho\sigma} \frac{\partial x^{\nu'}}{\partial x^{\gamma}} \frac{\partial x^{\alpha'}}{\partial x^{\delta}} g^{\gamma\delta} \\
&= \frac{\partial x^{\rho}}{\partial x^{\mu'}} \underbrace{\left(\frac{\partial x^{\sigma}}{\partial x^{\nu'}} \frac{\partial x^{\nu'}}{\partial x^{\gamma}} \right)}_{\delta^{\sigma}_{\gamma}} \frac{\partial x^{\alpha'}}{\partial x^{\delta}} g_{\rho\sigma} \, g^{\gamma\delta} \\
&= \frac{\partial x^{\rho}}{\partial x^{\mu'}} \frac{\partial x^{\alpha'}}{\partial x^{\delta}} \underbrace{g_{\rho\gamma} \, g^{\gamma\delta}}_{\delta^{\delta}_{\rho}} \\
&= \frac{\partial x^{\rho}}{\partial x^{\mu'}} \frac{\partial x^{\alpha'}}{\partial x^{\rho}} = \delta^{\alpha'}_{\mu'}.
\end{aligned}
$$

The equation $g_{\mu'\nu'} g^{\nu'\alpha'} = \delta^{\alpha'}_{\mu'}$ in the coordinate system $\left\{ x^{\mu'} \right\}$ defines uniquely the inverse of the matrix $\left(g_{\mu'\nu'} \right)$.

The metric tensor allows one to define the algebraic tensor operations consisting of *raising* and *lowering* tensor indices.

Definition 4.20 Let A^{μ} be a contravariant vector; we define

$$
A_{\mu} \equiv g_{\mu\nu} A^{\nu} \quad (\text{"lowering" of } \mu). \tag{4.78}
$$

If B_{μ} is a covariant vector we define

$$
B^{\mu} \equiv g^{\mu\nu} B_{\nu} \quad (\text{"raising" of } \mu). \tag{4.79}
$$

The raising of a lowered index and the lowering of a raised index give back the original object:

$$
A^{\mu} = g^{\mu\nu} A_{\nu} = g^{\mu\nu} \left(g_{\nu\alpha} A^{\alpha} \right) = \delta^{\mu}_{\alpha} \, A^{\alpha} = A^{\mu},
$$

$$
B_{\mu} = g_{\mu\nu} B^{\nu} = g_{\mu\nu} \left(g^{\nu\beta} B_{\beta} \right) = \delta^{\beta}_{\mu} \, B_{\beta} = B_{\mu}.
$$

The operations of raising or lowering apply to any tensor index, for example

$$T^{\alpha}{}_{\beta\gamma} = g^{\alpha\mu} T_{\mu\beta\gamma}, \qquad Q^{\mu\nu} = g^{\mu\alpha} g^{\nu\beta} Q_{\alpha\beta}.$$

A^{μ} and A_{μ} are just different descriptions of the same object because it is always possible to obtain one from the other by lowering or raising indices with the metric $g_{\mu\nu}$ or the inverse metric $g^{\mu\nu}$. In particular, Eq. (4.77) expresses the fact that the Kronecker delta is the mixed form of the metric tensor $g_{\mu\nu}$.

The scalar product of two vectors A^{μ} and B^{μ} can now be written as[5]

$$g_{\mu\nu} A^{\mu} B^{\nu} = g^{\mu\nu} A_{\mu} B_{\nu} = A_{\mu} B^{\mu} = A^{\mu} B_{\mu} \tag{4.80}$$

and, for example, $A_{\mu\nu} B^{\mu\nu} = g_{\mu\alpha} g_{\nu\beta} A^{\alpha\beta} B^{\mu\nu}$.

The symmetries of a tensor are conserved by raising/lowering its indices:

$$T_{\nu\mu} = \pm T_{\mu\nu}$$

if and only if

$$T^{\nu\mu} = \pm T^{\mu\nu}$$

(the proof is left as an exercise).

Example 4.21 Consider, in three dimensions, a deformable medium in a state of rest and let $x^i = \left(x^1, x^2, x^3\right)$ be the position of a generic point P in it. If the body is subject to stresses, it deforms so that the point P moves to $x^i + \mathrm{d}x^i$. Let $s_i\left(x^1, x^2, x^3\right)$ be the displacement vector of the point P. At a nearby point Q of position

$$\left(x^1 + \mathrm{d}x^1, x^2 + \mathrm{d}x^2, x^3 + \mathrm{d}x^3\right)$$

at rest, the displacement is

$$s_i\left(x^1, x^2, x^3\right) + \sum_{j=1}^{3} \frac{\partial s_i}{\partial x^j} \mathrm{d}x^j + \mathrm{O}(2). \tag{4.81}$$

The symmetric part of the tensor $\partial s_i / \partial x^j$, i.e.,

$$\frac{1}{2}\left(\frac{\partial s_i}{\partial x^j} + \frac{\partial s_j}{\partial x^i}\right), \tag{4.82}$$

is important in continuum mechanics. For the point Q,

$$\mathbf{dx}' = \mathbf{dx} + \frac{\partial \mathbf{s}}{\partial \mathbf{x}} \cdot \mathbf{dx} = \left(\hat{I}_d + \frac{\partial \mathbf{s}}{\partial \mathbf{x}}\right) \cdot \mathbf{dx},$$

[5] The rule $A_{\mu} B^{\mu} = A^{\mu} B_{\mu}$ is known as the *see-saw rule*.

or, in components,

$$\mathrm{d}x^{i'} = \mathrm{d}x^i + \sum_j \frac{\partial s_i}{\partial x^j}\,\mathrm{d}x^j = \sum_j \left(\delta_{ij} + \frac{\partial s_i}{\partial x^j}\right)\mathrm{d}x^j, \tag{4.83}$$

where $\left(\frac{\partial s_i}{\partial x^j}\right)$ is the displacement gradient and $\left(\delta_{ij} + \frac{\partial s_i}{\partial x^j}\right)$ is the deformation gradient. We have, for infinitesimal distances within the medium,

$$\frac{(\mathrm{d}l')^2 - \mathrm{d}l^2}{\mathrm{d}l^2} = \frac{(\mathrm{d}l')^2}{\mathrm{d}l^2} - 1 \neq 0 \tag{4.84}$$

(the deformation does not preserve distances between points, which makes it clear that we are discussing a non-rigid body). The deformation is described by

$$(\mathrm{d}l')^2 = \left[\left(\hat{I}_d + \frac{\partial \mathbf{s}}{\partial \mathbf{x}}\right)\cdot \mathrm{d}\mathbf{x}\right]^T \left[\left(\hat{I}_d + \frac{\partial \mathbf{s}}{\partial \mathbf{x}}\right)\cdot \mathrm{d}\mathbf{x}\right]$$

or, in components,

$$\begin{aligned}
g_{ij}\,\mathrm{d}x^{i'}\mathrm{d}x^{j'} &= g_{ij}\left(\delta^i_l + \frac{\partial s^i}{\partial x^l}\right)\mathrm{d}x^l \left(\delta^j_m + \frac{\partial s^j}{\partial x^m}\right)\mathrm{d}x^m \\
&= g_{ij}\left(\delta^i_l + \frac{\partial s^i}{\partial x^l}\right)\left(\delta^j_m + \frac{\partial s^j}{\partial x^m}\right)\mathrm{d}x^l\mathrm{d}x^m \\
&= g_{ij}\left(\delta^i_l\delta^j_m + \delta^i_l\frac{\partial s^j}{\partial x^m} + \delta^j_m\frac{\partial s^i}{\partial x^l} + \frac{\partial s^i}{\partial x^l}\frac{\partial s^j}{\partial x^m}\right)\mathrm{d}x^l\mathrm{d}x^m \\
&= \big(\text{in Cartesian coordinates in which } g_{ij} = \delta_{ij}\big) \\
&= \left(\delta_{ij}\delta^i_l\delta^j_m + \delta_{ij}\delta^i_l\frac{\partial s^j}{\partial x^m} + \delta_{ij}\delta^j_m\frac{\partial s^i}{\partial x^l} + \delta_{ij}\frac{\partial s^i}{\partial x^l}\frac{\partial s^j}{\partial x^m}\right)\mathrm{d}x^l\mathrm{d}x^m \\
&= \left(\delta_{lm} + \frac{\partial s_l}{\partial x^m} + \frac{\partial s_m}{\partial x^l} + \delta_{ij}\frac{\partial s^i}{\partial x^l}\frac{\partial s^j}{\partial x^m}\right)\mathrm{d}x^l\,\mathrm{d}x^m \\
&= (\delta_{lm} + 2L_{lm})\,\mathrm{d}x^l\mathrm{d}x^m,
\end{aligned}$$

where

$$L_{lm} = \frac{1}{2}\left(\frac{\partial s_l}{\partial x^m} + \frac{\partial s_m}{\partial x^l} + \delta_{ij}\frac{\partial s^i}{\partial x^l}\frac{\partial s^j}{\partial x^m}\right) \tag{4.85}$$

is the (symmetric) *deformation tensor*.

4.8.2 Metric Determinant

The metric tensor in $\mathbb{R}^n$ is represented by an $n \times n$ symmetric non-degenerate matrix and the *metric determinant* g is the determinant of this matrix

$$g \equiv \text{Det}\left(g_{\mu\nu}\right). \tag{4.86}$$

The requirement that the metric tensor be non-degenerate implies that the matrix associated with this tensor in any given basis is non-singular, i.e.,

$$g \neq 0.$$

This property guarantees the existence of the inverse (or contravariant) metric $g^{\mu\nu}$ represented, in any coordinate system by the inverse matrix $(g^{\mu\nu})$ of $\left(g_{\mu\nu}\right)$.

4.9 The Levi-Civita Symbol and Tensor Densities

Definition 4.21 The *Levi-Civita* (or *alternating*) *symbol* in three dimensions is defined as

$$\varepsilon_{\mu\nu\rho} \equiv \begin{cases} +1 & \text{if } \mu\nu\rho \text{ is an even permutation of } 123, \\ -1 & \text{if } \mu\nu\rho \text{ is an odd permutation of } 123, \\ 0 & \text{otherwise.} \end{cases} \tag{4.87}$$

Remember that a permutation of 123 is an ordering of the set $\{1, 2, 3\}$ obtained by starting with the natural order 123 and exchanging only two digits at a time, once or several times. An even (odd) permutation is one that requires an even (odd) number of such exchanges. It is useful to remember that there are $n!$ permutations of n numbers. The non-zero components of the Levi-Civita symbol $\varepsilon_{\mu\nu\rho}$ are, therefore,

$$\begin{aligned} &\varepsilon_{123} = +1, \ \varepsilon_{132} = -1, \ \varepsilon_{312} = +1, \\ &\varepsilon_{321} = -1, \ \varepsilon_{231} = +1, \ \varepsilon_{213} = -1. \end{aligned}$$

We are particularly interested in the case $n = 4$.

Definition 4.22 The Levi-Civita or alternating symbol in four dimensions is defined as

$$\varepsilon_{\mu\nu\rho\sigma} \equiv \begin{cases} +1 & \text{if } \mu\nu\rho\sigma \text{ is an even permutation of } 0123, \\ -1 & \text{if } \mu\nu\rho\sigma \text{ is an odd permutation of } 0123, \\ 0 & \text{otherwise}, \end{cases} \tag{4.88}$$

where a permutation of 0123 is an ordering of the set $\{0, 1, 2, 3\}$ obtained by starting with the natural order 0123 and exchanging only two digits at a time.

There are $4! = 24$ permutations of four indices, which give the 24 components[6]

$$\begin{aligned}
&\varepsilon_{0123} = +1, \ \varepsilon_{0132} = -1, \ \varepsilon_{0213} = -1, \\
&\varepsilon_{0231} = +1, \ \varepsilon_{0312} = +1, \ \varepsilon_{0321} = -1, \\
&\varepsilon_{1023} = -1, \ \varepsilon_{1032} = +1, \ \varepsilon_{1203} = +1, \\
&\varepsilon_{1230} = -1, \ \varepsilon_{1302} = -1, \ \varepsilon_{1320} = +1, \\
&\varepsilon_{2013} = +1, \ \varepsilon_{2031} = -1, \ \varepsilon_{2103} = -1, \\
&\varepsilon_{2130} = +1, \ \varepsilon_{2301} = +1, \ \varepsilon_{2310} = -1, \\
&\varepsilon_{3012} = -1, \ \varepsilon_{3021} = +1, \ \varepsilon_{3102} = +1, \\
&\varepsilon_{3120} = -1, \ \varepsilon_{3201} = -1, \ \varepsilon_{3210} = +1.
\end{aligned}$$

The Levi-Civita symbol transforms as a tensor only in $\mathbb{R}^n$ in Cartesian coordinates, but not in other coordinate systems (it is called a "symbol", not a tensor). In fact, it has the same components in any coordinate system, an unusual property. The Levi-Civita symbol is not a tensor, but a tensor density.

Definition 4.23 A *tensor density* in four dimensions[7] $\tau_{\mu_1\mu_2\mu_3\mu_4}$ is a quantity that transforms as

$$\tau_{\mu'_1\mu'_2\mu'_3\mu'_4} = \left[\text{Det}\left(\frac{\partial x^{\mu'}}{\partial x^{\mu}}\right)\right]^w \tau_{\mu_1\mu_2\mu_3\mu_4} \frac{\partial x^{\mu_1}}{\partial x^{\mu'_1}} \frac{\partial x^{\mu_2}}{\partial x^{\mu'_2}} \frac{\partial x^{\mu_3}}{\partial x^{\mu'_3}} \frac{\partial x^{\mu_4}}{\partial x^{\mu'_4}}, \tag{4.89}$$

where w is the *weight of the density*.

From a tensor density $\tau_{\mu\nu\rho\sigma}$ of weight w, one can obtain a tensor by multiplying it by the power of the absolute value of the metric determinant $|g|^{w/2}$;

$$t_{\mu\nu\rho\sigma} \equiv \left(\sqrt{|g|}\right)^w \tau_{\mu\nu\rho\sigma} \tag{4.90}$$

[6] For example, 0132 is an odd permutation and $\varepsilon_{0132} = -1$, while 0321 is obtained by $0123 \to 0213 \to 0231 \to 0321$, i.e., with 3 permutations; by an odd permutation, and therefore $\varepsilon_{0321} = -1$. And 0231 is obtained by $0123 \to 0213 \to 0231$, two permutations, therefore $\varepsilon_{0231} = +1$.

[7] The extension of this definition to n dimensions is obvious, although notationally more cumbersome.

is a tensor because $|g|$ is a scalar density of weight -2.

The Levi-Civita symbol is a tensor density of weight $+1$, *while the metric determinant* g *is a (scalar) density of weight* -2.

Proof Let M be an invertible 4×4 matrix, then

$$\varepsilon_{\mu'\nu'\rho'\sigma'}\text{Det}\,(M) = \varepsilon_{\mu\nu\rho\sigma}\, M^{\mu}_{\mu'}\, M^{\nu}_{\nu'}\, M^{\rho}_{\rho'}\, M^{\sigma}_{\sigma'}, \tag{4.91}$$

an expression valid for the determinant of any 4×4 matrix in linear algebra. In particular, we can choose M as the inverse matrix of the transformation $x^{\mu} \longrightarrow x^{\mu'}$, or $M^{\mu}_{\nu'} = \dfrac{\partial x^{\mu}}{\partial x^{\nu'}}$; then

$$\varepsilon_{\mu'\nu'\rho'\sigma'}\left|\frac{\partial x^{\alpha}}{\partial x^{\beta'}}\right| = \varepsilon_{\mu\nu\rho\sigma} M^{\mu}_{\mu'}\, M^{\nu}_{\nu'}\, M^{\rho}_{\rho'}\, M^{\sigma}_{\sigma'} \tag{4.92}$$

using the fact that $(\text{Det}\,(M))^{-1} = \left|\frac{\partial x^{\alpha}}{\partial x^{\beta'}}\right|^{-1} = \left|\frac{\partial x^{\beta'}}{\partial x^{\alpha}}\right| = \text{Det}\left(M^{-1}\right)$, we have

$$\varepsilon_{\mu'\nu'\rho'\sigma'} = \left|\frac{\partial x^{\beta'}}{\partial x^{\alpha}}\right| \frac{\partial x^{\mu}}{\partial x^{\mu'}}\frac{\partial x^{\nu}}{\partial x^{\nu'}}\frac{\partial x^{\rho}}{\partial x^{\rho'}}\frac{\partial x^{\sigma}}{\partial x^{\sigma'}}\,\varepsilon_{\mu\nu\rho\sigma}, \tag{4.93}$$

hence the levi-Civita symbol is a tensor density of weight 1.

The determinant of the metric tensor is also a tensor density; we have

$$g_{\mu'\nu'} = \frac{\partial x^{\mu}}{\partial x^{\mu'}}\frac{\partial x^{\nu}}{\partial x^{\nu'}}\, g_{\mu\nu} \tag{4.94}$$

and taking the determinant of both sides and using the fact that the determinant of a matrix product is the product of the determinants,

$$g' = \left|\frac{\partial x^{\mu}}{\partial x^{\mu'}}\right|^{2} g$$

or

$$g' = \left|\frac{\partial x^{\mu'}}{\partial x^{\mu}}\right|^{-2} g.$$

Hence, *g is a (scalar) density of weight -2.* □

Now, Eq. (4.92) tells us that the Levi-Civita symbol $\varepsilon_{\mu\nu\rho\sigma}$ transforms as a tensor density because of the factor $\left|\frac{\partial x^{\mu'}}{\partial x^{\nu}}\right|$ in the right hand side of this equation. However, if we consider the quantity

$$\tilde{\varepsilon}_{\mu\nu\rho\sigma} \equiv \sqrt{|g|}\,\varepsilon_{\mu\nu\rho\sigma}, \tag{4.95}$$

this transforms according to

$$\begin{aligned}
\tilde{\varepsilon}_{\mu'\nu'\rho'\sigma'} &= \sqrt{|g'|}\,\varepsilon_{\mu'\nu'\rho'\sigma'} \\
&= \left|\frac{\partial x^{\alpha'}}{\partial x^{\alpha}}\right|^{-1} \sqrt{|g|}\,\left|\frac{\partial x^{\alpha'}}{\partial x^{\alpha}}\right| \frac{\partial x^{\mu}}{\partial x^{\mu'}}\frac{\partial x^{\nu}}{\partial x^{\nu'}}\frac{\partial x^{\rho}}{\partial x^{\rho'}}\frac{\partial x^{\sigma}}{\partial x^{\sigma'}}\,\varepsilon_{\mu\nu\rho\sigma} \\
&= \sqrt{|g|}\,\frac{\partial x^{\mu}}{\partial x^{\mu'}}\frac{\partial x^{\nu}}{\partial x^{\nu'}}\frac{\partial x^{\rho}}{\partial x^{\rho'}}\frac{\partial x^{\sigma}}{\partial x^{\sigma'}}\,\varepsilon_{\mu\nu\rho\sigma} \\
&= \frac{\partial x^{\mu}}{\partial x^{\mu'}}\frac{\partial x^{\nu}}{\partial x^{\nu'}}\frac{\partial x^{\rho}}{\partial x^{\rho'}}\frac{\partial x^{\sigma}}{\partial x^{\sigma'}}\,\tilde{\varepsilon}_{\mu\nu\rho\sigma},
\end{aligned}$$

i.e., as a true tensor. In general, we can obtain a tensor from a tensor density by multiplying by the appropriate power of $|g|$. If $\tau_{\mu_1\mu_2\mu_3\mu_4}$ is a tensor density of weight w then

$$\tilde{\tau}_{\mu_1\mu_2\mu_3\mu_4} = \left(\sqrt{|g|}\right)^{w} \tau_{\mu_1\mu_2\mu_3\mu_4}$$

is a tensor.

4.9.1 Properties of the Levi-Civita Symbol in Four Dimensions

By raising indices with the inverse metric we can define $\varepsilon^{\mu\nu\rho\sigma}$ and

$$\tilde{\varepsilon}^{\,\mu\nu\rho\sigma} = \frac{1}{\sqrt{|g|}}\,\varepsilon^{\mu\nu\rho\sigma}. \tag{4.96}$$

Contraction gives

$$\tilde{\varepsilon}^{\,\mu\nu\rho\sigma}\,\tilde{\varepsilon}_{\mu\nu\rho\sigma} = -4! = -24, \tag{4.97}$$

$$\tilde{\varepsilon}^{\,\mu\nu\rho\alpha}\,\tilde{\varepsilon}_{\mu\nu\rho\beta} = -3!\,1!\,\delta^{\alpha}_{\beta} = -6\,\delta^{\alpha}_{\beta}, \tag{4.98}$$

$$\tilde{\varepsilon}^{\,\mu\nu\alpha\beta}\,\tilde{\varepsilon}_{\mu\nu\gamma\delta} = -2!\,2!\,\delta^{[\alpha}_{\gamma}\,\delta^{\beta]}_{\delta} = -2\left(\delta^{\alpha}_{\gamma}\,\delta^{\beta}_{\delta} - \delta^{\beta}_{\gamma}\,\delta^{\alpha}_{\delta}\right), \tag{4.99}$$

$$\tilde{\varepsilon}^{\,\mu\nu\alpha\beta}\tilde{\varepsilon}_{\mu\gamma\delta\varphi} = -3!\,1!\,\delta^{[\nu}_{\gamma}\,\delta^{\alpha}_{\delta}\,\delta^{\beta]}_{\varphi}. \tag{4.100}$$

Finally, if $A_{\mu\nu}$ is an antisymmetric 2-index tensor, its *dual* is the tensor density

$$^{*}A_{\mu\nu} \equiv \frac{1}{2}\,\varepsilon_{\mu\nu}{}^{\rho\sigma}\,A_{\rho\sigma}\,. \tag{4.101}$$

The Levi-Civita symbol in three dimensions with the Euclidean metric can be used to define the cross-product of 3-vectors and the curl of 3-vector fields, as shown by the following examples.

Example 4.22 The cross-product of two vectors **A** and **B** in 3 dimensions in Cartesian coordinates can be expressed as $\mathbf{C} = \mathbf{A} \times \mathbf{B}$, where

$$C_i = \varepsilon_{ijk} A^j B^k\,. \tag{4.102}$$

In fact, $C_1 = \varepsilon_{123} A^2 B^3 + \varepsilon_{132} A^3 B^2 = A^2 B^3 - A^3 B^2$ or

$$C_x = A_y B_z - A_z B_y\,,$$

$C_2 = \varepsilon_{231} A^3 B^1 + \varepsilon_{213} A^1 B^3 = A^3 B^1 - A^1 B^3$ or

$$C_y = A_z B_x - A_x B_z\,,$$

and $C_3 = \varepsilon_{312} A^1 B^2 + \varepsilon_{321} A^2 B^1 = A^1 B^2 - A^2 B^1$, or

$$C_z = A_x B_y - A_y B_x\,,$$

so the rule (4.102) coincides with the rule "$C_i = A_j B_k - A_k B_j$ with *i,j,k cyclic*", or with the empirical rule

$$\mathbf{C} = \begin{vmatrix} \hat{i} & \hat{j} & \hat{k} \\ A_x & A_y & A_z \\ B_x & B_y & B_z \end{vmatrix} \tag{4.103}$$

based on a pseudo-determinant.

Example 4.23 The 3-dimensional curl of a vector field **A** is defined as the pseudo-determinant

$$\nabla \times \mathbf{A} = \begin{vmatrix} \hat{i} & \hat{j} & \hat{k} \\ \partial_x & \partial_y & \partial_z \\ A_x & A_y & A_z \end{vmatrix}\,.$$

$$= \hat{i}\left(\partial_y A_z - \partial_z A_y\right) - \hat{j}\left(\partial_x A_z - \partial_z A_x\right) + \hat{k}\left(\partial_x A_y - \partial_y A_x\right)$$

The curl can be defined as

$$(\nabla \times \mathbf{A})^i = \varepsilon^{ijk} \frac{\partial A_j}{\partial x^k}.$$

In fact, we have

$$\begin{aligned}
\varepsilon^{1jk} \frac{\partial A_j}{\partial x^k} &= \varepsilon^{123} \frac{\partial A_2}{\partial x^3} + \varepsilon^{132} \frac{\partial A_3}{\partial x^2} = \frac{\partial A_2}{\partial x^3} - \frac{\partial A_3}{\partial x^2} = \frac{\partial A_y}{\partial z} - \frac{\partial A_z}{\partial y} \\
&= (\nabla \times \mathbf{A})^x, \\
\varepsilon^{2jk} \frac{\partial A_j}{\partial x^k} &= \varepsilon^{213} \frac{\partial A_1}{\partial x^3} + \varepsilon^{231} \frac{\partial A_3}{\partial x^1} = -\frac{\partial A_1}{\partial x^3} + \frac{\partial A_3}{\partial x^1} = \frac{\partial A_z}{\partial x} - \frac{\partial A_x}{\partial z} \\
&= (\nabla \times \mathbf{A})^y, \\
\varepsilon^{3jk} \frac{\partial A_j}{\partial x^k} &= \varepsilon^{312} \frac{\partial A_1}{\partial x^2} + \varepsilon^{321} \frac{\partial A_2}{\partial x^1} = \frac{\partial A_1}{\partial x^2} - \frac{\partial A_2}{\partial x^1} = \frac{\partial A_x}{\partial y} - \frac{\partial A_y}{\partial x} \\
&= (\nabla \times \mathbf{A})^z.
\end{aligned}$$

4.9.2 Volume Element

The volume element $\mathrm{d}^n x$ in n dimensions does not transform as a tensor under coordinate transformations but as a tensor density because it involves the Jacobian of the transformation:

$$\mathrm{d}^n x' = \mathrm{Det}\left(\frac{\partial \mathrm{x}^{\alpha'}}{\partial \mathrm{x}^{\beta}} \right) \mathrm{d}^{\mathrm{n}}\mathrm{x}. \tag{4.104}$$

Using the fact that the determinant g of the metric tensor is a tensor density of weight -2, it is easy to realize that the quantity $\sqrt{|g|}$ can be used to construct a covariant volume element. The quantity $\sqrt{|g|}\,\mathrm{d}^n x$ is invariant:

$$\sqrt{|g'|}\,\mathrm{d}^n x' = \sqrt{|g|}\,\mathrm{d}^n x \tag{4.105}$$

and this is the "correct" volume element to use in relativity.[8] Therefore, a covariant expression of the integral of a function $f(x^\mu)$ over a region Ω of $\mathbb{R}^n$ is

[8] There is more to the volume element than what we present here: a full discussion (see, e.g., Refs. [1–3]) involves differential forms and the Levi-Civita symbol, but the above will suffice for our purposes.

$$\int_{\Omega} \mathrm{d}^n x \sqrt{|g|}\, f \tag{4.106}$$

and not simply $\int_{\Omega} \mathrm{d}^n x\, f$. Similarly, the volume element $\sqrt{|g|}\,\mathrm{d}^n x$ will appear in the integral of tensorial quantities:

$$\int_{\Omega} \mathrm{d}^n x \sqrt{|g|}\, T^{\mu_1 \mu_2 \dots \mu_k}{}_{\nu_1 \nu_2 \dots \nu_l}\,. \tag{4.107}$$

This property is not new: consider for example, the integral of a scalar function f over a region $\Omega \subseteq \mathbb{R}^3$ in spherical coordinates $\{r, \theta, \varphi\}$. The metric components in these coordinates are given by the line element

$$\mathrm{d}l_{(3)}^2 = \mathrm{d}r^2 + r^2 \left(\mathrm{d}\theta^2 + \sin^2\theta\, \mathrm{d}\varphi^2\right),$$

or

$$g_{\mu\nu} = \mathrm{diag}\left(1, \mathrm{r}^2, \mathrm{r}^2 \sin^2\theta\right)$$

with $g = r^4 \sin^2\theta$. The invariant volume element is $\sqrt{g}\,\mathrm{d}^3\mathbf{x} = r^2 \sin\theta\, \mathrm{d}r \mathrm{d}\theta \mathrm{d}\varphi$ and the integral of the scalar $f(r, \theta, \varphi)$ on a region $\Omega \subseteq \mathbb{R}^3$ is

$$\int_{\Omega} \mathrm{d}r \mathrm{d}\theta \mathrm{d}\varphi\, r^2 \sin\theta\, f\,(r, \theta, \varphi)\,,$$

an expression familiar from elementary calculus.

4.10 Conclusion

The formalism of tensors has many applications to mathematics and physics and to science and technology in general. For us, its main advantage is that it allows the equations of physics to be formulated in a manner that is independent of particular coordinate systems (*covariance*). The invariance of the theory under changes of inertial frame is the essence of the Principle of Relativity, one of the only two pillars of the theory, and then it is clear that covariance is essential for Special Relativity and its equations must be formulated as tensor equations. In this sense, the formalism of tensors is particularly suited for Special Relativity. The restriction to inertial observers and inertial frames is eliminated in General Relativity, for which arbitrary coordinate transformations corresponding to arbitrary observers are possible, and the formalism of tensors becomes even more necessary there.

Problems

4.1. Is

$$x' = x + 2y,$$
$$y' = 3x + 6y,$$

an admissible coordinate transformation in $\mathbb{R}^2$?

4.2. Find the "new" components of $v^\mu = (1, 5, 11)$ after the linear homogeneous coordinate transformation

$$x' = ax + by + z,$$
$$y' = 3x + y + 5z,$$
$$z' = \frac{y}{3} + z.$$

Under what condition on the constants a and b is this an admissible coordinate transformation?

4.3. In $\mathbb{R}^2$, let $\{x^\mu\} = \{x, y\}$ be Cartesian coordinates and $\left\{x^{\mu'}\right\} = \{r, \varphi\}$ be polar coordinates.
(a) If the vector field A^μ has Cartesian components $A^\mu = \left(x^2, 2y^2\right)$, find the polar components $A^{\mu'}$.
(b) If $f(x, y)$ is a scalar function, find the components of the gradient $\nabla_\alpha f$ in polar coordinates.

4.4. Given the Cartesian components (A^x, A^y, A^z) of a vector **A** in 3-dimensional Euclidean space $\mathbb{R}^3$, find its "new" components in spherical coordinates $\{r, \theta, \varphi\}$ in terms of the "old" components.

4.5. Prove Eq. (4.17).

4.6. Let $R^\mu{}_\nu$ describe the infinitesimal transformations which can be connected continuously to the identity and preserve the scalar product in Minkowski spacetime, i.e.,

$$R^\mu{}_\nu = \delta^\mu_\nu + \varepsilon\, l^\mu{}_\nu$$

(where ε is a smallness parameter) and, for any 4-vector V^μ,

$$V^{\mu'} V_{\mu'} = V^\mu V_\mu$$

where $V^{\mu'} = R^{\mu'}{}_\nu V^\nu$. Show that there are only six such transformations $R^\mu{}_\nu$ and interpret them physically.

4.7. Compute $\dfrac{\partial^2 x^\mu}{\partial x^\alpha \partial x^\beta}$.

4.8. Show that, if A^μ and B^μ are contravariant vectors, then $C^{\mu\nu} \equiv A^\mu B^\nu$ is a contravariant tensor of rank 2.

4.9. If A^μ and B_ν are a contravariant and a covariant vector, respectively, show that $A^\mu B_\nu$ is a mixed tensor of type $\begin{pmatrix} 1 \\ 1 \end{pmatrix}$.

4.10. Show that, if X^μ is a contravariant vector field and f is a scalar function, then $X^\alpha \partial_\alpha f$ is a scalar function.

4.11. Show that the Kronecker delta defined by Eq. (4.15) describes a true tensor of type $\begin{pmatrix} 1 \\ 1 \end{pmatrix}$ and that it is left unchanged by coordinate transformations.

4.12. Prove the see-saw rule $A_\mu B^\mu = A^\mu B_\mu$.

4.13. Show that, if $F_{\mu\nu} = F_{[\mu\nu]}$, then $F_{\mu\nu} A^\mu A^\nu = 0$.

4.14. Compute

$$X_{\mu(\nu} Y_{\rho)\sigma} A^{[\nu} B^{\rho]}$$

and

$$A_{\mu(\nu} B_{\rho)}{}^{\sigma} C_{\sigma(\tau} D_{\alpha)\beta} E^{[\alpha} E^{\beta]}$$

by knowing that $D_{\alpha\beta}$ is symmetric.

4.15. Let $A^\mu = B^{\mu\nu} C_\nu$ and $D^\mu = A^\mu A^\sigma A_\sigma$. Compute $F^{[\mu\nu]}$, where $F^{\mu\nu} \equiv A^\mu B^{\nu\alpha} E_{\alpha\beta} D^\beta$.

4.16. Compute

$$X_{\alpha[\beta} Y_{\gamma]}{}^{(\delta} A^{\mu)} A^\alpha A^\beta + \frac{1}{4} X_{\alpha\gamma} A^\alpha A^\beta Y_\beta{}^\mu A^\delta - \frac{1}{4} X_{\alpha\beta} A^\alpha A^\beta Y_\gamma{}^\delta A^\mu.$$

4.17. Compute the trace T of $A_{[\mu} B_{\nu]} A^\mu C^\alpha$ by knowing that $A^\mu A_\mu = -3$ and $B_\mu A^\mu = 4$. If $B_\mu C^\mu = a$, $B_\mu B^\mu = 0$, and $A_\mu C^\mu = b$, determine the constants a and b so that $T = 0$ and $\left(A_\mu + B_\mu + C_\mu\right)(A^\mu + B^\mu) = 0$.

4.18. Show that

$$\left(A_{\mu\nu} + A_{\nu\mu}\right) B^\mu B^\nu = 2 A_{\mu\nu} B^\mu B^\nu$$

and that

$$\left(A_{\mu\nu\rho} + A_{\nu\rho\mu} + A_{\rho\mu\nu}\right) B^\mu B^\nu B^\rho = 3 A_{\mu\nu\rho} B^\mu B^\nu B^\rho.$$

4.19. Given an antisymmetric tensor $A^{\mu\nu}$, find explicitly all the components $A^{\mu'\nu'}$ in terms of the components $A^{\mu\nu}$ when a Lorentz transformation in the 4-dimensional Minkowski spacetime is applied.

4.20. Verify directly that the property of two vectors A^μ and B^μ of being orthogonal is coordinate-independent (i.e., if the two vectors are orthogonal in a coordinate system, they are orthogonal in *any* coordinate system).

4.21. Prove that $T_{\nu\mu} = \pm T_{\mu\nu}$ if and only if $T^{\nu\mu} = \pm T^{\mu\nu}$.

4.22. Prove that the decomposition (4.43)–(4.45) is unique.

4.23. The decomposition of a 2-index tensor into a symmetric and an antisymmetric part extends to *pairs* of indices in a tensor of rank $k > 2$, but not to triples or larger sets of indices. For example, show that for any tensor $A_{\mu\nu\rho}$ of type $\binom{0}{3}$,

$$A_{\mu\nu\rho} = A_{(\mu\nu)\rho} + A_{[\mu\nu]\rho},$$

but

$$A_{\mu\nu\rho} \neq A_{(\mu\nu\rho)} + A_{[\mu\nu\rho]}.$$

4.24. Show that if $A^{\mu\nu}$ is symmetric and $B^{\mu\nu}$ is antisymmetric, then $A_{\mu\nu}B^{\mu\nu} = 0$.
4.25. Show that $A_{\mu\nu}B^{\mu}C^{\nu} = A_{(\mu\nu)}B^{\mu}C^{\nu}$.
4.26. Show that the inverse $g^{\mu\nu}$ of the metric tensor $g_{\mu\nu}$ is symmetric.
4.27. Consider the Euclidean 3-dimensional space in Cartesian coordinates and, using the Levi-Civita symbol, show that $\mathbf{C} \equiv \mathbf{A} \times \mathbf{B}$ is perpendicular to both **A** and **B**.
4.28. *Prove the properties (4.70)–(4.72) of the commutator.

References

1. C.W. Misner, K.S. Thorne, J.A. Wheeler, *Gravitation* (Freeman, New York, 1973)
2. R.M. Wald, *General Relativity* (Chicago University Press, Chicago, 1984)
3. S.M. Carroll, *Spacetime and Geometry, An Introduction to General Relativity* (Addison-Wesley, San Francisco, 2004)
4. R. d'Inverno, *Introducing Einstein's Relativity* (Clarendon Press, Oxford, 2002)
5. B. Lesche, I. Alfocorado, M.L. Bedran, Am. J. Phys. **60**, 545 (1992)

Chapter 5
Tensors in Minkowski Spacetime

Make everything as simple as possible, but no simpler.
—Albert Einstein

5.1 Introduction

After studying the general formalism of tensors, we can now apply it specifically to the 4-dimensional spacetime arena. A fundamental addition to the general baggage of tensors is the causal character of 4-vectors in Minkowski spacetime, which is due to the Lorentzian signature of the Minkowski metric. As a consequence of this signature, the line element is not positive-definite and we already know that this feature is linked to the existence of light cones, which play a crucial role because they determine the causal structure of the theory. Moreover, some specific notation which was not used in the general discussion of tensors applies to the Minkowski spacetime of Special Relativity.

5.2 Vectors and Tensors in Minkowski Spacetime

Let us consider now Minkowski spacetime and label the coordinates with the indices 0, 1, 2, and 3. The index 0 refers to the time coordinate. For example, in Cartesian coordinates, it is

$$x^0 = ct, \quad x^1 = x, \quad x^2 = y, \quad x^3 = z. \tag{5.1}$$

We adopt the convention that Greek indices assume the values 0, 1, 2, 3 and label spacetime quantities while Latin indices assume the values 1, 2, 3 and label spatial quantities. A contravariant vector in Minkowski spacetime is simply called a *4-vector*

V. Faraoni, *Special Relativity*, Undergraduate Lecture Notes in Physics,

DOI: 10.1007/978-3-319-01107-3_5,

and has components

$$A^{\mu} = (A^0, \mathbf{A}).$$

A^0 is the *time component* while A^1, A^2, A^3 are the *space components* which, together, form a 3-vector $\mathbf{A} = \left(A^1, A^2, A^3\right)$ with respect to purely spatial coordinate transformations $x^i \longrightarrow x^i(x^j)$ (here $i, j = 1, 2, 3$). However, $\mathbf{A}$ does not transform as a vector under 4-dimensional coordinate transformations.

Example 5.1 The position 4-vector is $x^{\mu} = (ct, \mathbf{x})$, and the gradient of a scalar function f is $\partial_{\mu} f = \left(\frac{\partial f}{c\partial t}, \nabla f\right)$.

A contravariant 4-tensor of rank 2 is a set of $4^2 = 16$ quantities that transform like the product of components of two 4-vectors

$$T^{\mu'\nu'} = \frac{\partial x^{\mu'}}{\partial x^{\alpha}} \frac{\partial x^{\nu'}}{\partial x^{\beta}} T^{\alpha\beta},$$

etc. The coordinate changes that we are interested in are mostly the Lorentz transformations between inertial frames with constant relative velocity (in standard configuration or otherwise), but one could transform from an inertial to an accelerated or rotating frame as well.

5.3 The Minkowski Metric

Let $\{x^{\mu}\} = \{ct, x, y, z\}$ be Cartesian coordinates in Minkowski spacetime. The spacetime interval between nearby points of coordinates $x^{\mu} = (ct, x, y, z)$ and $x^{\mu} + \mathrm{d}x^{\mu} = (ct + c\mathrm{d}t, x + \mathrm{d}x, y + \mathrm{d}y, z + \mathrm{d}z)$ is

$$\mathrm{d}s^2 = -c^2\mathrm{d}t^2 + \mathrm{d}x^2 + \mathrm{d}y^2 + \mathrm{d}z^2. \tag{5.2}$$

We stick to this quantity because we know that it is left invariant by Lorentz transformations, which we have adopted as the fundamental symmetries of Special Relativity following the lesson coming from Maxwell's electromagnetism. By contrast, the usual Euclidean distance (squared) between two spatial points $\mathrm{d}l_{(3)}^2 = \mathrm{d}x^2 + \mathrm{d}y^2 + \mathrm{d}z^2$, or their time separation (squared) $c^2\mathrm{d}t^2$, are not invariant under Lorentz transformations.

The infinitesimal interval, or *line element* of Minkowski spacetime $\mathrm{d}s^2$ can be obtained by introducing the metric tensor which, in Cartesian coordinates, has the components. [1]

[1] The symbol $\doteq$ denotes equality in a particular coordinate system.

$$(\eta_{\mu\nu}) \doteq \begin{pmatrix} -1 & 0 & 0 & 0 \\ 0 & 1 & 0 & 0 \\ 0 & 0 & 1 & 0 \\ 0 & 0 & 0 & 1 \end{pmatrix} = \text{diag}\,(-1, 1, 1, 1) \tag{5.3}$$

(*Minkowski metric in Cartesian coordinates*) and contracting it twice with the coordinate differentials $\mathrm{d}x^\mu$:

$$\begin{aligned} \eta_{\mu\nu}\mathrm{d}x^\mu\mathrm{d}x^\nu &= \begin{pmatrix} \mathrm{d}x^0 \\ \mathrm{d}x^1 \\ \mathrm{d}x^2 \\ \mathrm{d}x^3 \end{pmatrix}^T \begin{pmatrix} -1 & 0 & 0 & 0 \\ 0 & 1 & 0 & 0 \\ 0 & 0 & 1 & 0 \\ 0 & 0 & 0 & 1 \end{pmatrix} \begin{pmatrix} \mathrm{d}x^0 \\ \mathrm{d}x^1 \\ \mathrm{d}x^2 \\ \mathrm{d}x^3 \end{pmatrix} \\ &= -\left(\mathrm{d}x^0\right)^2 + \left(\mathrm{d}x^1\right)^2 + \left(\mathrm{d}x^2\right)^2 + \left(\mathrm{d}x^3\right)^2 \\ &= -c^2\mathrm{d}t^2 + \mathrm{d}x^2 + \mathrm{d}y^2 + \mathrm{d}z^2, \end{aligned}$$

thus

$$\mathrm{d}s^2 = \eta_{\mu\nu}\mathrm{d}x^\mu\mathrm{d}x^\nu \doteq -c^2\mathrm{d}t^2 + \mathrm{d}x^2 + \mathrm{d}y^2 + \mathrm{d}z^2. \tag{5.4}$$

In this sense, the Minkowski metric introduces the notion of distance between two points in spacetime. If the two points $x^\mu_{(1,2)}$ are not at infinitesimal distance, the finite spacetime interval in Cartesian coordinates is given by

$$\Delta s^2 = -c^2\Delta t^2 + \Delta x^2 + \Delta y^2 + \Delta z^2, \tag{5.5}$$

where $\Delta x^\mu \equiv x^\mu_{(2)} - x^\mu_{(1)}$.

The line element is not positive-definite because of the negative sign in the time-time component η_{00} of the Minkowski metric (5.3). This feature is absolutely necessary in order to introduce the notion of causality in Special Relativity and the notion of a limiting speed c. As already remarked, two points separated by a null interval $\mathrm{d}s^2 = 0$ can be related by a signal traveling at the speed of light:

$$\mathrm{d}s^2 = \eta_{\mu\nu}\mathrm{d}x^\mu\mathrm{d}x^\nu = -c^2\mathrm{d}t^2 + \mathrm{d}\mathbf{x}^2 = 0.$$

Considering, for simplicity, a configuration such that $\mathrm{d}y = \mathrm{d}z = 0$ (points on the x-axis), it is $c^2\mathrm{d}t^2 = \mathrm{d}\mathbf{x}^2$ for this signal, or

$$\frac{dx}{dt} = \pm c$$

i.e., a signal travelling along the positive or negative x-axis and connecting two points $(ct, x, 0, 0)$ and $(ct + c\mathrm{d}t, x + \mathrm{d}x, 0, 0)$ with $\mathrm{d}s^2 = 0$ must necessarily travel at speed c, and vice-versa.

The Lorentzian signature $-+++$ of the metric makes it clear that time is treated differently from space, although they both concur to build spacetime.
A metric tensor is positive-definite (or a *Riemannian metric*) if $g_{\mu\nu}x^\mu x^\nu > 0$ for any non-zero vector x^μ. Note that the convention on the metric signature is not unique. Several textbooks use the opposite signature $+---$ for relativity, the physics being, of course, unchanged.

Example 5.2 Thus far, we have reasoned in Cartesian coordinates. However, it is clear that the components of a metric tensor will change if we change coordinates. Consider, as an example, the 3-dimensional space $\mathbb{R}^3$ with the Euclidean metric e_{ij} and the change from Cartesian to cylindrical coordinates $\{x, y, z\} \longrightarrow \{r, \varphi, z\}$

$$\begin{aligned} x &= r\,\cos\varphi, \\ y &= r\,\sin\varphi, \\ z &= z. \end{aligned}$$

We want to compute the components of the Euclidean metric tensor e_{ij} in these cylindrical coordinates. We begin by writing the Euclidean line element in $\mathbb{R}^3$

$$\mathrm{d}l^2_{(3)} = e_{ij}\,\mathrm{d}x^i\mathrm{d}x^j = \mathrm{d}x^2 + \mathrm{d}y^2 + \mathrm{d}z^2$$

and by differentiating the inverse coordinate transformation $\{r, \varphi, z\} \longrightarrow \{x, y, z\}$,

$$\begin{aligned} \mathrm{d}x &= \mathrm{d}r\,\cos\varphi - r\,\sin\varphi\mathrm{d}\varphi, \\ \mathrm{d}y &= \mathrm{d}r\,\sin\varphi + r\,\cos\varphi\mathrm{d}\varphi, \\ \mathrm{d}z &= \mathrm{d}z. \end{aligned}$$

Substituting into the expression of $\mathrm{d}l^2_{(3)}$, we obtain

$$\begin{aligned} \mathrm{d}l^2_{(3)} &= \mathrm{d}x^2 + \mathrm{d}y^2 + \mathrm{d}z^2 = (\mathrm{d}r\cos\varphi - r\sin\varphi\mathrm{d}\varphi)^2 \\ &\quad + (\mathrm{d}r\sin\varphi + r\cos\varphi\mathrm{d}\varphi)^2 + \mathrm{d}z^2 \\ &= \mathrm{d}r^2\left(\cos^2\varphi + \sin^2\varphi\right) + r^2\left(\sin^2\varphi + \cos^2\varphi\right)\mathrm{d}\varphi^2 \\ &\quad -2r\,\sin\varphi\cos\varphi\,\mathrm{d}r\,\mathrm{d}\varphi + 2r\,\cos\varphi\sin\varphi\,\mathrm{d}r\,\mathrm{d}\varphi + dz^2 \\ &= \mathrm{d}r^2 + r^2\mathrm{d}\varphi^2 + \mathrm{d}z^2. \end{aligned}$$

Since $\mathrm{d}l^2_{(3)} = e_{ij}\,\mathrm{d}x^i\mathrm{d}x^j$ is a tensor equation and is true in any coordinate system, we have that

$$\mathrm{d}l^2_{(3)} = \underbrace{e_{ij}\,\mathrm{d}x^i\mathrm{d}x^j}_{\mathrm{d}x^2+\mathrm{d}y^2+\mathrm{d}z^2} = \underbrace{e_{i'j'}\,\mathrm{d}x^{i'}\mathrm{d}x^{j'}}_{\mathrm{d}r^2+r^2\mathrm{d}\varphi^2+\mathrm{d}z^2}$$

and, therefore, the components $e_{i'j'}$ of the Euclidean metric in cylindrical coordinates $\{r, \varphi, z\}$ can be read off as

$$\left(e_{i'j'}\right) \doteq \begin{pmatrix} 1 & 0 & 0 \\ 0 & r^2 & 0 \\ 0 & 0 & 1 \end{pmatrix} = \mathrm{diag}\left(1, r^2, 1\right). \tag{5.6}$$

Example 5.3 Consider the surface of a sphere of radius R and centre in the origin of $\mathbb{R}^3$. We want to compute the metric induced by $\mathbb{R}^3$ on this 2-dimensional sphere using spherical coordinates $\{\theta, \varphi\}$. The coordinate transformation $\{x, y, z\} \longrightarrow \{r, \theta, \varphi\}$ has inverse

$$\begin{aligned} x &= r\sin\theta\cos\varphi, \\ y &= r\sin\theta\sin\varphi, \\ z &= r\cos\theta, \end{aligned}$$

with $r = R =$ const. on the surface of the given sphere. The Euclidean 3-dimensional metric in Cartesian coordinates has components $e_{ij} = \mathrm{diag}\,(1, 1, 1)$ and produces the line element between two infinitesimally nearby points

$$\mathrm{d}l^2_{(3)} = \delta_{ij}\,\mathrm{d}x^i\mathrm{d}x^j = \mathrm{d}x^2 + \mathrm{d}y^2 + \mathrm{d}z^2.$$

Now express the differentials $\mathrm{d}x^i = (\mathrm{d}x, \mathrm{d}y, \mathrm{d}z)$ on the surface of the sphere in terms of the differentials of θ and φ:

$$\begin{aligned} \mathrm{d}x &= R\left(\cos\theta\cos\varphi\,\mathrm{d}\theta - \sin\theta\sin\varphi\,\mathrm{d}\varphi\right), \\ \mathrm{d}y &= R\left(\cos\theta\sin\varphi\,\mathrm{d}\theta + \sin\theta\cos\varphi\,\mathrm{d}\varphi\right), \\ \mathrm{d}z &= -R\sin\theta\,\mathrm{d}\theta, \end{aligned}$$

hence,

$$\begin{aligned}
\mathrm{d}l^2_{(3)} &= R^2\Big[(\cos\theta\cos\varphi\,\mathrm{d}\theta - \sin\theta\sin\varphi\,\mathrm{d}\varphi)^2 \\
&\quad + (\cos\theta\sin\varphi\,\mathrm{d}\theta + \sin\theta\cos\varphi\,\mathrm{d}\varphi)^2 + \sin^2\theta\,\mathrm{d}\theta^2\Big] \\
&= R^2\Big[\cos^2\theta\left(\cos^2\varphi + \sin^2\varphi\right)\mathrm{d}\theta^2 + \sin^2\theta\left(\sin^2\varphi + \cos^2\varphi\right)\mathrm{d}\varphi^2 \\
&\quad + \sin^2\theta d\theta^2\Big] = R^2\left(d\theta^2 + \sin^2\theta\,\mathrm{d}\varphi^2\right).
\end{aligned}$$

By calling $e^{(s)}_{ij}$ the restriction of the Euclidean metric e_{ij} to the 2-sphere of radius R, we can write the line element on this 2-sphere as

$$\mathrm{d}l^2_{(2)} = e^{(s)}_{i'j'}\,\mathrm{d}x^{i'}\mathrm{d}x^{j'} = R^2 d\theta^2 + R^2\sin^2\theta\,\mathrm{d}\varphi^2 \quad \left(i', j' = 1, 2\right), \tag{5.7}$$

or

$$\left(e^{(s)}_{i'j'}\right) = \begin{pmatrix} R^2 & 0 \\ 0 & R^2\sin^2\theta \end{pmatrix} \tag{5.8}$$

in coordinates $\{\theta, \varphi\}$. A widely used notation for the line element on the unit 2-sphere is

$$\mathrm{d}\Omega^2_{(2)} \equiv \mathrm{d}\theta^2 + \sin^2\theta\,\mathrm{d}\varphi^2. \tag{5.9}$$

* * *

In Cartesian coordinates, the inverse Minkowski metric $\eta^{\mu\nu}$ is equal to $\eta_{\mu\nu}$:

$$\eta^{\mu\nu} \doteq \mathrm{diag}\,(-1, 1, 1, 1)\,. \tag{5.10}$$

Let us update our definition of Minkowski spacetime: the *Minkowski spacetime* is the pair $\left(\mathbb{R}^4, \eta_{\mu\nu}\right)$, the set of all spacetime events endowed with the Minkowski metric $\eta_{\mu\nu}$. This definition reflects the fact that it is not only the set of events $\mathbb{R}^4$ that matters, but also the metric with which it is endowed. The relations between spacetime points determined by this metric are more important than the events themselves.

The *Minkowski metric in cylindrical coordinates* $\{ct, r, \theta, z\}$ is given by

$$ds^2 \doteq -c^2 dt^2 + dr^2 + r^2 d\theta^2 + dz^2, \tag{5.11}$$

or

$$g_{\mu\nu} \doteq \text{diag}\left(-1, 1, r^2, 1\right) \tag{5.12}$$

with

$$g^{\mu\nu} \doteq \text{diag}\left(-1, 1, 1/r^2, 1\right). \tag{5.13}$$

The *Minkowski metric in spherical coordinates* $\{ct, r, \theta, \varphi\}$ is given by

$$ds^2 \doteq -c^2 dt^2 + dr^2 + r^2 d\Omega^2_{(2)}, \tag{5.14}$$

or

$$g_{\mu\nu} \doteq \text{diag}\left(-1, 1, r^2, r^2 \sin^2\theta\right), \tag{5.15}$$

$$g^{\mu\nu} \doteq \text{diag}\left(-1, 1, \frac{1}{r^2}, \frac{1}{r^2 \sin^2\theta}\right). \tag{5.16}$$

In Minkowski spacetime the Minkowski metric $\eta_{\mu\nu}$ satisfies the requirement of non-singularity and the inverse metric $\eta^{\mu\nu}$ is well-defined. In Cartesian coordinates $\{ct, x, y, z\}$ we have $\eta_{\mu\nu} = \text{diag}\,(-1, 1, 1, 1) = \eta^{\mu\nu}$,

$$\text{Det}\left(\eta_{\mu\nu}\right) = -1,$$

and

$$\eta_{\mu\nu}\,\eta^{\nu\alpha} = \delta^{\alpha}_{\mu}.$$

5.4 Scalar Product and Length of a Vector in Minkowski Spacetime

In addition to the distance between spacetime points, the metric tensor provides the notion of scalar product between 4-vectors and that of length of a 4-vector in Minkowski spacetime. Remember that the scalar product between two vectors X^α and Y^β is $< \mathsf{X}, \mathsf{Y} > \equiv g_{\alpha\beta}\, X^\alpha Y^\beta$ and that the length squared of a vector X^α is the scalar product of X^α with itself $< \mathsf{X}, \mathsf{X} > \equiv g_{\alpha\beta}\, X^\alpha X^\beta$. In Minkowski spacetime, the length of a vector is not positive-definite because of the Lorentzian signature of the Minkowski metric. In Cartesian coordinates we have

$$\eta_{\mu\nu} X^\mu X^\nu = -\left(X^0\right)^2 + \left(X^1\right)^2 + \left(X^2\right)^2 + \left(X^3\right)^2.$$

This means that a vector can have zero length even if its components are not all zero. For example, the vectors with components

$$l^{\mu} = \left(l^0, \pm l^0, 0, 0\right) \qquad (l^0 \neq 0)$$

in Cartesian coordinates have length

$$\eta_{\mu\nu}\, l^{\mu}\, l^{\nu} = -\left(l^0\right)^2 + \left(l^0\right)^2 = 0.$$

These vectors are orthogonal to themselves but are not identically zero. Again, two vectors X^{α}, Y^{β} are orthogonal if $g_{\alpha\beta} X^{\alpha} Y^{\beta} = 0$.

Definition 5.1 If two vectors X^{α}, Y^{β} have non-zero lengths, the *angle* θ *between* X^{α} *and* Y^{β} is defined by

$$\cos\theta \equiv \frac{g_{\mu\nu} X^{\mu} Y^{\nu}}{\sqrt{|g_{\alpha\beta} X^{\alpha} X^{\beta}| \cdot |g_{\rho\sigma} Y^{\rho} Y^{\sigma}|}}. \tag{5.17}$$

Example 5.4 In $\mathbb{R}^3$, the scalar product defined by the Euclidean metric e_{ij} coincides with the ordinary dot product of vectors. In Cartesian coordinates, for two vectors $\mathbf{a} = \left(a^x, a^y, a^z\right)$, $\mathbf{b} = \left(b^x, b^y, b^z\right)$,

$$e_{ij}\, a^i b^j = \delta_{ij} a^i b^j = a^x b^x + a^y b^y + a^z b^z \equiv \mathbf{a} \cdot \mathbf{b}; \tag{5.18}$$

the length squared of a vector **a** is

$$e_{ij}\, a^i a^j = \delta_{ij}\, a^i a^j = a^x a^x + a^y a^y + a^z a^z = \mathbf{a} \cdot \mathbf{a} = ||\mathbf{a}||^2. \tag{5.19}$$

In cylindrical coordinates $\{r, \varphi, z\}$, using the expression (5.12) we obtain

$$e_{i'j'}\, a^{i'} a^{j'} = \begin{pmatrix} a^r \\ a^{\varphi} \\ a^z \end{pmatrix}^T \begin{pmatrix} 1 & 0 & 0 \\ 0 & r^2 & 0 \\ 0 & 0 & 1 \end{pmatrix} \begin{pmatrix} a^r \\ a^{\varphi} \\ a^z \end{pmatrix} = \left(a^r\right)^2 + r^2 \left(a^{\varphi}\right)^2 + \left(a^z\right)^2.$$

* * *

In the space $\mathbb{R}^n$ *with Euclidean metric* and *in Cartesian coordinates*, vector components with upper and lower indices are the same because the metric reduces to the Kronecker delta.[2] For example,

[2] This is the reason why, so far, we have not distinguished between contravariant components A^i and covariant components A_i of a vector **A** in $\mathbb{R}^3$ with Cartesian coordinates.

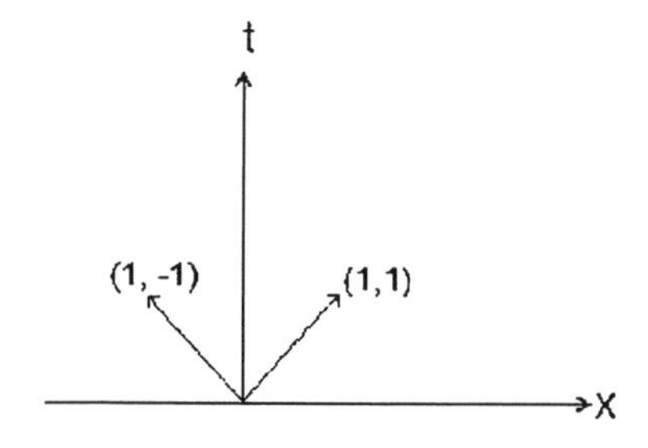

Fig. 5.1 The null vectors $(1, \pm 1)$ defining a cone (*light cone*) in 2-dimensional Minkowski spacetime

$$A_i = g_{ij} A^j = \delta_{ij} A^j = A^i,$$

$$A^k = g^{kl} A_l = \delta^{kl} A_l = A_k,$$

$$T_{ij} = g_{il}\, g_{jm}\, T^{lm} = \delta_{il}\, \delta_{jm}\, T^{lm} = T^{ij}.$$

Of course, this property is no longer true when non-Cartesian coordinates are used and $g_{ij} \neq \delta_{ij}$, or when a Lorentzian metric is used instead of the Euclidean one.

Example 5.5 Consider the 2-dimensional Minkowski spacetime $\mathbb{R}^2$ with the Minkowski metric

$$(\eta_{\mu\nu}) = \begin{pmatrix} -1 & 0 \\ 0 & 1 \end{pmatrix}$$

in coordinates $\{ct, x\}$. The vector $X^\mu = (1, 1)$ is not identically zero but

$$\eta_{\mu\nu} X^\mu X^\nu = -(1)^2 + (1)^2 = 0;$$

the same is true for the vector $Y^\mu = (1, -1)$. These two vectors identify the future light cone through the origin of this Minkowski space (Fig. 5.1).

Vectors which appear to have the same length in a (x, t) Minkowski diagram (suppressing two dimensions) do not necessarily have the same length according to the Minkowski metric. Similarly, vectors which appear to be orthogonal in the (x, t) spacetime diagram are not necessarily orthogonal in the sense of the Minkowski metric. Consider, for example $A^\mu = (1, 1, 0, 0)$ and $B^\mu = (1, -1, 0, 0)$; it is

$$A_\mu B^\mu = \eta_{\mu\nu} A^\mu B^\nu = -A^0 B^0 + A^1 B^1 + A^2 B^2 + A^3 B^3 = -2 \neq 0$$

but these two vectors appear orthogonal in a Minkowski diagram (Fig. 5.1). The representation in the spacetime diagram distorts four-dimensional lengths and angles. However, the sum of two vectors in the diagram corresponds to their sum in the sense of the Minkowski metric and the graphical notion of parallelism of two vectors (i.e., $A^\mu = \lambda B^\mu$ for some scalar $\lambda > 0$ defines "A^μ is parallel to B^μ") corresponds to parallelism in the sense of the Minkowski metric.

5.5 Raising and Lowering Tensor Indices

As we have seen, the metric $g_{\mu\nu}$ and the inverse metric $g^{\mu\nu}$ can be used to lower or raise tensor indices.

- For any contravariant vector X^μ we can define a corresponding covariant vector $X_\mu \equiv g_{\mu\nu} X^\nu$.
- For any covariant vector Y_α we can define a corresponding contravariant vector $Y^\alpha \equiv g^{\alpha\beta} Y_\beta$.

Similarly, any tensor index can be raised or lowered using $g^{\alpha\beta}$ or $g_{\alpha\beta}$. For example, we can associate to the tensor $T^\alpha{}_\beta$ both $T_{\alpha\beta} \equiv g_{\alpha\gamma} T^\gamma{}_\beta$ and $T^{\alpha\beta} \equiv g^{\beta\gamma} T^\alpha{}_\gamma$. If we want to lower the index α_j of the tensor $T^{\alpha_1 \dots \alpha_j \dots \alpha_k}{}_{\beta_1 \dots \beta_l}$, we use $g_{\alpha_j \gamma}$ obtaining

$$T^{\alpha_1 \dots \alpha_{j-1} \alpha_{j+1} \dots \alpha_n}{}_{\alpha_j \beta_1 \dots \beta_l} \equiv g_{\alpha_j \gamma}\, T^{\alpha_1 \dots \alpha_{j-1} \gamma\, \alpha_{j+1} \dots \alpha_k}{}_{\beta_1 \dots \beta_l} . \tag{5.20}$$

X^α and X_α are two different representations of the same vector provided by $g_{\alpha\beta}$, $g^{\alpha\beta}$, and δ^α_β.

In Minkowski spacetime a contravariant 4-vector in Cartesian coordinates is written as

$$A^\mu = \left(A^0, \mathbf{A}\right) \equiv \left(A^0, A^x, A^y, A^z\right)$$

and the corresponding covariant 4-vector (dual vector) is

$$A_\mu = \eta_{\mu\nu} A^\nu = \left(-A^0, \mathbf{A}\right) = \left(-A^0, A^x, A^y, A^z\right) = \left(A_0, A_x, A_y, A_z\right) .$$

Example 5.6 In Minkowski spacetime in Cartesian coordinates, find the components T_{00} and $T_0{}^0$ of a tensor $T^{\mu\nu}$.
We have

$$T_{00} = \eta_{0\mu}\, \eta_{0\nu}\, T^{\mu\nu} \underset{\substack{\uparrow \\ \text{in} \\ \text{Cartesian} \\ \text{coordinates}}}{\doteq} \left(-\delta_{0\mu}\right)\left(-\delta_{0\nu}\right) T^{\mu\nu} = T^{00}$$

and

$$T_0{}^\nu = \eta_{0\mu} T^{\mu\nu} = -\delta_{0\mu} T^{\mu\nu} = -T^{0\nu} ;$$

in particular, for $\nu = 0$ one has $T_0{}^0 = -T^{00}$.

* * *

The scalar product can also be used to define the divergence of a vector field **A** in the 3-dimensional space with Euclidean metric and in Cartesian coordinates:

$$\nabla \cdot \mathbf{A} \equiv g^{ij} \frac{\partial A_i}{\partial x^j} = \delta^{ij} \frac{\partial A_i}{\partial x^j} = \frac{\partial A_1}{\partial x^1} + \frac{\partial A_2}{\partial x^2} + \frac{\partial A_3}{\partial x^3},$$

which matches the familiar expression

$$\nabla \cdot \mathbf{A} \equiv \frac{\partial A^x}{\partial x} + \frac{\partial A^y}{\partial y} + \frac{\partial A^z}{\partial z}.$$

Definition 5.2 The *divergence of a 4-vector field* $A^\mu = \left(A^0, \mathbf{A}\right)$ in Minkowski spacetime in Cartesian coordinates is

$$\begin{aligned} \partial_\mu A^\mu &= \delta^\mu_\nu \partial_\mu A^\nu = \delta^{\mu\nu} \partial_\mu A_\nu \\ &= \frac{\partial A^0}{\partial (ct)} + \frac{\partial A^x}{\partial x} + \frac{\partial A^y}{\partial y} + \frac{\partial A^z}{\partial z} \end{aligned} \tag{5.21}$$

so

$$\partial_\mu A^\mu = \frac{\partial A^0}{\partial (ct)} + \nabla \cdot \mathbf{A}. \tag{5.22}$$

Definition 5.3 The *d'Alembertian* of a scalar field ϕ is

$$\Box \phi \doteq \partial^\mu \partial_\mu \phi \doteq \eta^{\mu\nu} \partial_\mu \partial_\nu \phi \tag{5.23}$$

in Cartesian coordinates,[3] for which

$$\Box \phi \doteq \eta^{\mu\nu} \partial_\mu \partial_\nu \phi = -\frac{1}{c^2} \frac{\partial^2 \phi}{\partial t^2} + \frac{\partial^2 \phi}{\partial x^2} + \frac{\partial^2 \phi}{\partial y^2} + \frac{\partial^2 \phi}{\partial z^2} = -\frac{1}{c^2} \frac{\partial^2 \phi}{\partial t^2} + \nabla^2 \phi, \tag{5.24}$$

where $\nabla^2 \equiv \frac{\partial^2}{\partial x^2} + \frac{\partial^2}{\partial y^2} + \frac{\partial^2}{\partial z^2}$ is the usual Laplace operator in Cartesian coordinates.

[3] The definition of the d'Alembertian $\nabla^\mu \nabla_\mu \phi$ in general coordinates requires the notion of covariant derivative ∇_α introduced in Chap. 10.

The d'Alembertian used in wave mechanics (when the waves propagate at the speed of light c) appears to be a straightforward generalization of the Laplacian to four dimensions with the Lorentzian signature.

The d'Alembertian of ϕ coincides with the divergence of the gradient $\partial_\mu \phi$, or $\Box\phi \doteq \partial^\mu \left(\partial_\mu \phi\right)$. This is analogous to the situation in three dimensions in which the Laplacian is the divergence of the gradient: $\nabla^2\phi = \nabla \cdot (\nabla\phi)$.

5.5.1 Working with Tensors in Minkowski Spacetime

From a 4-vector $A^\mu = \left(A^0, \mathbf{A}\right) = \left(A^0, A^i\right)$ $(i = 1, 2, 3)$ one can obtain covariant components by lowering the indices with the Minkowski metric. In Cartesian coordinates $\left\{x^\mu\right\} = \{ct, x, y, z\}$, it is

$$A_\mu = \eta_{\mu\nu} A^\nu = (A_0, A_i) = \left(-A^0, A^i\right). \tag{5.25}$$

If A^μ, B^μ are two 4-vectors then

$$A_\mu B^\mu = \eta_{\mu\nu} A^\mu B^\nu = -A^0 B^0 + A^1 B^1 + A^2 B^2 + A^3 B^3 = -A^0 B^0 + \mathbf{A} \cdot \mathbf{B}.$$

We have also

$$A_\mu A^\mu = -\left(A^0\right)^2 + (\mathbf{A})^2.$$

The transformation property of the components of a 4-vector A^μ under Lorentz transformations in standard configuration is $A^\mu \to A^{\mu'}$ with

$$A^{0'} = \gamma\left(A^0 - \frac{v}{c} A^1\right), \tag{5.26}$$

$$A^{1'} = \gamma\left(A^1 - \frac{v}{c} A^0\right), \tag{5.27}$$

$$A^{2'} = A^2, \tag{5.28}$$

$$A^{3'} = A^3, \tag{5.29}$$

as follows from $A^{\mu'} = \dfrac{\partial x^{\mu'}}{\partial x^\alpha} A^\alpha \equiv L_{(v)}{}^{\mu'}{}_\alpha A^\alpha$. For a covariant vector B_μ, we have

$$B_{0'} = \gamma\left(B_0 + \frac{v}{c} B_1\right), \tag{5.30}$$

$$B_{1'} = \gamma\left(B_1 + \frac{v}{c} B_0\right), \tag{5.31}$$

$$B_{2'} = B_2, \tag{5.32}$$

$$B_{3'} = B_3, \tag{5.33}$$

according to $B_{\mu'} = \dfrac{\partial x^\alpha}{\partial x^{\mu'}} B_\alpha = L_{(v)\,\mu'}{}^{\alpha} B_\alpha$.

For higher rank tensors, one applies repeatedly the Lorentz matrix $L_\alpha{}^{\beta'} = \dfrac{\partial x^{\beta'}}{\partial x^\alpha}$ and its inverse, for example

$$T^{\alpha'\beta'}{}_{\gamma'} = L^{\alpha'}{}_{\rho}\, L^{\beta'}{}_{\sigma}\, L^{\delta}{}_{\gamma'}\, T^{\rho\sigma}{}_{\delta}.$$

5.6 Causal Nature of 4-Vectors

A vector X^μ in Minkowski spacetime is

- *timelike* if $X_\mu X^\mu < 0$,
- *null* or *lightlike* if $X_\mu X^\mu = 0$,
- *spacelike* if $X_\mu X^\mu > 0$.

A timelike or null vector is called a *causal vector* (Fig. 5.3).

The *light cone* (or *null cone*) at a spacetime point P is the set of null vectors at P. This is a vector space of dimension 2 and a surface in Minkowski space (and, as will be clear later, is generated by the tangents to ingoing and outgoing radial null rays at P).

In Cartesian coordinates, a null vector satisfies

$$\eta_{\mu\nu}\, X^\mu X^\nu \doteq 0$$

or

$$-\left(X^0\right)^2 + \left(X^1\right)^2 + \left(X^2\right)^2 + \left(X^3\right)^2 = 0.$$

If X^μ coincides with the Cartesian position 4-vector x^μ, this is the equation of the double cone

$$x^0 = \pm\sqrt{\left(x^1\right)^2 + \left(x^2\right)^2 + \left(x^3\right)^2}$$

(Fig. 5.2). Let $t^\mu = (1, 0, 0, 0) = \delta^{0\mu}$; this unit vector points in the "direction of time" (the direction of the time axis) and has unit norm.

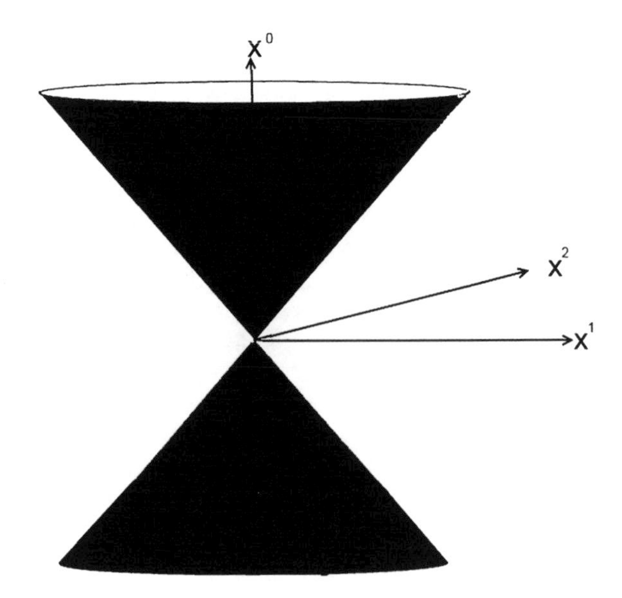

Fig. 5.2 The light cone through the spacetime point $O = (0, 0, 0, 0)$

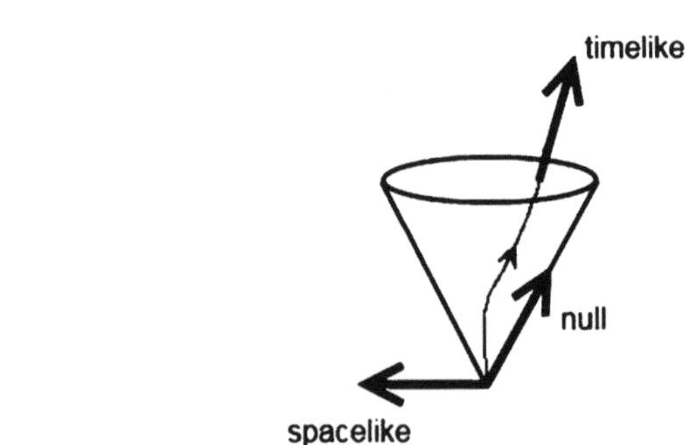

Fig. 5.3 A timelike vector at a spacetime point P points *inside* the light cone at P; a null vector points *along* the light cone, and a spacelike vector points *outside* of it. The tangent to the worldline of a massive particle always points inside the light cone

A timelike or null vector X^μ is

- *future-pointing* if $t_\mu X^\mu < 0$, or
- *past-pointing* if $t_\mu X^\mu > 0$.

Note that this definition involves a scalar product, therefore it is independent of the coordinate system used.

Let X^μ be a timelike or null vector and let $\left(X^0, X^1, X^2, X^3\right)$ be its components in a coordinate system $\{x^\mu\} = \left\{x^0, \mathbf{x}\right\}$, with time (multiplied by c) as the zeroth component. If $X^0 > 0$, then X^μ is future-pointing, while if $X^0 < 0$, X^μ is past-pointing, and $\eta_{\mu\nu} t^\mu X^\nu \equiv t_\mu X^\mu$ is the projection of X^α on the time direction.

Two causal vectors are called *isochronous* if they are both future-pointing or both past-pointing. This means that, if $A^\mu = \left(A^0, \mathbf{A}\right)$ and $B^\mu = \left(B^0, \mathbf{B}\right)$, it is $A^0 B^0 > 0$.

Example 5.7 The 4-vector of components $A^\mu = (1, 0, 3, 1)$ is spacelike since

$$A_\mu A^\mu = -(A^0)^2 + (A^1)^2 + (A^2)^2 + (A^3)^2 = -1 + 9 + 1 = 9 > 0.$$

The 4-vector $l^\mu = \left(\frac{1}{2}, 0, \frac{1}{2}, 0\right)$ is null since

$$l_\mu l^\mu = -(l^0)^2 + (l^1)^2 + (l^2)^2 + (l^3)^2 = -\frac{1}{4} + \frac{1}{4} = 0.$$

The 4-vector $B^\mu = (3, 1, 1, 0)$ is timelike since

$$B_\mu B^\mu = -(B^0)^2 + (B^1)^2 + (B^2)^2 + (B^3)^2 = -9 + 1 + 1 = -7 < 0.$$

* * *

It is often convenient to choose an inertial frame which simplifies the calculations. The following results are useful to this regard:

- if $A^\mu = \left(A^0, \mathbf{A}\right)$ is a timelike 4-vector, it is always possible to find an inertial frame in which $A^{\mu'} = \left(A^{0'}, 0, 0, 0\right)$. This frame is unique.
- For a *spacelike* 4-vector B^μ, it is always possible to find an inertial frame in which the components reduce to $B^{\mu'} = \left(0, B^{1'}, 0, 0\right)$. This frame is unique.
- (*Zero component lemma*) If a 4-vector has the same component (for example the time component) equal to zero in *all* inertial frames, it must be the zero vector $(0, 0, 0, 0)$.

The proof of these statements is left as an exercise.

In a Minkowski diagram, the frame S has its time axis vertical and its x-axis horizontal and simultaneities (events occurring at the same time in this frame) form a horizontal straight line. All other inertial frames have apparently non-orthogonal t'- and x'- axes and simultaneities of these other frames are represented by oblique lines in the (x, t) plane. The apparent orthogonality has no physical significance because the physical metric is the Minkowski one, not the Euclidean metric upon which our intuition is built and which suggests orthogonality in the Euclidean sense. Taking different inertial frames corresponds to taking different time slices of Minkowski spacetime (which have different time axes) and with hyperplanes inclined with respect to the 3-spaces t =const. of S (Fig. 5.4).

A null vector $l^\mu = \left(l^0, \mathbf{l}\right)$ can always be reduced to the form

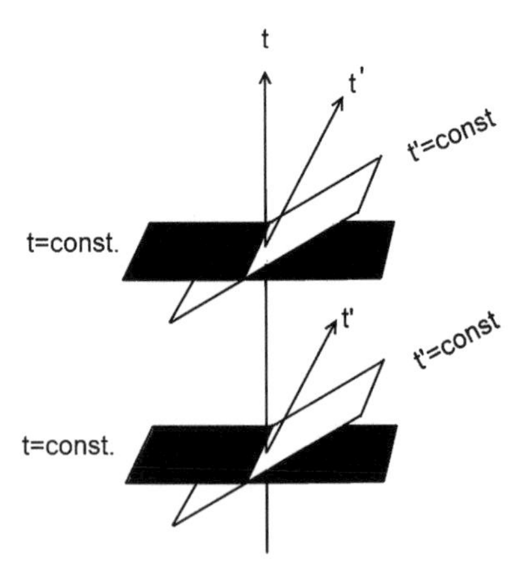

Fig. 5.4 Two slicings of Minkowski spacetime with two different times and 3-dimensional spaces

$$l^{\mu'} = \left(l^{0'}, \pm l^{0'}, 0, 0 \right)$$

(this is trivial to show: simply align the x'-axis with $\pm \mathbf{l}$). Note also that a null vector is defined up to a constant without affecting its normalization. If l^{μ} is such that $l^{\mu} l_{\mu} = 0$, then $\forall \alpha \neq 0$, $m^{\mu} = \alpha \, l^{\mu}$ is parallel to l^{μ} and null: $m_{\mu} m^{\mu} = \alpha^2 \, l^{\mu} l_{\mu} = 0$. If l^{μ} is future- or past-pointing one must choose the constant α positive in order for m^{μ} to remain future- or past-pointing.

The following statements hold true, assuming that none of the null 4-vectors involved coincide with the trivial vector $(0, 0, 0, 0)$:

- The sum of two isochronous timelike 4-vectors is a timelike 4-vector isochronous with them.
- The sum of a timelike 4-vector and an isochronous null 4-vector is a timelike 4-vector isochronous with both.
- The sum of two isochronous null 4-vectors is a timelike 4-vector unless the two 4-vectors are parallel, in which case their sum is a null 4-vector.
- The difference of two isochronous null 4-vectors is a spacelike 4-vector unless the two null 4-vectors are parallel, in which case their difference is a null 4-vector.
- The sum of any number of isochronous null or timelike 4-vectors is a timelike or null 4-vector isochronous with them and it is null if and only if all the 4-vectors added are null and parallel.
- Any timelike 4-vector can be expressed as the sum of two isochronous null 4-vectors.
- Any spacelike 4-vector can be expressed as the difference of two isochronous null 4-vectors.
- A timelike 4-vector cannot be orthogonal to a causal 4-vector.
- A 4-vector orthogonal to a null 4-vector A^{μ} must be spacelike, or else it is null and parallel or antiparallel to A^{μ}.
- Any 4-vector orthogonal to a causal 4-vector A^{μ} is spacelike, or else it is a null 4-vector parallel to A^{μ}.

- The scalar product of two isochronous timelike 4-vectors is negative.
- The scalar product of two isochronous null 4-vectors is negative unless they are parallel (in which case their product vanishes).
- The scalar product of a timelike 4-vector and an isochronous null 4-vector is negative.

The proofs of these statements are left as exercises.

Definition 5.4 A curve $x^\mu(\lambda)$ in Minkowski spacetime is a timelike/null/spacelike curve at a point if its 4-tangent $u^\mu = dx^\mu/d\lambda$ is timelike/null/spacelike, respectively, at that point. A *timelike curve* is one whose 4-tangent is everywhere timelike, etc., i.e., the causal character of a spacetime curve is the causal character of its tangent.

The worldline of a massive particle is a timelike curve, while a null ray (the spacetime trajectory of a photon) is a null curve.

Example 5.8 Consider the geometric curve $x^\mu(\lambda) = \left(3\lambda, 6\lambda^2, 0, 0\right)$ in Minkowski spacetime in Cartesian coordinates. The 4-tangent to this curve is

$$u^\mu = \frac{dx^\mu}{d\lambda} = (3, 12\lambda, 0, 0)$$

and its square is

$$u_\mu u^\mu = \eta_{\mu\nu} u^\mu u^\nu = -(u^0)^2 + (u^1)^2 + (u^2)^2 + (u^3)^2 = -9 + 144\lambda^2.$$

We have $u_\mu u^\mu < 0$ if $|\lambda| < 1/4$ and the curve is spacelike for $\lambda < -1/4$, null at $\lambda = 1/4$, timelike for $-1/4 < \lambda < 1/4$, null again at $\lambda = 1/4$, and then spacelike again for $\lambda > 1/4$ (this curve cannot be the worldline of a physical particle).

Example 5.9 Consider the curve with parametric representation

$$x^\mu(\lambda) = \frac{1}{\sqrt{7}}\left(3\lambda, \lambda, 3, \lambda\right)$$

in Minkowski spacetime in Cartesian coordinates. The 4-tangent is

$$u^\mu = \frac{dx^\mu}{d\lambda} = \frac{1}{\sqrt{7}}\left(3, 1, 0, 1\right)$$

and its square is $u_\mu u^\mu = \eta_{\mu\nu} u^\mu u^\nu = -9 + 1 + 1 = -1 < 0$. This curve is timelike.

Example 5.10 Consider the curve of parametric representation

$$x^\mu(\lambda) = \left(3\lambda^2, 1, 13, 3\lambda^2\right)$$

in Minkowski spacetime in Cartesian coordinates. The 4-tangent is $u^\mu = \frac{dx^\mu}{d\lambda} = (6\lambda, 0, 0, 6\lambda)$ and its square is $u_\mu u^\mu = \eta_{\mu\nu} u^\mu u^\nu = -36\lambda^2 + 36\lambda^2 = 0$. This curve is always null (and, therefore, it could represent the worldline of a photon).

5.7 Hypersurfaces

Definition 5.5 A *hypersurface* in an n-dimensional space is a surface of dimension $n-1$. A hypersurface is

- *timelike* if its normal n^μ is spacelike, $n_\mu n^\mu > 0$;
- *null* if its normal n^μ is null, $n_\mu n^\mu = 0$;
- *spacelike* if its normal n^μ is timelike, $n_\mu n^\mu < 0$.

Example 5.11 Any hypersurface t =constant is spacelike. In fact, the equation of the hypersurface is

$$f(t) = t - \text{const.} = 0.$$

The normal to this surface has the direction of the gradient of f,

$$n_\mu = \nabla_\mu f = \nabla_\mu t = (1, 0, 0, 0)$$

and is already normalized:

$$n_\mu n^\mu = -1.$$

Example 5.12 Any hypersurface x^1 = constant is timelike. In fact, the equation of this surface is $f(x) = x^1 - \text{const.} = 0$. The normal has the direction of the gradient,

$$n_\mu = \nabla_\mu f = (0, 1, 0, 0)$$

and is normalized, $n_\mu n^\mu = 1$ and spacelike, hence $x^1 = $ const. is a timelike hypersurface.

* * *

The null cone through any point of Minkowski spacetime is a null surface.

Proof Let $x^\mu_{(0)} = (ct_0, x_0, y_0, z_0)$ in Cartesian coordinates (the result, however, will not depend on the coordinates adopted). The light cone through $x^\mu_{(0)}$ has equation

$$f(t, x, y, z) \equiv -c^2 (t - t_0)^2 + (x - x_0)^2 + (y - y_0)^2 + (z - z_0)^2 = 0.$$

The normal to this surface is

$$n_\mu = \nabla_\mu f \bigg|_{f=0}$$

$$= -2c\,(t-t_0)\,\delta_{0\mu} + 2\,(x-x_0)\,\delta_{1\mu} + 2\,(y-y_0)\,\delta_{2\mu}$$

$$+2\,(z-z_0)\,\delta_{3\mu}\Big|_{f=0}$$

or

$$n_\mu = 2\Big(-c\,(t-t_0)\,,x-x_0,y-y_0,z-z_0\Big),$$

while

$$n^\mu = 2\Big(c\,(t-t_0)\,,x-x_0,y-y_0,z-z_0\Big)$$

so that

$$n^\mu n_\mu = 4\Big[-c^2\,(t-t_0)^2 + (x-x_0)^2 + (y-y_0)^2 + (z-z_0)^2\Big]_{f=0}$$

$$= 4f\,(t,x,y,z)\Big|_{f=0} = 0\,.$$

□

The null cone is a 2-dimensional surface.

Proof The null cone through any point P of Minkowski spacetime is generated by only two linearly independent null vectors at that point. If $l^\mu = \left(l^0, \mathbf{l}\right)$ is a null vector at P, one can align the x-axis with $\mathbf{l}$ and then, in these coordinates, it is $l^\mu = \left(l^0, l^1, 0, 0\right)$. The normalization $l_\mu l^\mu = -(l^0)^2 + (l^1)^2 = 0$ yields $l^1 = \pm l^0$ and $l^\mu_{(1,2)} = \left(l^0, \pm l^0, 0, 0\right)$. These are all the null vectors at P. It is easy to see that $l^\mu_{(1)}$ and $l^\mu_{(2)}$ are linearly independent and, therefore, the null cone generated by them has dimension two.

5.8 Gauss' Theorem

Let Ω be a 4-dimensional simply connected region of Minkowski spacetime which has volume element $d\Omega \equiv \sqrt{|g|}\,\mathrm{d}^4x \doteq \mathrm{d}x^0\mathrm{d}x^1\mathrm{d}x^2\mathrm{d}x^3$ in Cartesian coordinates in which $g \doteq \mathrm{Det}\left(\eta_{\mu\nu}\right) = -1$. Let $\partial\Omega \equiv S$ be its 3-dimensional boundary with its own 3-dimensional volume element ("surface element") $\mathrm{d}S$ associated with an outward-pointing normal n^μ. This normal is assumed to be normalized, i.e., it has norm squared +1 if the surface is timelike, −1 if the surface is spacelike and, of course, zero if the surface is null. Let V^μ be a 4-vector field defined in Ω. The *Gauss theorem* states that

$$\int_\Omega \mathrm{d}^4x\, \partial_\mu V^\mu = \int_{\partial\Omega} \mathrm{d}S\, V^\mu n_\mu . \tag{5.34}$$

Example 5.13 Consider the spacetime region

$$\Omega = \{(ct, r, \vartheta, \varphi) : \quad 0 \le r \le r_0\} ,$$

which is a"tube" formed by the region of 3-dimensional space enclosed by the sphere of radius r_0, with time spanning the entire t-axis. The boundary of this regions is

$$\partial\Omega = \{(ct, r, \vartheta, \varphi) : \quad r = r_0\} ,$$

and the unit normal to $\partial\Omega$ is

$$n^\mu = (0, 1, 0, 0) , \qquad n_\mu = (0, 1, 0, 0) .$$

Consider the vector field

$$V^\mu = \left(0, \frac{r}{1+\alpha t^2}, 0, 0, \right) = \left(V^0, \mathbf{V}\right) ,$$

where $\alpha > 0$ is a constant with the dimensions of an inverse time squared. The 4-divergence of V^μ is

$$\partial_\mu V^\mu = \frac{\partial V^0}{\partial(ct)} + \nabla \cdot \mathbf{V} = 0 + \frac{1}{r^2}\frac{\partial}{\partial r}\left(r^2 V^r\right) + 0 + 0 = \frac{1}{r^2}\frac{\partial}{\partial r}\left(\frac{r^3}{1+\alpha t^2}\right) = \frac{3}{1+\alpha t^2}$$

and its integral over the 4-region Ω is

$$\begin{aligned} \int_\Omega \mathrm{d}^4x\, \partial_\mu V^\mu &= \int_\Omega \mathrm{d}^4x\, \frac{3}{1+\alpha t^2} = \int_{-\infty}^{+\infty} \mathrm{d}t \int_0^{r_0} \mathrm{d}r \int_0^\pi \mathrm{d}\vartheta \int_0^{2\pi} \mathrm{d}\varphi\, r^2 \sin\vartheta\, \frac{3}{1+\alpha t^2} \\ &= 3\,\frac{4\pi}{3}\, r_0^3 \int_{-\infty}^{+\infty} \frac{\mathrm{d}t}{1+\alpha t^2} = 4\pi r_0^3 \left[\frac{1}{\sqrt{\alpha}} \tan^{-1}\left(\sqrt{\alpha}\, t\right)\right]_{-\infty}^{+\infty} = \frac{4\pi^2 r_0^3}{\sqrt{\alpha}} . \end{aligned}$$

Now compute the surface integral of the vector field V^μ itself on the boundary $\partial\Omega$:

$$\int_{\partial\Omega} \mathrm{d}S\, n_\mu V^\mu = \int_{-\infty}^{+\infty} \mathrm{d}t \int_0^\pi \mathrm{d}\vartheta \int_0^{2\pi} \mathrm{d}\varphi\, r^2 \sin\vartheta\, \frac{r}{1+\alpha t^2}\Big|_{r=r_0}$$

$$= 4\pi r_0^3 \int_{-\infty}^{+\infty} \mathrm{d}t \, \frac{1}{1+\alpha t^2} = \frac{4\pi^2 r_0^3}{\sqrt{\alpha}} \, .$$

Therefore, Gauss' theorem $\int_{\Omega} d^4x \, \partial_\mu V^\mu = \int_{\partial\Omega} \mathrm{d}S \, n_\mu V^\mu$ is verified.

5.9 Conclusion

Thus far, we have seen that a 4-dimensional world view is convenient and even necessary because a change of inertial frame mixes space and time coordinates, similar to the way in which spatial rotations in three dimensions mix different spatial coordinates. We have studied the geometry of Minkowski spacetime and it is now time to do physics in this spacetime. Physics must be given a relativistic (i.e., Lorentz-invariant) formulation. Beginning with mechanics, we know that Newton's second law is Galilei- but not Lorentz-invariant, and it must be modified. Maxwell's theory is already Lorentz-invariant and does not need to be modified, but only rewritten in the 4-dimensional formalism.

A physical theory will be expressed by basic physical laws which must be theoretically and experimentally consistent with our (limited) knowledge and must make new predictions which are falsifiable.[4] Further, these laws must be expressed in a covariant way by tensor equations. We will only consider physics without gravity (gravity is included in General Relativity but not in Special Relativity) and with the stipulation that there exists a preferred class of reference frames, the inertial frames. We begin our study of physics in Minkowski spacetime by reformulating the mechanics of point particles (we know that Newtonian mechanics is not invariant under Lorentz boosts) and then moving on to geometric optics, fluid physics, and the physics of scalar and electromagnetic fields, while giving some general prescriptions applicable to any branch of physics whenever possible.

The relativistic corrections to Newtonian mechanics and their predictions were studied theoretically long before their experimental verification. Today, relativistic mechanics is the basis for the working of particle physics accelerators, nuclear power generation, the Global Positioning System, positron annihilation spectroscopy, and various tools used in medicine and the industry. Newtonian mechanics is adequate in the limit of small velocities $|v| \ll c$. In particle physics experiments, instead, γ-factors of order 10^4 have been achieved and factors $\gamma \sim 10^{11}$ are common in cosmic rays.

[4] It is an old Popperian adage that a theory cannot be verified: it can only be falsified.

Problems

5.1 Find all the future-oriented and all the past-oriented null vectors of the 2-dimensional Minkowski spacetime (ct, x) with the Minkowski metric

$$\eta_{\mu\nu} = \text{diag}\,(-1, 1)\,.$$

5.2 Let A^μ be a 4-vector in Minkowski spacetime. Prove directly, using the transformation properties, that $g_{\mu\nu} A^\mu A^\nu$ is invariant under arbitrary coordinate transformations $x^\mu \longrightarrow x^{\mu'}$.

5.3 Are the following 4-vectors (with components given in Cartesian coordinates) orthogonal to each other in Minkowski spacetime?

$$A^\mu = (1, 0, 0, 1)\,,$$
$$B^\mu = (1, 0, 0, 0)\,,$$
$$C^\mu = (0, 1, 0, 0)\,.$$

5.4 Determine the timelike, spacelike, or null character of the 4-vectors

$$u^\mu = (1, 0, 0, 0)\,, \qquad v^\mu = (1, 1, 1, 1)\,,$$
$$w^\mu = (1, 0, 0, 1)\,, \qquad x^\mu = \left(1, 0, 3, \sqrt{3}\right),$$
$$y^\mu = (1, 0, -1, 1)\,, \qquad z^\mu = (100, 3, 4, 17)\,,$$
$$q^\mu = (0, 1, 0, 0)\,, \qquad t^\mu = \left(5, 0, 0, \sqrt{7}\right),$$
$$r^\mu = (0, 1, 0, 1)\,, \qquad s^\mu = \left(\sqrt{5}, \pi, \sqrt{11}, \mathrm{e}\right),$$

in Minkowski spacetime in Cartesian coordinates.

5.5 Determine the spacelike, null, or timelike character of the 4-vectors given, in Cartesian coordinates, by

$$A^\mu = \left(1, 0, \frac{1}{\sqrt{2}}, \frac{1}{\sqrt{2}}\right),$$
$$B^\mu = \left(1, 0, \frac{1}{2}, \frac{1}{2}\right),$$
$$C^\mu = (0, 0, 1, 0)\,,$$
$$D^\nu = A^{\mu\nu} E_\mu\,, \quad \text{where } \mathrm{A}^{\mu\nu} = \delta^{\mu 0}\delta^{\nu 1} \text{ and } \mathrm{E}^\mu = \left(\frac{1}{2}, 1, 0, 1\right).$$

5.6 If u^μ is timelike and s^μ is spacelike, is it true that $u_\mu v^\mu = 0$?

5.7 Prove the following statements or disprove them with counterexamples. In general,

(a) is the sum of two null 4-vectors a null 4-vector?
(b) Is the sum of two spacelike 4-vectors a spacelike 4-vector?
(c) Is the sum of two timelike 4-vectors a timelike 4-vector?

5.8 (a) Let X^μ and Y^μ be two spacelike 4-vectors in Minkowski spacetime. Is $X_\mu Y^\mu \geq 0$?
(b) Let X^μ and Y^μ be two timelike 4-vectors in Minkowski spacetime. Is $X_\mu Y^\mu \leq 0$?
(c) Let X^μ and Y^μ be a timelike and a spacelike 4-vector, respectively. Is $X_\mu Y^\mu = 0$?

5.9 Show that if A^μ is a timelike 4-vector, it is always possible to find a unique inertial frame in which $A^{\mu'} = \left(A^{0'}, 0, 0, 0\right)$.

5.10 Show that if A^μ is a spacelike 4-vector, it is always possible to find a unique inertial frame in which $A^{\mu'} = (0, \mathbf{A}')$.

5.11 Show that, if $l^\mu = \left(l^0, \mathbf{l}\right)$ (in Cartesian coordinates) is a null vector, l^0 has the same sign in all inertial frames.

5.12 Prove the zero-component lemma for 4-vectors in Minkowski spacetime.

5.13 Let $X^\mu = (A, B, C, D)$ be a 4-vector in Minkowski spacetime in Cartesian coordinates. Under what conditions on the constants A, B, C, and D is X^μ null and orthogonal to $Y^\mu = (1, 0, 1, 1)$, $Z^\mu = (0, 1, 2, 0)$, $W^\mu = (0, 0, 1, 0)$ and future-pointing?

5.14 Show that, if $A_{\mu\nu} X^\mu X^\nu = 0$ for all 4-vectors X^μ, then $A_{\mu\nu}$ is antisymmetric.

5.15 Show that the sum of two isochronous timelike 4-vectors is a timelike 4-vector isochronous with them.

5.16 Show that the sum of a timelike 4-vector and an isochronous null (non-trivial) 4-vector is a timelike 4-vector isochronous with both.

5.17 Show that the sum of two isochronous null (non-trivial) 4-vectors is a timelike 4-vector unless the two 4-vectors are parallel, in which case their sum is a null 4-vector.

5.18 Show that the difference of two isochronous null (non-trivial) 4-vectors is a spacelike 4-vector unless the two null 4-vectors are parallel, in which case their difference is a null 4-vector.

5.19 Show that the sum of any number of isochronous null (non-trivial) or timelike 4-vectors is a timelike or null 4-vector isochronous with them and it is null if and only if all the 4-vectors added are null and parallel.

5.20 Show that any timelike 4-vector can be expressed as the sum of two isochronous null 4-vectors.

5.21 Show that any spacelike 4-vector can be expressed as the difference of two isochronous null 4-vectors.

5.22 Show that a timelike 4-vector cannot be orthogonal to a causal non-trivial 4-vector.

5.23 Show that any (non-trivial) 4-vector orthogonal to a (non-trivial) causal 4-vector A^μ is spacelike, or else it is a null 4-vector parallel to A^μ.

5.24 Show that the scalar product of two isochronous timelike 4-vectors is negative.

5.25 Show that the scalar product of two isochronous null (non-trivial) 4-vectors is negative unless they are parallel (in which case their product vanishes).

5.26 Show that the scalar product of a timelike 4-vector and an isochronous null (non-trivial) 4-vector is negative.

5.27 Write the Minkowski metric $g_{\mu\nu}$ using the *null coordinates*

$$u \equiv \frac{ct - x}{\sqrt{2}} \qquad \textit{(retarded time)},$$

$$v \equiv \frac{ct + x}{\sqrt{2}} \qquad \textit{(advanced time)}.$$

Compute $g_{\mu\nu}$, $\sqrt{|g|}$, and $g^{\mu\nu}$ in these coordinates. Write down the wave equation $\Box\phi = 0$ for a scalar field $\phi = \phi\,(u, v)$. Compute the normals n_μ and m_μ to the surfaces $u =$ const. and $v =$ const., show that they are null 4-vectors, and compute their scalar product $n^\mu m_\mu$. Draw the surfaces $u =$ const. and $v =$ const. in an (x, t) spacetime diagram: what do these surfaces represent?

5.28 Consider a slicing of Minkowski spacetime with hypersurfaces of constant time Σ_t. On each slice Σ_t, consider a 2-sphere

$$\mathscr{S} = \left\{ (ct, r, \theta, \varphi) : \quad t = \text{const.}, \quad \text{r} = \text{const.} \right\}.$$

(a) Show that $\mathcal{S}$ is spacelike (could it be otherwise, considering that $\mathcal{S} \subseteq \Sigma_t$?).
(b) Let s^μ be the outward-directed unit normal to $\mathcal{S}$ in Σ_t and let n^μ be the future-pointing timelike unit normal to Σ_t. What is the causal character of $l^\mu \equiv n^\mu + s^\mu$ and $m^\mu \equiv n^\mu - s^\mu$? Normalize l^μ and m^μ so that $l_\mu m^\mu = -1$.

5.29 Given a 2-index tensor $T_{\mu\nu}$, we say that a vector v^μ is an *eigenvector* of $T_{\mu\nu}$ if there exists a scalar λ (*eigenvalue*) such that $T_{\mu\nu}v^\nu = \lambda v_\mu$.
(a) Find all the eigenvectors of the Minkowski metric $\eta_{\mu\nu}$.
(b) Let $T_{\mu\nu}$ be symmetric; what is the maximum number of independent eigenvectors of $T_{\mu\nu}$ in 4-dimensional Minkowski spacetime?
(c) In general, one cannot diagonalize simultaneously the Minkowski metric and a symmetric 2-tensor with a coordinate transformation in Minkowski spacetime. This fact is linked to the existence of null vectors. Take $T^{\mu\nu} = k^\mu k^\nu$, where $k^\mu \doteq (1, 1, 0, 0)$ in Cartesian coordinates and prove that no Lorentz transformation can diagonalize $T^{\mu\nu}$.
(d) Let $F_{\mu\nu}$ be antisymmetric and let v^μ be an eigenvector of $F_{\mu\nu}$ with eigenvalue λ. What can you say about v^μ and/or λ?

5.30 Show that, given two 4-vectors $A^\mu = \left(A^0, \mathbf{A}\right)$ and $B^\mu = \left(B^0, \mathbf{B}\right)$ in Cartesian coordinates in Minkowski spacetime, the quantity

$$\mathscr{I} \equiv \frac{(A^0 - A^1)(B^0 + B^1)}{(A^0 + A^1)(B^0 - B^1)}$$

is Lorentz-invariant.

5.31 In an (x, t) spacetime diagram, draw
(a) an hypersurface which is asymptotically null as $|x| \longrightarrow +\infty$;
(b) an hypersurface which is null in the far past.

5.32 Find the form of the Minkowski line element in the coordinate system $\{ct, r, \theta, \varphi\}$ related to Cartesian coordinates by

$$x = \sqrt{r^2 + a^2}\, \sin\theta \cos\varphi,$$
$$y = \sqrt{r^2 + a^2}\, \sin\theta \sin\varphi,$$
$$z = r\cos\theta.$$

In General Relativity, the spacetime outside a rotating stationary black hole of mass M and angular momentum per unit mass a is given by the Kerr metric (here expressed in Boyer-Lindquist coordinates and in units in which Newton's constant and c are unity [1–4])

$$\mathrm{d}s^2 = -\left(1 - \frac{2Mr}{\Sigma}\right)\mathrm{d}t^2 - \frac{4aMr\sin^2\theta}{\Sigma}\,\mathrm{d}\theta\mathrm{d}\varphi + \frac{\Sigma}{\Delta}\,\mathrm{d}r^2 + \Sigma\mathrm{d}\theta^2$$
$$+ \left(r^2 + a^2 + \frac{2Mra^2}{\Sigma}\sin^2\theta\right)\sin^2\theta\,\mathrm{d}\varphi^2,$$

where

$$\Delta = r^2 - 2Mr + a^2, \qquad \Sigma = r^2 + a^2\cos^2\theta.$$

In the limit $M \to 0$, gravity disappears and General Relativity reduces to Special Relativity, therefore the Kerr spacetime must reduce to the Minkowski spacetime. Check that this is indeed the case.

References

1. L.D. Landau, E. Lifschitz, *The Classical Theory of Fields* (Pergamon Press, Oxford, 1989)
2. R.M. Wald, *General Relativity* (Chicago University Press, Chicago, 1984)
3. S.M. Carroll, *Spacetime and Geometry, An Introduction to General Relativity* (Addison-Wesley, San Francisco, 2004)
4. R. d'Inverno, *Introducing Einstein's Relativity* (Clarendon Press, Oxford, 2002)

Chapter 6
Relativistic Mechanics

I know not with what weapons World War III will be fought, but World War IV will be fought with sticks and stones.
—Albert Einstein

6.1 Introduction

Consider a particle of position x^μ in Minkowski spacetime. Instead of describing its motion using time t as $\mathbf{x}(t)$ in Newtonian mechanics, we use a 4-dimensional scalar parameter λ and the spacetime coordinates (including time) are functions of this parameter, $x^\mu = x^\mu(\lambda)$. This description is more suitable for a 4-dimensional world view. An extended object composed of discrete particles, or a portion of a continuum, will describe a worldtube enclosing the worldlines of its constituent particles.

The requirement that the 3-speed $|d\mathbf{x}/dt|$ of a particle be less than the speed of light c has the geometrical meaning that the tangent to the worldline cannot make an angle larger than 45 ° (in units in which $c = 1$) with the vertical. At all times, the tangent

$$u^\mu \equiv \frac{dx^\mu}{d\lambda} \tag{6.1}$$

must not point outside of the light cone at that point (Fig. 5.3). If the particle is massless and moves at the speed of light, the tangent u^μ will lie along the local light cone while, if the particle is massive, u^μ will point inside the light cone. Recall that

- A *timelike curve* is one whose tangent $u^\mu \equiv dx^\mu/d\lambda$ is always timelike, $u^\mu u_\mu < 0$ everywhere. The worldlines of massive particles are timelike and they are always contained inside the light cone.

V. Faraoni, *Special Relativity*, Undergraduate Lecture Notes in Physics,
DOI: 10.1007/978-3-319-01107-3_6, © Springer International Publishing Switzerland 2013

- A *lightlike* (or *null*) *curve* has tangent u^μ that is everywhere null, $u^\mu u_\mu = 0$ (these curves are tangent to the local light cone). The worldlines of massless particles (e.g., photons) are null.
- A *spacelike curve* has $u^\mu u_\mu > 0$ everywhere. A curve that has $u^\mu u_\mu > 0$ even at a single point cannot represent a physical particle.

Timelike/lightlike curves are *future-* [*past-*] *oriented* if their tangent u^μ is future- [past-] oriented.

For timelike observers, the notions of "before" and "after" are universal: not so for spacelike observers, for which the notion of cause and effect cannot be defined because what is cause and what is effect depend on the reference frame adopted. The division into causal future, causal past, and elsewhere is invariant for inertial timelike observers because $\eta_{\mu\nu}$ is invariant under Lorentz transformations.

6.1.1 Massive Particles

The tangent $u^\mu \equiv dx^\mu/d\tau$ to the worldline of a particle, where the proper time τ of the particle is chosen as parameter, is called *4-velocity*. The components of the 4-velocity can be expressed in terms of the 3-dimensional velocity $\mathbf{v} \equiv d\mathbf{x}/dt$,

$$u^0 = \frac{cdt}{d\tau} = \frac{c}{\sqrt{1-\frac{v^2}{c^2}}} \equiv \gamma c \tag{6.2}$$

using $\mathrm{dt} = \gamma\, \mathrm{d}\tau$ and

$$v^i \equiv \frac{dx^i}{dt} = \frac{dx^i}{d\tau}\frac{d\tau}{dt} \equiv \frac{u^i}{\gamma} = u^i\sqrt{1-\frac{v^2}{c^2}}, \tag{6.3}$$

so that

$$u^i = \gamma v^i$$

and the 4-velocity has components

$$u^\mu = (\gamma\, c, \gamma\, \mathbf{v})\,. \tag{6.4}$$

Consider a massive particle (or an observer, which we describe as a particle) with 4-velocity u^μ; the 4-velocity of a timelike particle is normalized to

$$u^{\mu} u_{\mu} = -c^2 \quad \text{(massive particles)}. \tag{6.5}$$

In fact, consider the inertial reference frame in which the particle is at rest and its 3-dimensional velocity vanishes, $x^{\mu} = (ct, 0, 0, 0)$ (this rest frame always exists for a timelike 4-velocity). Use as a parameter the proper time τ (remember that this is the time measured by a clock carried by the particle in the frame in which the particle is at rest). Clearly, in this frame it is $\mathrm{d}s^2 = -c^2 \mathrm{d}\tau^2$, $x^{\mu}(\tau) = (c\tau, \text{const.}, \text{const.}, \text{const.})$, and $u^{\mu} \equiv dx^{\mu}/d\tau = (c, 0, 0, 0)$ so, in Cartesian coordinates, $\eta_{\mu\nu} u^{\mu} u^{\nu} = \eta_{00} c^2 = -c^2$. Since $\eta_{\mu\nu} u^{\mu} u^{\nu}$ is a world scalar, it assumes the same value $-c^2$ in any other coordinate system. Alternatively, in any other inertial frame it is $u^{\mu} = (\gamma\, c, \gamma\, \mathbf{v})$, which implies that $u^{\mu} u_{\mu} = -\left(u^0\right)^2 + \mathbf{u}^2 = -c^2\gamma^2 + \gamma^2 v^2 = -c^2\gamma^2 \left(1 - \frac{v^2}{c^2}\right) = -c^2$.

The time dilation formula is contained implicitly in the expression of the Minkowski line element. In the rest frame of the particle, it is $\mathrm{d}s^2 = -c^2 \mathrm{d}\tau^2$. In another inertial frame moving with velocity $\mathbf{v} = d\mathbf{x}/dt$ with respect to the rest frame of the particle, the line element is

$$\begin{aligned} \mathrm{d}s^2 &= -c^2 \mathrm{d}t^2 + \mathrm{d}x^2 + \mathrm{d}y^2 + \mathrm{d}z^2 \\ &= -c^2 \mathrm{d}t^2 \left\{ 1 - \frac{1}{c^2} \left[\left(\frac{dx}{dt} \right)^2 + \left(\frac{dy}{dt} \right)^2 + \left(\frac{dz}{dt} \right)^2 \right] \right\} \\ &= -c^2 \mathrm{d}t^2 \left(1 - \frac{\mathbf{v}^2}{c^2} \right). \end{aligned} \tag{6.6}$$

Combining this equation with $\mathrm{d}s^2 = -c^2 \mathrm{d}\tau^2$, we obtain

$$\mathrm{d}\tau = \sqrt{1 - \frac{v^2}{c^2}}\, \mathrm{d}t, \tag{6.7}$$

or $\mathrm{d}t = \gamma \mathrm{d}\tau$, the time dilation formula. For events that are at finite separations we have

$$\tau = \int_{t_1}^{t_2} \frac{\mathrm{d}t}{\gamma} = \int_{t_1}^{t_2} \mathrm{d}t \sqrt{1 - \frac{v^2}{c^2}}. \tag{6.8}$$

Example 6.1 Consider the worldline describing the motion of a particle accelerated along the x-axis with the law

$$t(\lambda) = \frac{c}{a}\sinh\left(\frac{a\lambda}{c^2}\right), \tag{6.9}$$

$$x(\lambda) = \frac{c^2}{a}\cosh\left(\frac{a\lambda}{c^2}\right), \tag{6.10}$$

$$y = 0, \tag{6.11}$$

$$z = 0, \tag{6.12}$$

where a is a positive constant with the dimensions of an acceleration and $\lambda \in (-\infty, +\infty)$ is a parameter with the dimensions of a length. The worldline is the hyperbola of equation $x^2 - c^2t^2 = c^4/a^2$ in the (x, t) plane or, if you prefer,

$$t(x) = \frac{\pm 1}{c}\sqrt{x^2 - \frac{c^4}{a^2}}. \tag{6.13}$$

This worldline is curved, meaning that the particle is accelerated. The proper time along the worldline is given by

$$\begin{aligned} c^2 d\tau^2 &= -ds^2 = c^2 dt^2 - dx^2 \\ &= \frac{c^4}{a^2}\left[\left(\cosh\left(\frac{a\lambda}{c^2}\right)\frac{a}{c^2}d\lambda\right)^2 - \left(\sinh\left(\frac{a\lambda}{c^2}\right)\frac{a}{c^2}d\lambda\right)^2\right] \\ &= d\lambda^2, \end{aligned} \tag{6.14}$$

hence λ/c coincides[1] with the proper time τ. The 4-velocity along this worldline is

$$u^\mu \equiv \frac{dx^\mu}{d\tau} = \left(c\cosh\left(\frac{a\tau}{c}\right), c\sinh\left(\frac{a\tau}{c}\right), 0, 0\right). \tag{6.15}$$

Of course, u^μ is normalized correctly:

$$u^\mu u_\mu = \eta_{\mu\nu}u^\mu u^\nu = -\left(u^0\right)^2 + \left(u^1\right)^2 = -c^2\cosh^2(a\tau) + c^2\sinh^2(a\tau) = -c^2.$$

The particle's 3-dimensional velocity is $\mathbf{v} = (v^x, 0, 0)$ with

$$v^x = \frac{dx}{dt} = \frac{dx}{d\tau}\frac{d\tau}{dt} = \frac{u^1}{u^0/c} = c\,\frac{\sinh\left(\frac{a\tau}{c}\right)}{\cosh\left(\frac{a\tau}{c}\right)} = c\tanh\left(\frac{a\tau}{c}\right).$$

[1] Strictly speaking $\lambda = \pm c\tau$ but the worldline should be future-oriented, hence we choose the positive sign. In any case, the worldline (6.13) is symmetric with respect to time reflection $t \to -t$.

Note that $|v^x| < c \ \ \forall \tau \in (-\infty, +\infty)$ but $v^x \rightarrow \pm c$ as $\tau \rightarrow \pm\infty$. The hyperbola (6.13) is asymptotic to the light cone $t = \pm x/c$ as $\tau \rightarrow \pm\infty$ (the worldline is asymptotically null but never exactly null).

6.2 Relativistic Dynamics of Massive Particles

We have already seen that Newtonian mechanics is not invariant under Lorentz transformations and, therefore, it must be corrected. We can now begin this program.

The *mass of a particle m* is defined as the mass in the frame in which the particle is at rest and the 4-velocity of the particle is

$$u^\mu \equiv \frac{dx^\mu}{d\tau} = (\gamma\, c, \gamma\, \mathbf{v}) \,, \tag{6.16}$$

where τ is the proper time of the particle and $x^\mu(\tau)$ is its 4-position (u^μ has the dimensions of a velocity).
The *4-acceleration* of the particle is defined analogously to Newtonian mechanics, but with the 4-dimensional view:

$$a^\mu \equiv \frac{du^\mu}{d\tau} = \frac{d^2 x^\mu}{d\tau^2}. \tag{6.17}$$

A property of the 4-acceleration is that it is always orthogonal to the 4-velocity:

$$a_\mu u^\mu = 0. \tag{6.18}$$

To prove Eq. (6.18), differentiate the normalization relation $u_\mu u^\mu = -c^2$ with respect to proper time:

$$0 = \frac{d}{d\tau}\left(-c^2\right) = \frac{d}{d\tau}\left(u_\mu u^\mu\right) = 2\, u_\mu \frac{du^\mu}{d\tau} = 2\, a^\mu u_\mu. \tag{6.19}$$

All inertial observers agree on the fact that a particle is accelerated but they disagree on the value of the components a^μ of the 4-acceleration. We have

$$a^\mu = \frac{du^\mu}{d\tau} = \gamma \frac{du^\mu}{dt} = \gamma \frac{d}{dt}\left(\gamma\, c, \gamma\, \mathbf{v}\right) = \gamma \left(c\frac{d\gamma}{dt}, \frac{d\gamma}{dt}\mathbf{v} + \gamma \frac{d\mathbf{v}}{dt} \right), \tag{6.20}$$

where $\gamma = \gamma(\mathbf{v})$. In the instantaneous rest frame of the particle it is

$$u^\mu \doteq (c, \mathbf{0}), \qquad a^\mu \doteq \left(0, \frac{d\mathbf{v}}{dt}\right), \tag{6.21}$$

therefore, $a^\mu = 0$ only if the proper acceleration (the magnitude of the 3-acceleration in the rest frame) is equal to zero. By contrast, the 4-velocity can never vanish (in the rest frame of the particle $u^0 = c \neq 0$, which means that time always goes on and cannot be stopped, so $u^\mu \doteq (c, 0, 0, 0)$ in this frame).

One can give a physical meaning to the scalar product of two 4-velocities $u^\mu = \left(u^0, \mathbf{u}\right)$ and $\tilde{u}^\mu = \left(\tilde{u}^0, \tilde{\mathbf{u}}\right)$. Since $u^\mu \tilde{u}_\mu$ is a scalar, we can evaluate it in the rest frame of one of the two particles, say of the particle with 4-velocity u^μ. Then in this frame it is $u^\mu \doteq (c, \mathbf{0})$ and

$$u^\mu \tilde{u}_\mu = -u^0 \tilde{u}^0 + \mathbf{u} \cdot \tilde{\mathbf{u}} = -c\left(\gamma_{\tilde{u}}\, c\right) = -c^2 \gamma_{\tilde{u}} \tag{6.22}$$

(this equation reduces to $u_\mu u^\mu = -c^2$ if $\tilde{u}^\mu = u^\mu$), so $-u^\mu \tilde{u}_\mu$ *is* c^2 *times the Lorentz factor of the relative velocity of the two particles.* One can define a spacelike 4-vector v^α which, in the rest frame of u^μ, has components $v^\alpha \doteq (0, \mathbf{v})$ and such that

$$\tilde{u}^\mu = \gamma\left(u^\mu + v^\mu\right), \tag{6.23}$$

$$\gamma = \frac{1}{\sqrt{1 - \frac{v^\alpha v_\alpha}{c^2}}} \doteq \frac{1}{\sqrt{1 - \frac{v^2}{c^2}}}, \tag{6.24}$$

$$u^\mu v_\mu = 0. \tag{6.25}$$

This formalism is useful, for example, in the analysis of particle scattering or when one must examine a distribution of matter in a reference frame which is in motion with respect to the rest frame of that matter.

Similarly to Newtonian mechanics, we define the *4-momentum* of a particle of mass m as

$$p^\mu \equiv m u^\mu. \tag{6.26}$$

From $u^\mu u_\mu = -c^2$ it follows immediately that

$$p_\mu p^\mu = -m^2 c^2. \tag{6.27}$$

The components of the particle 4-momentum are

$$p^\mu = (\gamma\, mc, \gamma\, m\, \mathbf{v}) \equiv \left(\frac{E}{c}, \mathbf{p}\right) = \left(p^0, \mathbf{p}\right), \tag{6.28}$$

where E is the *relativistic energy*

$$E = \gamma\, mc^2 \tag{6.29}$$

and

$$\mathbf{p} = \gamma\, m\, \mathbf{v} \tag{6.30}$$

is the *relativistic 3-momentum*.[2]

For low speeds $|v|/c \ll 1$ it is $\gamma \equiv \frac{1}{\sqrt{1-\frac{v^2}{c^2}}} = 1 + \frac{v^2}{2c^2} + \cdots$ and

$$p^0 \equiv \frac{E}{c} = mc + \frac{mv^2}{2c} + \cdots \simeq \frac{\text{rest energy}}{c} + \frac{\text{Newtonian kinetic energy}}{c}, \tag{6.31}$$

where mc^2 is the *rest energy* of the particle. We also have

$$\mathbf{p} \equiv \gamma\, m\, \mathbf{v} = m\, \mathbf{v} + \cdots, \tag{6.32}$$

which reproduces the Newtonian 3-momentum in the slow-motion approximation. The component p^0 of the 4-momentum is the relativistic energy of the particle: for this

[2] To avoid confusion, pay attention to the rather unfortunate, but common, notation. The 3-velocity $d\mathbf{x}/dt$ is denoted with the familiar symbol $\mathbf{v}$ and $u^\mu = \left(u^0, \mathbf{u}\right)$ with $\mathbf{u} = \gamma\, \mathbf{v}$, but the other familiar symbol $\mathbf{p}$ denotes $\gamma\, m\mathbf{v}$, not $m\mathbf{v}$, with $p^\mu = \left(p^0, \mathbf{p}\right)$.

reason p^μ is also called the *energy-momentum 4-vector*. Its zero component can be regarded as a relativistic energy while its spatial part $\mathbf{p}$ is a relativistic 3-dimensional momentum.

Since the position 4-vector is $x^\mu = (ct, \mathbf{x})$, the 0-component being just time rescaled by c, identifying p^μ with $\left(\frac{E}{c}, \mathbf{p}\right)$ corresponds to thinking of energy as a quantity associated with 3-momentum in the same way that time is associated with spatial position. So, if the 3-momentum $\mathbf{p}$ is canonically conjugated with the position $\mathbf{x}$ in classical mechanics, also energy is canonically conjugated with time (if you have studied Hamiltonian mechanics this sounds familiar, doesn't it?).

6.2.1 Relativistic Energy

Having defined $p^\mu = (\gamma\, mc, \gamma\, m\mathbf{v}) \equiv \left(\frac{E}{c}, \mathbf{p}\right)$, the normalization $p_\mu p^\mu = -m^2c^2$ now gives the famous relation between relativistic energy and momentum:

$$E = \sqrt{p^2c^2 + m^2c^4} \tag{6.33}$$

(*free particle Hamiltonian*). In the rest frame of the particle, where $\mathbf{v} = 0$, this equation reduces to

$$E \doteq mc^2 \quad \text{(rest frame)}, \tag{6.34}$$

the most famous equation of physics! One may have expected that this renowned formula would require a lengthy and arcane derivation but, once the reality of the 4-world is accepted, this formula drops out without effort. It is natural to introduce $u^\mu = dx^\mu/d\tau$ and $p^\mu = mu^\mu$ and the formula follows immediately from the inescapable normalization $u^\mu u_\mu = -c^2$. It's really that simple, once the 4-dimensional world is accepted.

Equation (6.34) expresses the fact that a free particle possesses energy just because it has mass, and that a small mass can free up an enormous amount of energy because the factor c^2 is large in ordinary units (as demonstrated in nuclear reactions and nuclear bombs): this is the *equivalence of mass and energy*. 1 gram of mass stores an equivalent 9×10^{13} joules, or approximately 20 ktons, the energy of the Hiroshima bomb.[3]

---

[3] 1 ton (of TNT) is equivalent to 4.2×10^9 J.

At low speeds $\frac{|v|}{c} \ll 1$ (or $\frac{|p|}{mc} \ll 1$), Eq. (6.33) becomes

$$E = mc^2\sqrt{1 + \frac{p^2}{m^2c^2}} = mc^2\left(1 + \frac{p^2}{2m^2c^2} + \cdots\right). \tag{6.35}$$

Equation (6.33) is valid in any inertial frame because it is nothing but the covariant normalization of the momentum $p_\mu p^\mu = -m^2c^2$. E and $\mathbf{p}$ depend on the inertial frame since they are components of $p^\mu = \left(\frac{E}{c}, \mathbf{p}\right)$ but the *relation* between energy and momentum does not. $E^2 - p^2c^2 = m^2c^4 =$ const. in all inertial frames, in other words $c^2\, \eta_{\mu\nu}\, p^\mu p^\nu = -E^2 + p^2c^2 = -m^2c^4$ is a Lorentz invariant or world scalar.

It is customary to express the energy of subatomic particles in electronvolts. Remember that 1 eV is the energy of an electron (with charge $e = 1.6 \times 10^{-19}$ Coulomb) accelerated by the potential difference of 1 V:

$$1\,\text{eV} = 1.60 \times 10^{-19}\,\text{J} \tag{6.36}$$

and it is standard practice to *measure particle masses in electronvolts/c^2* using the equivalence between mass and energy (6.34). For example, the mass of the electron is $m_e = 9.11 \times 10^{-31}$ kg and its rest energy is, using (6.36),

$$m_ec^2 = 511\ \text{keV} = 8.20 \times 10^{-14}\,\text{J},$$

or

$$m_e = 0.511\ \frac{\text{MeV}}{c^2}. \tag{6.37}$$

In practice, physicists exchange units of mass and energy freely and refer to the mass of the electron as 0.511 MeV omitting the factor c^2 (in other words, using units in which $c = 1$). Similarly, the proton mass is

$$m_p = 1.673 \times 10^{-27}\text{kg} = 938.3\,\text{MeV/c}^2.$$

The equivalence between mass and energy manifests itself also in the fact that light behaves as if it had weight. A light ray is deflected by a gravitational field, resulting in the phenomenon of *gravitational lensing* studied in the context of General Relativity [1], or as energy loss when climbing out of a gravitational potential well (*gravitational redshift*). And electromagnetic radiation gravitates, in fact it is believed that it was the dominant source of gravity in the universe at some early time when the latter was extremely hot and dense [2–6].

Example 6.2 Compare the relations

$$-(ct)^2 + \mathbf{x}^2 = s^2 \quad \text{Lorentz invariant}$$

$$-\left(\frac{E}{c}\right)^2 + \mathbf{p}^2 = -m^2c^4 \quad \text{Lorentz invariant.}$$

This parallel suggests that $\left(\frac{E}{c}, \mathbf{p}\right)$ transforms in the same way as $(ct, \mathbf{x})$ under Lorentz transformations, i.e.,

$$\frac{E'}{c} = \gamma\left(\frac{E}{c} - \frac{v}{c}p_x\right),$$

$$p^{x'} = \gamma\left(p^x - \frac{v}{c^2}E\right),$$

$$p^{y'} = p^y,$$

$$p^{z'} = p^z;$$

this is true because p^μ transforms as a 4-vector, like x^μ.

6.2.2 *Pair Production and Annihilation*

The equivalence between mass and energy is demonstrated dramatically by the annihilation of pairs of matter-antimatter particles, for example electrons and positrons, into γ-ray photons, and by the phenomenon of *pair creation* in which γ-ray photons disappear and an electron-positron pair appears. Annihilation and creation of particle-antiparticle pairs occur for any kind of particle admitting a distinct antiparticle, not only for electrons and positrons. Since the work of Dirac in the 1920s we know that to every fundamental particle there corresponds an antiparticle with the same mass, spin, and opposite charge.[4] The association of matter with antimatter is a fundamental symmetry of nature. With the electron e^- is associated the positron e^+ and vice-versa; with the neutron n the antineutron $\bar{n}$; with the proton p^+ the antiproton p^-, etc. In *pair production* two γ-photons with energies $E_\gamma \geq m_e c^2 = 0.511$ MeV can materialize into an electron-positron pair e^-- e^+:

[4] Some neutral particles coincide with their own antiparticle.

$$\gamma + \gamma \longrightarrow e^- + e^+.$$

The total charge, momentum, angular momentum (and other quantum numbers such as, e.g., the lepton number)[5] are conserved.

Pair annihilation occurs when a particle and its antiparticle collide; they then disappear generating two γ-photons,

$$e^- + e^+ \longrightarrow \gamma + \gamma.$$

Again the total charge, momentum, angular momentum, and lepton number are conserved. The annihilation of an electron-positron pair produces a characteristic spectral line at 511 keV. This spactral line has several applications: for example, it is used in medical physics in Positron Emission Tomography (PET), a standard diagnostic technique in which a radioactive chemical that emits positrons is injected in the body. The characteristic annihilation γ line is emitted from regions of the body where the chemical has concentrated due to physiological processes. This concentration, visualized in PET scans, allows doctors to measure blood flow in the brain or other tissues following strokes, brain tumours, or brain injuries. Another application is in astrophysics, where the detection of the 511 keV line by γ-ray detectors in space allows one to infer the presence of processes at these energies, or the temperature near a compact object such as a black hole.

Another application is positron annihilation spectroscopy (PAS), in which the lifetime of positrons emitted by a radioactive source depends on the differential density of electrons. PAS is used to study the structure of materials in solid state physics, for example to detect dislocations or gaps in a crystal.

We live in a universe composed almost entirely of matter, with essentially no antimatter. Given that the early universe must have been a very energetic place with constant production and annihilation of matter and antimatter it is surprising that, when it cooled, there was a predominance of matter versus antimatter left over, instead of a simple bath of gamma photons. This matter-antimatter asymmetry is an unsolved puzzle of cosmology and particle physics.

6.2.3 *Positron Emission Tomography

Positron Emission Tomography (PET) is used extensively in clinical oncology for the imaging of tumors and metastases and for diagnosing certain brain diseases, as well as in pure medical research (e.g., [7]). A short-lived isotope of a tracer is injected into a living body by binding it to a biologically reactive molecule (typically, glucose, water, or ammonia). This tracer is carried around the body and concentrates

[5] The known leptons are divided into three families: the electron e^- and the electron neutrino ν_e; the muon μ^- and the muon neutrino ν_μ; and the τ^- muon and its neutrino ν_τ, plus their respective antiparticles e^+, $\bar{\nu}_e$, μ^+, $\bar{\nu}_\mu$, τ^+, and $\bar{\nu}_\tau$. A lepton number $+1$ is assigned to the electron and the electron neutrino, a lepton number -1 is carried by their respective antiparticles e^+ and $\bar{\nu}_e$, and all other particles have lepton number zero.

in the tissues under study, where the radioisotope decays emitting positrons. When a positron annihilates with an electron, it emits a pair of 511 keV (γ) photons traveling in approximately opposite directions. These photons are detected by scintillators in the PET scanning machine, in which bursts of light arrive to photomultipliers. The near-simultaneous (within less than 10 ns from each other) detection of two γ photons moving in approximately opposite directions is a signature event. Within the experimental error, the reconstruction technique identifies the source of the γ photons along the line joining the two detection events and at a certain position along this line determined by the arrival time of each photon. In this way, using millions of data points, images of the tissue can be constructed.

The radiotracers employed are typically ^{11}C, ^{18}F, ^{13}N, and a few other radioactive elements; their half-lives are 20, 110, and 10 min, respectively. Because they are so short-lived, these radiotracers must be produced next to the PET scanning facility or reasonably nearby using a cyclotron, and then quickly bound to a biomolecule using standard chemistry but in a "hot lab" equipped for handling radioactive reagents. A significant amount of data processing is necessary. Modern PET technology combines PET scans with computerized tomography (CT) or magnetic resonance imaging (MRI) in order to improve the amount and quality of the information obtained.

6.3 The Relativistic Force

Let us consider now the notion of force in Special Relativity. This concept was fundamental in Newtonian mechanics and we need to discuss both the 4-dimensional and the 3-dimensional notions of force in the new special-relativistic mechanics.

6.3.1 The Relativistic 4-Force

Definition 6.1 The *4-force* acting on a particle of mass m is defined in a way reminiscent of Newton's second law but with a 4-dimensional flavour:

$$f^{\mu} \equiv \frac{dp^{\mu}}{d\tau}. \tag{6.38}$$

For a particle of constant mass m this definition reduces to

$$f^{\mu} \equiv ma^{\mu}, \tag{6.39}$$

Equation (6.38) is the 4-dimensional analogue of Newton's second law. It is not derived from more fundamental laws: simply, the definitions given are consistent with it. It is covariant, i.e., it takes the same form in all inertial frames, and it reduces to $dp^{\mu}/d\tau = 0$ when the particle is free (*conservation of 4-momentum*). Moreover, from Eq. (6.18) it follows immediately that

$$f^{\mu} u_{\mu} = 0 \tag{6.40}$$

for particles of constant mass m. However, if the mass m of the particle is not constant, the particle 4-momentum is not conserved: to begin with, the 4-force is

$$f^{\mu} = \frac{dp^{\mu}}{d\tau} = \frac{dm}{d\tau} u^{\mu} + m \frac{du^{\mu}}{d\tau} \equiv \dot{m}\, u^{\mu} + m a^{\mu}$$

and has a component $\dot{m} u^{\mu}$ along the time direction, in addition to the usual purely spatial component $m a^{\mu}$. The projection of the 4-force onto the time direction u^{μ} is then

$$f^{\mu} u_{\mu} = \left(\dot{m}\, u^{\mu} + m a^{\mu}\right) u_{\mu} = -\dot{m} c^2 + m a^{\mu} u_{\mu} = -\dot{m} c^2 \neq 0. \tag{6.41}$$

In this case $p^{\mu} p_{\mu} = -m^2 c^2$ is not constant. By differentiating this normalization relation and accounting for a non-vanishing $\dot{m} \equiv dm/d\tau$, one obtains again

$$f^{\mu} u_{\mu} = -\dot{m} c^2 = -\gamma \frac{dm}{dt} c^2, \tag{6.42}$$

where the last equality follows from the time dilation formula $dt = \gamma d\tau$.

Example 6.3 Compute the 4-force required to accelerate a particle of mass m along the worldline

$$t(\tau) = \frac{c}{a} \sinh\left(\frac{a\tau}{c}\right),$$

$$x(\tau) = \frac{c^2}{a} \cosh\left(\frac{a\tau}{c}\right),$$

$$y = 0,$$

$$z = 0.$$

We have

$$u^{\mu} \equiv \frac{dx^{\mu}}{d\tau} = \left(c \cosh\left(\frac{a\tau}{c}\right), c \sinh\left(\frac{a\tau}{c}\right), 0, 0 \right), \tag{6.43}$$

$$a^{\mu} \equiv \frac{du^{\mu}}{d\tau} = \left(a \sinh\left(\frac{a\tau}{c}\right), a \cosh\left(\frac{a\tau}{c}\right), 0, 0 \right) \tag{6.44}$$

and

$$a_{\mu}a^{\mu} = -a^2 \sinh^2\left(\frac{a\tau}{c}\right) + a^2 \cosh^2\left(\frac{a\tau}{c}\right) = a^2, \tag{6.45}$$

$$f^{\mu} = ma^{\mu} = ma\left(\sinh\left(\frac{a\tau}{c}\right), \cosh\left(\frac{a\tau}{c}\right), 0, 0 \right). \tag{6.46}$$

Of course, f^{μ} is orthogonal to the 4-velocity:

$$\begin{aligned} f_{\mu}u^{\mu} &= -f^0u^0 + f^1u^1 + f^2u^2 + f^3u^3 \\ &= -mac \sinh\left(\frac{a\tau}{c}\right)\cosh\left(\frac{a\tau}{c}\right) + mac \cosh\left(\frac{a\tau}{c}\right)\sinh\left(\frac{a\tau}{c}\right) = 0. \end{aligned}$$

6.3.2 The Relativistic 3-Force

Let us write the equation for the 4-force (6.38) using a derivative with respect to the coordinate time t instead of the proper time:

$$f^{\mu} = \frac{dp^{\mu}}{d\tau} = \frac{dp^{\mu}}{dt}\frac{dt}{d\tau} = \gamma\frac{dp^{\mu}}{dt}. \tag{6.47}$$

Defining the *relativistic 3-dimensional force* as

$$\mathbf{F} = \frac{d\mathbf{p}}{dt}, \tag{6.48}$$

where $\mathbf{p} = \gamma m \mathbf{v}$ is the relativistic 3-momentum, we have

$$f^{\mu} = \left(f^0, \mathbf{f}\right) = \left(\frac{\gamma}{c}\frac{dE}{dt}, \gamma\mathbf{F}\right), \tag{6.49}$$

where

$$\mathbf{f} = \gamma\,\mathbf{F} = \gamma\,\frac{d}{dt}\,(\gamma m \mathbf{v})\,, \tag{6.50}$$

and where $\mathbf{F}$ is constructed with the 3-dimensional components of the 4-force. dE/dt is the power, i.e., the rate at which the force transfers energy to the particle.

Since $f_\mu u^\mu = 0$ for a particle of constant mass, we have

$$-f^0 u^0 + \mathbf{f}\cdot\mathbf{u} = 0,$$

$$f^0 \gamma c = \mathbf{f}\cdot\mathbf{u},$$

$$f^0 = \frac{f\cdot\mathbf{u}}{\gamma c} = \mathbf{F}\cdot\frac{\mathbf{u}}{c} = \mathbf{F}\cdot\gamma\,\frac{\mathbf{v}}{c}$$

and

$$f^\mu = \left(\gamma\,\mathbf{F}\cdot\frac{\mathbf{v}}{c},\,\gamma\,\mathbf{F}\right), \tag{6.51}$$

where $\mathbf{v} = d\mathbf{x}/dt$ is the usual 3-dimensional velocity. The time component of the equation of motion $f^\mu = dp^\mu/d\tau$ becomes $f^0 = dp^0/d\tau$ or $\gamma\,\mathbf{F}\cdot\dfrac{\mathbf{v}}{c} = \dfrac{\gamma}{c}\,dE/dt$:

$$\frac{dE}{dt} = \mathbf{F}\cdot\mathbf{v}, \tag{6.52}$$

which expresses the energy loss of the particle as work done against the 3-dimensional relativistic force $\mathbf{F}$. This equation is a consequence of the other three equations of motion $\mathbf{F} = d\mathbf{p}/dt$ in the limit $|v|/c \ll 1$, $\mathbf{f} = \gamma\,\mathbf{F} \approx \mathbf{F} \equiv d\mathbf{p}/dt$.

Note the parallel between the expression of the 4-force and that of the 4-velocity:

$$f^\mu = \left(\gamma\,\frac{dE}{dt},\,\gamma\,\mathbf{F}\right), \qquad u^\mu = (\gamma\,c,\,\gamma\,\mathbf{v})\,. \tag{6.53}$$

For a particle subject to a constant 3-force (cause) the 3-acceleration (effect) is smaller than the corresponding acceleration in Newtonian mechanics and, at higher and higher speeds, it is harder and harder to accelerate the particle. As a result, it is impossible to accelerate the massive particle to speed c. In fact,

$$\mathbf{F} = \frac{d\mathbf{p}}{dt} = \frac{d}{dt}\,(\gamma m\mathbf{v})\,. \tag{6.54}$$

Considering, for simplicity, motion in one dimension, we have

$$F = m\frac{d}{dt}\left(\frac{v}{\sqrt{1-\frac{v^2}{c^2}}}\right) = m\left[\frac{\frac{dv}{dt}}{\sqrt{1-\frac{v^2}{c^2}}} + v\frac{-\left(-\frac{2v}{c^2}\right)\frac{dv}{dt}}{2\left(1-\frac{v^2}{c^2}\right)^{3/2}}\right]$$

$$= \frac{m}{\sqrt{1-\frac{v^2}{c^2}}}\frac{dv}{dt}\left[1+\frac{\frac{v^2}{c^2}}{1-\frac{v^2}{c^2}}\right]$$

$$= \frac{m}{\sqrt{1-\frac{v^2}{c^2}}}\frac{dv}{dt}\frac{1-\frac{v^2}{c^2}+\frac{v^2}{c^2}}{\left(1-\frac{v^2}{c^2}\right)} = \frac{m}{\left(1-\frac{v^2}{c^2}\right)^{3/2}}\frac{dv}{dt}$$

and

$$\frac{dv}{dt} = \frac{F}{m}\left(1-\frac{v^2}{c^2}\right)^{3/2} < \frac{F}{m}, \quad \text{not} \quad \frac{\mathrm{dv}}{\mathrm{dt}} = \frac{\mathrm{F}}{\mathrm{m}}. \tag{6.55}$$

As $v \to c$ the 3-acceleration dv/dt caused by any finite constant force tends to zero. It is impossible to accelerate a massive particle to speed c. As we have seen in Chap. 2, the possibility of breaking this barrier would jeopardize causality, but causality turns out to be protected by the impossibility of reaching c for a massive particle. The following calculation shows that this goal would require an infinite amount of work done on the particle.

6.3.3 *Relativistic Kinetic Energy*

The (Newtonian) work done by a relativistic 3-force **F** parallel to the x-axis is

$$W = \int_{x_1}^{x_2} F(x)\,\mathrm{d}x \equiv \int_{x_1}^{x_2} \mathrm{d}x\,\frac{dp}{dt}; \tag{6.56}$$

now, in relativity, we have

$$\frac{dp}{dt} = \frac{d}{dt}\left(\frac{mv}{\sqrt{1-\frac{v^2}{c^2}}}\right) = \frac{m}{\left(1-\frac{v^2}{c^2}\right)^{3/2}}\frac{dv}{dt}$$

(which we have already computed to derive Eq. (6.55)), so that

$$W = \int_{x_1}^{x_2} \mathrm{d}x \, \frac{m}{\left(1 - \frac{v^2}{c^2}\right)^{3/2}} \frac{dv}{dt}.$$

Formally, $\mathrm{d}x \, \frac{dv}{dt} = v\mathrm{d}v$ and

$$W = \int_0^v \mathrm{d}v' \frac{mv'}{\left(1 - \frac{v'^2}{c^2}\right)^{3/2}}$$

assuming that the particle is initially at rest. Then we have

$$W = -\frac{mc^2}{2} \int_0^v \frac{\mathrm{d}\left(1 - \frac{v'^2}{c^2}\right)}{\left(1 - \frac{v'^2}{c^2}\right)^{3/2}} = -\frac{mc^2}{2} \int_1^{1-\frac{v^2}{c^2}} \frac{\mathrm{d}s}{s^{3/2}}$$

$$= \frac{mc^2}{2} \int_{\left(1-\frac{v^2}{c^2}\right)}^{0} \mathrm{d}s \, s^{-3/2} = \frac{mc^2}{2} \left[-2s^{-1/2}\right]_{1-\frac{v^2}{c^2}}^{1}$$

and finally

$$W = mc^2 \left(\frac{1}{\sqrt{1 - \frac{v^2}{c^2}}} - 1 \right). \tag{6.57}$$

We have thus established the *work-energy theorem*

$$W = \Delta T \tag{6.58}$$

where

$$T = \frac{mc^2}{\sqrt{1 - \frac{v^2}{c^2}}} - mc^2 = (\gamma - 1) \, mc^2 \tag{6.59}$$

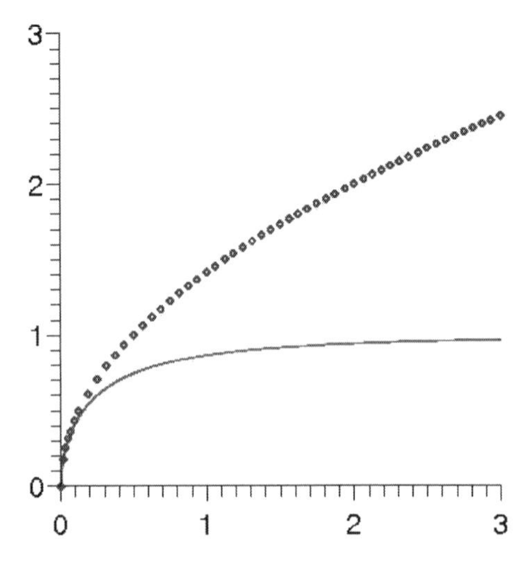

Fig. 6.1 The speed of a relativistic free particle v/c as a function of the rescaled kinetic energy T/mc^2 (*solid curve*), compared with its Newtonian counterpart (*dotted curve*)

is the *relativistic kinetic energy* (the relativistic energy of a free particle γmc^2 minus its rest energy mc^2). The work done on a particle is equal to the variation of its kinetic energy.

Equations (6.58) and (6.59) are confirmed every day in accelerators. In the limit $|v|/c \ll 1$ the quantity T reduces to the Newtonian kinetic energy,

$$T = mc^2 \left(\frac{1}{\sqrt{1 - \frac{v^2}{c^2}}} - 1 \right) = mc^2 \left(1 + \frac{v^2}{2c^2} + \cdots - 1 \right) \approx \frac{mv^2}{2}. \tag{6.60}$$

The speed v as a function of the kinetic energy T of a free particle is

$$v(T) = c\sqrt{1 - \left(\frac{T}{mc^2} + 1 \right)^{-2}}, \tag{6.61}$$

and it is easy to see that $v \longrightarrow c$ as $T \rightarrow +\infty$. The function $\frac{v}{c}$ of the ratio $\left(\frac{T}{mc^2} \right)$ is plotted in Fig. 6.1, where it is compared with its Newtonian counterpart $v_{Newtonian} = \sqrt{2T/m}$.

The *total energy of the free particle* is

$$E = T + mc^2 = \gamma\, mc^2. \tag{6.62}$$

This formula is, of course, consistent with the relation (6.33) between energy and momentum. In fact, remember that

$$p^{\mu} = \left(\frac{E}{c}, \gamma m \,\mathbf{v}\right) \equiv \left(p^0, \mathbf{p}\right)$$

and, therefore,

$$\begin{aligned} p^2c^2 + m^2c^4 &= m^2c^4 + \frac{m^2c^2v^2}{1-\frac{v^2}{c^2}} = m^2c^4\left(1+\frac{\frac{v^2}{c^2}}{1-\frac{v^2}{c^2}}\right) \\ &= \frac{m^2c^4}{1-\frac{v^2}{c^2}} = E^2. \end{aligned}$$

A particle with energy much larger than its rest energy, $E \gg mc^2$, is said to be *ultrarelativistic*.

6.3.4 Motion with Constant Acceleration

In Newton's theory a constant acceleration of a massive particle with velocity $u = dx/dt$ is defined by du/dt =const.$\equiv a$, which implies that $u(t) = u_0 + at \longrightarrow +\infty$ as $t \to +\infty$ if $a > 0$. This is not possible in Special Relativity where $|u| < c$. One needs a different definition of "constant acceleration".

Definition 6.2 The acceleration of a particle is constant if and only if it has the same value at each instant in any inertial frame instantaneously comoving with the particle (this frame keeps changing at every instant of time).[6]

At different t-times there are different inertial frames which are instantaneously comoving with the particle, but they all measure the same acceleration a. The definition can be stated by saying that the *proper* acceleration of the particle is constant.

If $\mathbf{u} = \mathbf{u}\,(t)$ is the velocity of the particle in a frame S, in any comoving frame (i.e., in any frame moving at speed $\mathbf{v} = \mathbf{u}$), the velocity of the particle is $u' \equiv 0$ and $du'/dt' =$ const. $\equiv a$. Then, remembering the transformation law of accelerations (2.38)–(2.40) we have, setting $v = u^x = u(t)$ and $\gamma(v) = \gamma(u(t))$,

$$\frac{du}{dt} = \frac{1}{\gamma^3\left(1+\frac{vu'}{c^2}\right)^3}\frac{du'}{dt'} = \left(1-\frac{u^2}{c^2}\right)^{3/2} a. \tag{6.63}$$

[6] In other words, the proper 3-acceleration of a particle is not invariant under Lorentz transformations, while it is invariant under Galilei transformations.

Integrate the equation

$$\left(1-\frac{u^2}{c^2}\right)^{-3/2}\frac{du}{dt}=a$$

between t_0 and t and use the fact that

$$\int\frac{\mathrm{d}\zeta}{\sqrt{\left(A^2-\zeta^2\right)^3}}=\frac{\zeta}{A^2\sqrt{A^2-\zeta^2}}$$

to obtain

$$\int \mathrm{d}u\left(1-\frac{u^2}{c^2}\right)^{-3/2}=c^3\int\frac{\mathrm{d}u}{\sqrt{\left(c^2-u^2\right)^3}}=\frac{c^3u}{c^2\sqrt{c^2-u^2}}$$

and

$$\left.\frac{cu'}{\sqrt{c^2-u'^2}}\right|_{u_0}^{u}=a\,(t-t_0)\,.$$

Assume the initial condition $u_0=0$, then

$$\frac{u}{\sqrt{1-\frac{u^2}{c^2}}}=a\,(t-t_0)\,.$$

This is a transcendental equation for u which can be solved exactly and yields

$$u^2=a^2\,(t-t_0)^2\left(1-\frac{u^2}{c^2}\right).$$

Isolating u^2 one obtains

$$u^2\left[1+\frac{a^2}{c^2}\,(t-t_0)^2\right]=a^2\,(t-t_0)^2\,,$$

and finally

$$u=\frac{a\,(t-t_0)}{\sqrt{1+\frac{a^2}{c^2}\,(t-t_0)^2}}=\frac{dx}{dt}.$$

Integrating again, we find

$$x\,(t)-x_0=\int_{t_0}^{t}\mathrm{d}t'\,\frac{a\left(t'-t_0\right)}{\sqrt{1+\frac{a^2}{c^2}\,(t'-t_0)^2}}.$$

Setting $\xi \equiv a\,(t-t_0)/c$ we have

$$
\begin{aligned}
x\,(t)-x_0 &= \frac{c^2}{a}\int_0^{\frac{a}{c}(t-t_0)} \mathrm{d}\xi' \,\frac{\xi'}{\sqrt{1+\xi'^2}} \\
&= \frac{c^2}{2a}\int_1^{1+\left[\frac{a}{c}(t-t_0)\right]^2} \left(1+\xi'^2\right)^{-1/2} \mathrm{d}\left(1+\xi'^2\right) \\
&= \frac{c^2}{a}\,[s]_1^{1+\left[\frac{a}{c}(t-t_0)\right]^2} \\
&= \frac{c^2}{a}\sqrt{1+\frac{a^2}{c^2}\,(t-t_0)^2}-\frac{c^2}{a} \\
&= \frac{c}{a}\sqrt{c^2+a^2\,(t-t_0)^2}-\frac{c^2}{a}.
\end{aligned}
$$

Re-arranging the terms, it is

$$
x-x_0+\frac{c^2}{a}=\frac{c^2}{a}\sqrt{1+\frac{a^2}{c^2}\,(t-t_0)^2}
$$

$$
\left(x-x_0+\frac{c^2}{a}\right)^2=\frac{c^4}{a}\left[1+\frac{a^2}{c^2}\,(t-t_0)^2\right]
$$

$$
\frac{\left(x-x_0+\frac{c^2}{a}\right)^2}{\left(\frac{c^2}{a}\right)^2}=1+\frac{1}{\frac{c^2}{a^2}}\,(t-t_0)^2
$$

and

$$
\frac{\left(x-x_0+\frac{c^2}{a}\right)^2}{\left(\frac{c^2}{a}\right)^2}-\frac{(ct-ct_0)^2}{\left(\frac{c^2}{a}\right)^2}=1. \tag{6.64}
$$

Adopting initial conditions $t_0=0$ and $x_0=c^2/a$ for simplicity, Eq. (6.64) reduces to

$$
x^2-(ct)^2=\left(\frac{c^2}{a}\right)^2, \tag{6.65}
$$

which is represented geometrically by a family of hyperbolae parametrized by a in the (x,t) plane (for this reason the uniformly accelerated motion is often called *hyperbolic motion*). These hyperbolae are the loci of points at constant spacetime distance from the origin and can be written as

$$ct = \pm\sqrt{x^2 - \left(\frac{c^2}{a}\right)^2}. \tag{6.66}$$

A parametrization of the hyperbolae is

$$ct\,(\tau) = \frac{c^2}{a}\,\sinh\left(\frac{a\tau}{c}\right), \tag{6.67}$$

$$x\,(\tau) = \frac{c^2}{a}\,\cosh\left(\frac{a\tau}{c}\right), \tag{6.68}$$

with y and z constant. It is easy to see, by substituting Eqs. (6.67) and (6.68), that Eq. (6.65) is satisfied. Each hyperbola is the worldline of a uniformly accelerated observer travelling in the x-direction. This observer O has an *event horizon*: light can only travel along 45° lines in a spacetime diagram and there is a region such that, if you send signals from it, they will never reach the observer (Fig. 6.2).

An event horizon is a barrier beyond which no information can be transmitted, or a causal barrier. In this respect, the acceleration horizon is similar to a black hole horizon, a surface which lets objects and radiation go through only in one direction. The horizon caused by uniform acceleration is referred to as a *Rindler* or *acceleration horizon* and is a null surface. Uniformly accelerated observers are important in quantum field theory; a uniformly accelerated observer in Minkowski spacetime will detect a thermal bath of particles of a quantum field at temperature $T = \frac{\hbar}{K_B c}\frac{a}{2\pi}$, where $\hbar$ is the reduced Planck constant and K_B is the Boltzmann constant (*Unruh temperature*), while a stationary inertial observer detects zero quanta of the field (*Unruh effect*) [8–10]. An accelerated observer moving along a constant acceleration path is called a *Rindler observer*. The Minkowski spacetime can be threaded by the worldlines of Rindler observers and their acceleration horizons are relative to these observers.

6.3.5 *Particle Accelerators

For a *given energy* E, the Lorentz factor of a particle is $\gamma = \frac{E}{mc^2}$, which means that light particles such as electrons and positrons can be made much more relativistic than heavier particles such as protons or nuclei, i.e., they can get closer to the speed of light or, equivalently, achieve larger γ-factors (the two statements are equivalent since $\beta = \left(1 - 1/\gamma^2\right)^{-1/2}$). When a particle is accelerated to speeds near the speed of light, further acceleration increases its kinetic energy $E = \gamma mc^2$ and its momentum $p = \gamma mv$ and, as a consequence, the greater the initial velocity the harder it is to

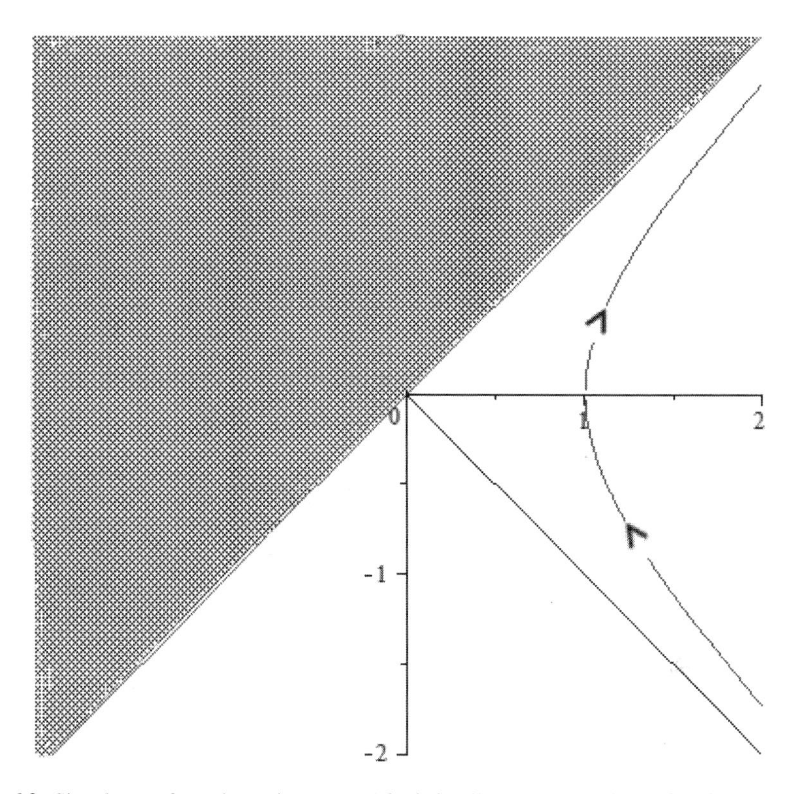

Fig. 6.2 Signals sent from the region $t \geq x$ (*shaded*) will never reach the uniformly accelerated observer with worldline described by the hyperbola. In order to do so these signals would have to travel faster than light and exit from the local light cone, with worldline tangent making an angle smaller than 45 ° with the horizontal

accelerate this particle further. It is useful to look at some numbers to get a feel for this aspect of relativistic dynamics.

An electron is easily accelerated; if it has energy $E = \gamma m_e c^2$ its speed is

$$v = c\sqrt{1 - \left(\frac{m_e c^2}{E}\right)^2}$$

in the laboratory frame. An 1 MeV electron has speed $v = 0.86c$. Accelerating it further to an energy of 10 MeV gives it the speed $v = 0.99869c$. Further acceleration to an energy of 50 MeV gives it the speed $v = 0.9995c$ and the energy of 1 GeV further brings it to $v = 9.999998c$. Therefore, electrons of vastly different energies (10 MeV or 1 GeV) do not differ much in speed.

Particle accelerators provide beams of high-energy charged particles which are scattered against fixed targets or other particles to study interations in particle physics. In the early 1900s Rutherford used natural radioactive samples as sources of

α-particles (helium nuclei, consisting of two protons and two neutrons) to probe the structure of the atom and later on cosmic rays were employed.[7] Over the years is turned out to be more practical to build accelerators as artificial sources of high-energy particles to use in scattering experiments. The first particle accelerators used static electric fields and, later, alternating electric fields were used, both in linear accelerator configurations (LINACs) and in cyclotrons. For heavier particles, linear drift tubes are limited by their size while, in a cyclotron, a particle beam can be circulated many times and accelerated repeteadly in a much smaller structure, achieving the same energy that would require a larger linear accelerator. In a cyclotron a strong magnetic field keeps charged particles on a circular orbit. The particle beam circulates inside two D-shaped cavities with a gap in between. An alternating electric field in the gap is responsible for accelerating the charged particles and must be kept synchronized with the particles because of the special-relativistic relation $E = \gamma mc^2$ replacing the Newtonian one $E = mv^2/2$ (hence the name "synchrotron").

While circular synchrotrons are very convenient for heavier particles, they have a very serious limitation for electrons and positrons. Accelerated charges emit electromagnetic radiation (*synchrotron radiation*). This effect is minimal for linear trajectories but is maximized for circular orbits. The energy lost to synchrotron radiation during each turn scales as γ^4/r, where r is the radius of the orbit and it is easy to see how very relativistic particles such as electrons and positrons will lose a much larger fraction of their energy than heavier particles like protons, antiprotons, or nuclei. The Large Hadron Collider (LHC) [11] at CERN in Geneva accelerates protons or lead ions to energies of 14 TeV. The energy lost by synchrotron emission by these particles is a negligible fraction of their total energy (although it must be resupplied). However, if the beams were composed of electrons or positrons with comparable energy, their energy loss would be much larger and the energy of the particle beams would be hopelessly depleted for the purpose of colliding high-energy particles and probe smaller and smaller spatial scales. Essentially, one would waste all the particle energy into synchrotron radiation. Acceleration rings for electrons or positrons become impractical for fundamental physics experiments at energies above $\sim$100 GeV and linear accelerators become useful again for these light particles because of the much smaller loss by syncrotron radiation in a linear trajectory. A possible future facility under design and intended to replace the LHC when it will go out of commission is a linear accelerator. The International Linear Collider (ILC) aims to accelerate beams of electrons at 0.5–1 TeV energies [12]. Circular rings are much more practical for accelerating heavy particles and are still useful for technical applications and synchrotron light imaging by accelerating electrons at moderate energies.

---

[7] High energy cosmic rays are much more energetic than the most energetic particle beams in accelerators. However, one cannot predict their location and trajectory, while particle beams in an accelerator can be aimed precisely at a target and their energies can be controlled.

6.4 Angular Momentum of a Particle

For a Newtonian particle of position $\mathbf{x}$ with respect to a given origin and of momentum $\mathbf{p}$, the angular momentum is $\mathbf{L} = \mathbf{x} \times \mathbf{p}$ and the torque of a force $\mathbf{f}$ is $\mathbf{N} = \mathbf{x} \times \mathbf{f}$. We know that $\mathbf{N} = \frac{d\mathbf{L}}{dt}$ and therefore, if the torque $\mathbf{N}$ vanishes, the angular momentum $\mathbf{L}$ is conserved. Let us see what the analogues of these Newtonian properties are in Special Relativity.

The definition of angular momentum carries over to Special Relativity, provided that $\mathbf{p}$ represents the relativistic 3-momentum.

Definition 6.3 For a particle of 4-momentum p^μ at position x^μ, the *angular momentum (pseudo-)tensor* is defined as

$$l^{\mu\nu} \equiv x^\mu p^\nu - x^\nu p^\mu \equiv 2x^{[\mu} p^{\nu]}. \tag{6.69}$$

This object is a pseudo-tensor because $l_{\mu\nu}$ can be written as

$$l_{\mu\nu} = \varepsilon_{\mu\nu\rho\sigma}\, x^\rho p^\sigma, \tag{6.70}$$

where $\varepsilon_{\mu\nu\rho\sigma}$ is the Levi-Civita symbol defined by Eq. (4.88). $l^{\mu\nu}$ is obviously antisymmetric, $l^{\mu\nu} = l^{[\mu\nu]}$.

The theorem of *conservation of angular momentum* states that the angular momentum of a particle which is not subject to forces is conserved,

$$f^\mu = 0 \quad \Rightarrow \quad \frac{dl^{\mu\nu}}{d\tau} = 0, \tag{6.71}$$

where τ is the proper time of the particle parametrizing its worldline. In fact, by differentiating Eq. (6.70) with respect to τ, one obtains

$$\begin{aligned}
\frac{dl_{\mu\nu}}{d\tau} &= \varepsilon_{\mu\nu\rho\sigma}\left(\frac{dx^\rho}{d\tau}\, p^\sigma + x^\rho\, \frac{dp^\sigma}{d\tau}\right) \\
&= \varepsilon_{\mu\nu\rho\sigma} u^\rho p^\sigma + \varepsilon_{\mu\nu\rho\sigma} x^\rho f^\sigma \\
&= m\, \varepsilon_{\mu\nu\rho\sigma} u^\rho u^\sigma + \varepsilon_{\mu\nu\rho\sigma} x^\rho f^\sigma .
\end{aligned}$$

The first term of the last line vanishes because the Levi-Civita symbol is antisymmetric in the indices ρ and σ while the product $u^\rho u^\sigma$ is instead symmetric. In the absence of forces, $f^\mu = 0$, or when the "torque" $N_{\mu\nu} \equiv \varepsilon_{\mu\nu\rho\sigma} x^\rho f^\sigma$ vanishes, it is $\frac{dl_{\mu\nu}}{d\tau} = 0$ and,

$$\frac{dl^{\alpha\beta}}{d\tau} = \frac{d\left(\eta^{\mu\alpha}\eta^{\nu\beta}l_{\mu\nu}\right)}{d\tau} = \eta^{\mu\alpha}\eta^{\nu\beta}\frac{dl_{\mu\nu}}{d\tau} = 0.$$

6.5 Particle Systems

Here we define briefly the momentum and angular momentum of discrete systems of point particles. Continuous systems such as elastic media and fluids will be discussed later in Chap. 9. The theory of point particle systems becomes a bit more specialized when applied to scattering processes and when discussing the conservation of angular momentum: here we omit the relevant discussion and refer the reader to more advanced textbooks (e.g., [5, 13]). It should be kept in mind that, in a discrete point-particle system, the inter-particle separation cannot be constant: rigid bodies correspond to infinite sound speed and are forbidden in Special Relativity.

For a system of N particles of masses $m_{(a)}$ and 4-momenta $p^\mu_{(a)}$, one defines the *total mass*

$$M \equiv \sum_{a=1}^{N} m_{(a)} \tag{6.72}$$

(where the subscript (a) denotes the particle, not a tensor component), the *total momentum*

$$P^\mu \equiv \sum_{a=1}^{N} p^\mu_{(a)} \tag{6.73}$$

and,[8] fixed an origin, the *total angular momentum* is

$$L^{\mu\nu} = \sum_{a=1}^{N} l^{\mu\nu}_{(a)}. \tag{6.74}$$

The law of conservation of momentum in Special Relativity is not derived but it is postulated because of its consistency with experiment, much as Newton's second law in classical mechanics. This law says that, for a system of particles in which there are no external forces and subject only to interactions (for example, collisions) between its particles,

$$P^\mu \text{ (before collision)} = P^\mu \text{ (after collision)}. \tag{6.75}$$

[8] It is not entirely trivial that P^μ is a 4-vector because it is obtained by adding 4-vectors at a single time, but different observers do not agree that these different 4-vectors at different locations are simultaneous. Nevertheless, it can be shown that P^μ is indeed a 4-vector [5, 13]. The same can be said of the angular momentum $L^{\mu\nu}$.

Since $P^{\mu} = \left(\frac{E}{c}, \mathbf{P}\right)$, we have the law of conservation of mass-energy

$$E\ (\text{before}) = E\ (\text{after})$$

and the 3-momentum conservation

$$\sum_{a=1}^{N} p^{\mu}_{(a)}\ (\text{before}) = \sum_{b=1}^{N} p^{\mu}_{(b)}\ (\text{after})\,. \tag{6.76}$$

One cannot have one law without the other because both energy and relativistic 3-momentum are part of the same 4-vector.

In the following section we discuss the conservation of mass-energy. Dynamical considerations on a system of point particles are less obvious than in Newtonian mechanics. For example, consider the concept of centre of mass, taking a system of two identical particles of equal mass and opposite velocities approaching each other. In Newtonian mechanics the centre of mass is halfway between the two particles. In Special Relativity an observer at rest with respect to one of the two identical particles will see the other particle (which has equal rest mass) approaching with the entire kinetic energy of the system. For this reason, the second particle is heavier than the first and the centre of mass will be closer to the second particle than to the first one at rest in this frame. Different inertial frames will assign different velocities and different kinetic energies to the two particles and the location of the centre of mass will differ in these frames. However, one can still define a 3-velocity

$$\mathbf{v}_{cm} \equiv \frac{\mathbf{P}}{M}, \tag{6.77}$$

where

$$\mathbf{P} \equiv \sum_{a=1}^{N} \mathbf{p}_{(a)}, \tag{6.78}$$

$\mathbf{p}_{(a)} = \gamma\left(v_{(a)}\right) m_{(a)} \mathbf{v}_{(a)}$ is the relativistic 3-momentum of each particle, and M is the total mass (6.72). In the inertial frame S' (*centre of momentum frame*) moving with velocity $\mathbf{v}_{cm}$ with respect to the original inertial frame S in which Eqs. (6.72) and (6.73) are defined, the total momentum (6.73) has no spatial components.

6.6 Conservation of Mass-Energy

For a particle at rest it is $E = mc^2$. Since in ordinary units $c^2 = 9 \times 10^{16}\,\text{m}^2/\text{s}^2$ is a very large number, a small amount of mass is equivalent to a very large energy. Dramatic verifications of this formula in nuclear physics include nuclear reactions, nuclear bombs, and nuclear power plants. In particle physics they include creation and annihilation of particle-antiparticle pairs. In these situations mass is converted into energy and energy into mass. Since mass can be converted into energy and vice-versa, conservation of mass and conservation of energy do not hold separately and one has *conservation of mass-energy* instead:

the total mass-energy of a system of particles is conserved.

Consider an isolated system of particles that are allowed to interact with each other (e.g., through scattering processes); the mass-energy of particle i is

$$E_i = \frac{m_i c^2}{\sqrt{1 - \frac{u_i^2}{c^2}}} \tag{6.79}$$

and the *total* mass-energy $E = \sum_{i=1}^{N} E_i$ is constant.

For example, consider the totally inelastic collision of two particles with equal masses $m_1 = m_2 \equiv m$, antiparallel velocities $u_1 = -u_2 \equiv u$, which come together and form a single particle of mass M. The total mass-energy before the collision is equal to the total mass-energy after it, i.e.,

$$\frac{mc^2}{\sqrt{1 - \frac{u^2}{c^2}}} + \frac{mc^2}{\sqrt{1 - \frac{u^2}{c^2}}} = Mc^2 \tag{6.80}$$

(because of the conservation of momentum the resulting particle is at rest with no kinetic energy). Then the mass M of the product particle is

$$M = \frac{2m}{\sqrt{1 - \frac{u^2}{c^2}}} \tag{6.81}$$

and is larger than the sum of the masses of the colliding particles, $M > 2m$. The *mass defect*

$$\Delta M \equiv M - 2m = \frac{T_1 + T_2}{c^2} \tag{6.82}$$

is defined so that $M = 2m + \Delta M$. We have

$$\Delta M \equiv M - 2m = \frac{2m}{\sqrt{1-\frac{u^2}{c^2}}} - 2m = 2m\left(\frac{1}{\sqrt{1-\frac{u^2}{c^2}}} - 1\right), \quad (6.83)$$

$$\frac{T_1+T_2}{c^2} = \frac{2}{c^2}\left(\frac{mc^2}{\sqrt{1-\frac{u^2}{c^2}}} - mc^2\right) = 2m\left(\frac{1}{\sqrt{1-\frac{u^2}{c^2}}} - 1\right), \quad (6.84)$$

which proves Eq. (6.82). We deduce that kinetic energy has been converted into mass.

In order to break down an atomic nucleus into its components, one must supply an amount of energy called *binding energy* (BE). The mass-energy of a bound system is

$$\underset{\substack{\uparrow \\ \text{bound} \\ \text{system's} \\ \text{mass}}}{Mc^2} + \underset{\substack{\uparrow \\ >0}}{BE} = \underset{\substack{\uparrow \\ \text{masses of the} \\ \text{free particle} \\ \text{components}}}{\sum_{i=1}^{N} m_i c^2}. \quad (6.85)$$

Note that the bound system is not a system of free particles: the binding energy is a potential energy of interaction between the particles. Since

$$Mc^2 = \sum_{i=1}^{N} m_i c^2 \underbrace{-BE}_{<0} < \sum_{i=1}^{N} m_i c^2,$$

the bound system is a configuration with less energy than the ensemble of free particles, hence it is energetically favoured and stable.

6.6.1 Nuclear Fusion

Atomic nuclei can fuse with each other and liberate energy. Nuclear fusion occurs spontaneously in the core of stars and also, in an uncontrolled way, in man-made hydrogen bombs. The search for controlled nuclear fusion holds the promise of clean nuclear power generation and remains the holy grail of research on power generation. In nuclear physics the strong force is much stronger than the electrical force and huge amounts of energy per particle mass are involved.

Read Eq. (6.85) from right to left, i.e., collide electrically charged nuclei (or other particles) with masses m_i until their Coulomb repulsion is overcome. Then the final particle has energy $Mc^2 < \sum_{i=1}^{N} m_i c^2$ and the energy BE is liberated.

The proton-proton chain reaction (*pp-chain*) is a nuclear fusion process providing the main source of pressure in main sequence stars (stars like our sun), in which hydrogen nuclei fuse into helium [14]. The first two steps of the pp-chain are

$$^1\mathrm{H} + {}^1\mathrm{H} \longrightarrow {}^2\mathrm{H} + \mathrm{e}^+ + \nu_\mathrm{e}$$

$$^2\mathrm{H} + {}^1\mathrm{H} \longrightarrow {}^3\mathrm{He} + \gamma$$

and from here the process can proceed with three possible reactions. In our sun the most important one is

$$^3\mathrm{He} + {}^3\mathrm{He} \longrightarrow {}^4\mathrm{He} + 2\,{}^1\mathrm{H}.$$

The net reaction liberates $\sim$26 MeV of energy in the form of photons and neutrinos (of which 0.3 MeV are carried away by neutrinos). Other cycles important for the evolution of stars more massive than the sun are the carbon-nitrogen-oxygen (CNO) and the triple-α cycles [14].

Nuclear masses can be measured very accurately using mass spectrometers and the conservation of mass-energy has been tested accurately.

6.6.2 Nuclear Fission

Consider an atomic nucleus of mass M at rest that decays into three particles with masses M_1, M_2, and M_3 and speeds u_1, u_2, and u_3, respectively. This process is called *nuclear fission*. The conservation of mass-energy yields

$$Mc^2 = \frac{M_1 c^2}{\sqrt{1 - \frac{u_1^2}{c^2}}} + \frac{M_2 c^2}{\sqrt{1 - \frac{u_2^2}{c^2}}} + \frac{M_3 c^2}{\sqrt{1 - \frac{u_3^2}{c^2}}}. \tag{6.86}$$

Clearly, it is

$$M > M_1 + M_2 + M_3$$

because all square roots in the denominators on the right hand side of Eq. (6.86) are less than unity. The mass loss is

$$\Delta M \equiv M - (M_1 + M_2 + M_3),$$

which goes into the kinetic energy of the three particles. The disintegration energy released per fission reaction is

$$Q = (\Delta M)\, c^2 = [M - (M_1 + M_2 + M_3)]\, c^2. \tag{6.87}$$

These considerations are important for atomic nuclei or for composite particles: when nucleons (protons and neutrons) bind together to form a nucleus the mass of the latter is *less* than the sum of the masses of the components. This means that the bound state is energetically favoured over a state corresponding to the isolated nucleons and a nucleus will form. Conversely, if sufficient energy is given the situation can be reversed and the bound state can decay into free nucleons, liberating an amount of energy $\Delta m\, c^2$ through fission. Fission processes were used in the first nuclear bombs and are used in nuclear reactors for power generation.

6.7 Conclusion

The basic notions of the kinematics and dynamics of massive particles are now known. As promised, the 4-dimensional world view unveils a deep relation between the energy and the momentum of a particle and the equivalence between mass and energy. Unfortunately, due to the vagaries of history, Einstein's famous formula $E = mc^2$ was applied to nuclear weapons before finding peaceful uses. Because of this early application, technology derived directly from fundamental physics has modified society profoundly during a large part of the twentieth century, a situation which seems now irreversible. The conversion of mass into energy explains the source of energy fuelling stars, which was a major puzzle before the development of nuclear physics. Leaving aside these most spectacular aspects of Special Relativity, a new view has emerged with the study of massive particles. We still have to see how massless particles behave in this theory, which provides more surprises.

Problems

6.1. A particle has 4-velocity u^μ in the direction of the vector $A^\mu = (3, 0, 0, 1)$, in Cartesian coordinates. Is it a massive or a massless particle? Write down the components of u^μ in these coordinates.

6.2. A massive particle has 4-velocity $u^\mu = (2, A, B, C)$ in units $c = 1$, where A, B, and C are unknown constants. Determine the possible values of its spatial components by knowing that

- u^μ is orthogonal to $A^\mu = (0, 1, 1, 1)$ and
- $u^\mu B_\mu = 3$, where $B^\mu = (0, 0, 0, 3)$.

6.3. In Minkowski spacetime in Cartesian coordinates, compute the tangents to the parametric curves

$$x^{\mu}_{(a)}(\lambda) = \left(l_0 \sin\left(\frac{\lambda}{l_0}\right), l_0 \cos\left(\frac{\lambda}{l_0}\right), \lambda, 0 \right),$$

$$x^{\mu}_{(b)}(\lambda) = \left(l_0 \cos\left(\frac{\lambda}{l_0}\right), l_0 \sin\left(\frac{\lambda}{l_0}\right), \lambda, 0 \right),$$

where l_0 is a constant and λ is a parameter, both with the dimensions of a length. Determine the causal character of these curves.

6.4. A particle is in uniform circular motion with linear speed v. Write the components of its 4-velocity u^μ in spherical coordinates $\{ct, r, \theta, \varphi\}$ (rotate the coordinate axes so that the motion takes place in the equatorial plane). Express u^μ in terms of the rapidity ϕ and check explicitly that u^μ is normalized correctly.

6.5. In a spacetime diagram (x, t), draw the worldline of a massive particle of 3-velocity

$$\mathbf{v}(t) = \left(\frac{v_0}{1 + \frac{t}{t_0}}, 0, 0 \right)$$

for $t \geq 0$, where v_0 and t_0 are positive constants. What is the proper time of this particle as a function of t? What is its 4-velocity?

6.6. A free massive particle is moving along the x-axis of an inertial frame with velocity $v = dx/dt$, passing through the origin at $t = 0$. Express the particle's worldline $x^\mu = x^\mu(\tau)$ parametrically in terms of the function $v(\tau)$ using the proper time τ as a parameter.

6.7. A point mass moves with uniform circular motion of angular frequency ω on a circle of radius R centered on the origin of the (x, y) plane.
(a) Draw the worldline of the particle in an (x, y, t) spacetime diagram with units $c = 1$.
(b) Given that R is fixed, can the value of ω be arbitrary?
(c) Provide a parametric representation of the particle worldline in terms of its proper time τ.

6.8. Use Eq. (6.21) to prove Eq. (6.18), making sure that the result holds in *any* inertial frame.

6.9. Compute[9] the percent change in mass $\delta m/m$ of a block of copper which is heated from 250 to 350 K. The molar specific heat of copper is $c_p = 2.6 \times 10^4 \, \frac{\text{J}}{\text{kmol}\cdot\text{K}}$ and its atomic weight is 29.

6.10. How fast must a particle travel in order for its kinetic energy to be equal to 15 times its rest energy?

---

[9] This classic exercise recurring in many thermodynamics and Special Relativity textbooks seems to be due originally to Pauli.

6.11. In 2-dimensional Minkowski spacetime the worldline (6.67) and (6.68) of a uniformly accelerated particle defines a hypersurface (except at $\tau = 0$, where the tangent is vertical). By computing its normal, show that this hypersurface is asymptotically null as $\tau \longrightarrow \pm\infty$ (as is clear from Fig. 6.2).

6.12. Is the hypersurface in Minkowski spacetime

$$\mathcal{S} = \left\{ (ct, x, y, z) : \quad -(ct-a)^2 + 2(x-b)^2 = -y^2 - z^2 \right\}$$

(where a and b are constants) timelike? Spacelike? Null? Find the unit normal to $\mathcal{S}$.

References

1. P. Schneider, J. Ehlers, E.E. Falco, *Gravitational Lenses*. (Springer, New York, 1999)
2. R.M. Wald, *General Relativity*. (Chicago University Press, Chicago, 1984)
3. S.M. Carroll, *Spacetime and Geometry: An Introduction to General Relativity*. (Addison-Wesley, San Francisco, 2004)
4. J.B. Hartle, *Gravity: An Introduction to Einstein's General Relativity*. (Addison-Wesley, San Francisco, 2003)
5. L.D. Landau, E. Lifschitz, *The Classical Theory of Fields*. (Pergamon Press, Oxford, 1989)
6. C.W. Misner, K.S. Thorne, J.A. Wheeler, *Gravitation*. (Freeman, New York, 1973)
7. D.L. Bailey, D.W. Townsend, P.E. Valk, M.N. Maisey, *Positron Emission Tomography: Basic Sciences*. (Springer, New York, 2005)
8. S.A. Fulling, Phys. Rev. D **7**, 2850 (1973)
9. P.C.W. Davies, J. Phys. A **8**, 609 (1975)
10. W.G. Unruh, Phys. Rev. D **14**, 870 (1976)
11. Large Hadron Collider website http://lhc.web.cern.ch/lhc/
12. International Linear Collider homepage http://www.linearcollider.org/
13. W. Rindler, *Introduction to Special Relativity*, 2nd edn. (Clarendon Press, Oxford, 1991)
14. D. Prialnik, *An Introduction to the Theory of Stellar Structure and Evolution*. (Cambridge University Press, Cambridge, 2000)

Chapter 7
Relativistic Optics

The eternal mystery of the world is its comprehensibility.
—Albert Einstein

7.1 Introduction

After studying the mechanics of massive test particles, it is now time to discuss massless particles. There are differences in the mechanics of massless and massive particles, as is evident from the fact that the Lorentz transformation, the formulae for time dilation and length contraction, and the concept of proper quantities, do not make sense in the limit in which the speed of an inertial observer goes to c—divergences occur. Moreover, there is no rest frame for massless particles. However, similarities remain in certain other formulae. Some of the effects presented here have applications in many areas of physics, astrophysics, and technology. Here we discuss geometric optics, the high-frequency limit of wave optics. Wave optics is discussed in Chap. 9.

7.2 Relativistic Optics: Null Rays

Massless particles (photons and gravitons) travel at the speed of light. In any inertial frame they satisfy the relation

$$\sqrt{\left(\frac{dx}{dt}\right)^2 + \left(\frac{dy}{dt}\right)^2 + \left(\frac{dz}{dt}\right)^2} = c \tag{7.1}$$

and

V. Faraoni, *Special Relativity*, Undergraduate Lecture Notes in Physics,

DOI: 10.1007/978-3-319-01107-3_7,

$$ds^2 = -c^2 dt^2 + d\mathbf{x}^2 = -dt^2 \left[c^2 - \left(\frac{dx}{dt}\right)^2 - \left(\frac{dy}{dt}\right)^2 - \left(\frac{dz}{dt}\right)^2 \right] = 0.$$

By introducing a parameter λ with the dimensions of a length along the null curve,[1]

$$\frac{ds^2}{d\lambda^2} = \eta_{\mu\nu} \frac{dx^\mu}{d\lambda} \frac{dx^\nu}{d\lambda} = \eta_{\mu\nu} u^\mu u^\nu = 0, \tag{7.2}$$

where $u^\mu \equiv dx^\mu / d\lambda$ is the 4-tangent to the particle worldline (which is also called a *null ray*). Therefore, for massless particles u^μ is dimensionless and its normalization is

$$u_\mu u^\mu = 0 \quad \text{(massless particles)}. \tag{7.3}$$

Do not be startled by the use of a parameter other than a time: for example, the worldline $x^\mu = (ct, ct, 0, 0)$ of a photon moving in the direction of the positive x-axis has equation $x = ct$ and can be described parametrically by

$$x^\mu = u^\mu \lambda, \tag{7.4}$$

where λ is the parameter and $u^\mu \doteq \delta^{0\mu} + \delta^{1\mu} = (1, 1, 0, 0)$. $u^\mu = dx^\mu / d\lambda$ is a null vector and, in addition, satisfies the equation

$$\frac{du^\mu}{d\lambda} = 0. \tag{7.5}$$

It is always possible to rotate the spatial axes so that the x-axis is aligned with the direction of the ray and u^μ assumes the form $u^\mu \doteq \delta^{\mu 0} + \delta^{\mu 1}$ in Cartesian coordinates. Therefore, Eq. (7.5) applies to all null rays in Cartesian coordinates. Since a null vector can always be reduced to the form $u^\mu = \alpha\,(1, 1, 0, 0)$ and the constant α is unimportant, the case $u^\mu = \delta^{0\mu} + \delta^{1\mu}$ is general.

The meaning of Eq. (7.5) is that light rays are not accelerated and follow straight lines in Minkowski spacetime; in fact, by integrating this equation one obtains the linear solution $x^\mu(\lambda) = \alpha^\mu \lambda + \beta^\mu$, where α^μ and β^μ are constants determined by the initial position and tangent of the null ray; this is the parametric representation of a straight line. Equations (7.3) and (7.5) can be expressed by saying that the massless particle does not accelerate (in Chap. 10 we will express this fact by saying that the

[1] This parameter is not the proper time, which is not defined because there is no rest frame for a particle travelling at light speed.

massless particle follows an affinely-parametrized null geodesic curve of Minkowski spacetime). Other parametrizations are possible for the null ray, for example one could use $\sigma = \lambda^5$ instead of λ. Then (forgetting about the odd dimensions of the new parameter σ) its 4-tangent would be

$$v^\mu \equiv \frac{dx^\mu}{d\sigma} = \frac{dx^\mu}{d\lambda}\frac{d\lambda}{d\sigma} = u^\mu \left(\frac{d\sigma}{d\lambda}\right)^{-1} = \frac{u^\mu}{5\lambda^4} = \frac{u^\mu}{5\sigma^{4/5}}$$

and v^μ would still be a null vector $\left(v^\mu v_\mu = 0\right)$ but

$$\frac{dv^\mu}{d\lambda} = \frac{-4\,v^\mu}{5\lambda^5} = -\frac{4v^\mu}{5\sigma} \neq 0.$$

A parameter λ such that

$$\frac{du^\mu}{d\lambda} = \frac{d^2x^\mu}{d\lambda^2} = 0$$

is called an *affine parameter*. The equation of motion of an affinely-parametrized null ray takes the same form as that for a massive particle with proper time as the parameter.

Let $u^\mu = \left(u^0, \mathbf{u}\right) = \left(\frac{dx^0}{d\lambda}, \frac{dx^i}{d\lambda}\right)$ $(i = 1, 2, 3)$, then the normalization (7.3) implies that $-\left(u^0\right)^2 + (\mathbf{u})^2 = 0$ and

$$u^\mu \equiv \frac{dx^\mu}{d\lambda} = \left(u^0, \mathbf{u}\right) \quad \text{with } u^0 = |\mathbf{u}|\,. \tag{7.6}$$

$u^0 \neq 0$ is a necessary condition for u^μ to be null and non-trivial and it must be $u^0 > 0$ for the 4-tangent u^μ to be future-pointing.

In the geometric optics approximation, which is the limit of Maxwell's wave optics when the wavelength of electromagnetic waves tends to zero, waves will propagate along null rays (which are normal to the wavefronts). These are often called *photon trajectories* although the terminology is not justified here since we are in a purely classical context while the photon is a quantum concept.

Let λ be a parameter along the null ray in 4-dimensional Minkowski spacetime in vacuo; this ray is a straight line. Let $u^\mu \equiv dx^\mu/d\lambda$ be the 4-vector tangent to the ray and define the *wave 4-vector*

$$k^{\mu} \equiv \frac{\omega}{c} u^{\mu} = \left(k^{0}, \mathbf{k}\right) = \left(\frac{\omega}{c}, \mathbf{k}\right) \quad \text{(null rays)} \tag{7.7}$$

where ω is the angular frequency, $|\mathbf{k}| = k^0$, and $\eta_{\mu\nu} k^{\mu} k^{\nu} = 0$. The dimensions of these quantities are $[u^{\mu}] = [0]$ and $[k^{\mu}] = \left[L^{-1}\right]$ (note that $k^0 \neq 0$ is a necessary condition for k^{μ} to be a null vector). Define the 3-dimensional "velocity" of the ray as

$$\begin{aligned} \mathbf{v} &\equiv \left(\frac{dx^1}{dx^0}, \frac{dx^2}{dx^0}, \frac{dx^3}{dx^0}\right) c = \left(\frac{dx^1/d\lambda}{dx^0/d\lambda}, \frac{dx^2/d\lambda}{dx^0/d\lambda}, \frac{dx^3/d\lambda}{dx^0/d\lambda}\right) c \\ &= \left(\frac{k^1}{k^0}, \frac{k^2}{k^0}, \frac{k^3}{k^0}\right) c = \frac{c^2}{\omega} \mathbf{k}. \end{aligned} \tag{7.8}$$

Then $k_{\mu} k^{\mu} = 0$ is equivalent to

$$-\left(k^0\right)^2 + |\mathbf{k}|^2 = 0,$$

$$-1 + \frac{|\mathbf{k}|^2}{\left(k^0\right)^2} = 0,$$

or

$$\frac{c^2}{\omega^2} \frac{\omega^2}{c^4} \mathbf{v}^2 = 1$$

and

$$\mathbf{v}^2 = c^2.$$

The equation

$$k^0 = \frac{\omega}{c} = |\mathbf{k}| \tag{7.9}$$

expresses the fact that the ray propagates in 3-dimensional space at the speed of light, so

$$k^{\mu} k_{\mu} = 0, \qquad \frac{du^{\mu}}{d\lambda} = \frac{d^2 x^{\mu}}{d\lambda^2} = 0. \tag{7.10}$$

In quantum mechanics the energy of a photon of angular frequency ω is

$$E = \hbar\omega, \tag{7.11}$$

where $\hbar \equiv \frac{h}{2\pi} = 1.05457266(63) \cdot 10^{-34}$ J·s is the reduced Planck constant, and the photon momentum is

$$\mathbf{p} = \hbar\,\mathbf{k}. \tag{7.12}$$

This relation leads one to define the *photon 4-momentum*

$$p^{\mu} \equiv \hbar k^{\mu} = \left(\frac{\hbar\omega}{c}, \hbar\,\mathbf{k}\right) = \left(\frac{E}{c}, \mathbf{p}\right). \tag{7.13}$$

The normalization $k^{\mu}k_{\mu} = 0$ (or $p_{\mu}p^{\mu} = 0$) can then be written as

$$E^2 = p^2c^4$$

or

$$E = pc, \tag{7.14}$$

i.e., as the limit for the rest mass $m \to 0$ of the familiar Eq. (6.34) valid for massive particles. The relation (7.14) is well known to be satisfied by classical electromagnetic waves in Maxwell's theory.

To summarize, in Special Relativity there exists a privileged class of observers (inertial observers) and the worldlines of these observers are privileged classes of curves in Minkowski spacetime such that:

- Ideal clocks travel along these timelike curves and measure the parameter τ (proper time) defined by $-c^2\mathrm{d}\tau^2 = \eta_{\mu\nu}\mathrm{d}x^{\mu}\mathrm{d}x^{\nu}$.
- Massive particles travel along *timelike* curves parametrized by proper time τ with 4-tangent u^{μ} normalized to $u_{\mu}u^{\mu} = -c^2$.
- Massless particles travel along *null* curves (light rays) with 4-tangents u^{μ} such that $u_{\mu}u^{\mu} = 0$ and the wave 4-vector $k^{\mu} = \omega u^{\mu}/c$ satisfies $k_{\mu}k^{\mu} = 0$.

7.3 The Drag Effect

Consider a flowing transparent liquid and light propagating through it: experiments performed by Fizeau in 1851, Michelson and Morley in 1886, and Zeeman in 1914 [1–3] indicate that the medium drags the light along with it. If the speed of light in the liquid medium *at rest* is c_m' and the liquid moves with speed v, the speed of light relative to an inertial frame outside the liquid is experimentally found to be

$$c_m = c'_m + \left(1 - \frac{1}{n^2}\right) v, \tag{7.15}$$

where $n \equiv c/c'_m$ is the refraction index of the medium and $(1 - 1/n^2)$ is known as the *Fresnel drag coefficient*. An explanation of this phenomenon based on the ether was given by Fresnel long before Einstein, but we now know that it cannot be correct. Instead, apply the (inverse) relativistic law of composition of velocities (2.31)–(2.33). The light travels with speed c'_m relative to the medium and the medium travels with speed v relative to an inertial frame S outside the liquid, hence

$$c_m = \frac{c'_m + v}{1 + \frac{c'_m v}{c^2}} = \left(c'_m + v\right)\left(1 - \frac{c'_m v}{c^2}\right) + \cdots$$

$$= c'_m - \frac{c'^2_m v}{c^2} + v + \cdots$$

$$= c'_m + \left(1 - \frac{1}{n^2}\right) v + \cdots$$

To lowest order, the drag effect is reproduced simply as the relativistic law of addition of velocities. This simple explanation was given by Laub and von Laue in 1907 [4, 5].

7.4 The Doppler Effect

The non-relativistic Doppler effect for sound waves can distinguish between motion of the source and motion of the observer. The angular frequency ω_{source} emitted by a source moving with velocity v_{source} is perceived by an observer moving with velocity $v_{observer}$ as

$$\omega = \omega_{source} \left(\frac{v \pm v_{observer}}{v \mp v_{source}}\right), \tag{7.16}$$

where the upper signs apply to motions of the source and observer toward each other, the lower signs to motions away from each other, and v is the speed of sound. Therefore, one can distinguish the absolute motion of frames with respect to the medium in which the longitudinal sound waves propagate (for example air). This distinction is possible because longitudinal sound waves require a medium to propagate: they cannot propagate in empty space and the medium provides an "absolute" frame of reference with respect to which motions are referred.

In the relativistic Doppler effect for electromagnetic waves in vacuo, instead, one cannot distinguish the motion of the source from the motion of the observer, according to the Principle of Relativity.[2] This situation is due to the fact that there is no medium (ether) and electromagnetic waves propagate in vacuo.

Consider, as in Fig. 7.1, a source of electromagnetic waves emitting at a single angular frequency ω_0 (the angular frequency in the rest frame of the source) and an observer who sees the source moving with velocity $\mathbf{v}$ and receives the angular frequency ω. Let α be the arrival angle of the light (the angle between the direction of propagation of the waves and the *negative* direction of the source (S') motion *in the observer's frame* S), let $\theta \equiv \pi - \alpha$, then $\alpha = \pi - \theta$. The variation in the angular frequency is obtained by writing the transformation law for the time component of the 4-vector

$$k^{\mu} = (|\mathbf{k}|, \mathbf{k}) = \left(\frac{\omega}{c}, \mathbf{k}\right) \tag{7.17}$$

or (the primed frame is the frame moving with the source)

$$k^{0'} = \gamma\left(k^0 - \frac{v}{c}k^1\right). \tag{7.18}$$

Since $k^1 = k\cos\theta = -\dfrac{\omega}{c}\cos\alpha$, using

$$\cos\theta = \cos(\pi - \alpha) = \cos\pi\cos\alpha + \sin\pi\sin\alpha = -\cos\alpha$$

in the last equality, it is

$$\frac{\omega_0}{c} \equiv \frac{\omega'}{c} = \gamma\left(\frac{\omega}{c} + \frac{v}{c}\frac{\omega}{c}\cos\alpha\right),$$

$$\omega_0 \equiv \omega' = \gamma\,\omega\left(1 + \frac{v}{c}\cos\alpha\right),$$

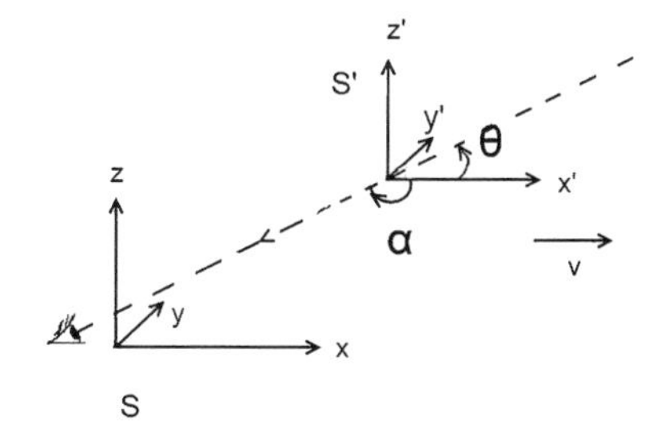

Fig. 7.1 A monochromatic source of light at rest in S' emits light at the angular frequency ω_0 toward an observer, who receives it at the angular frequency ω

[2] See Refs. [6, 7] for the Doppler formula for electromagnetic (or other) waves propagating in a refracting medium.

and

$$\omega = \frac{\omega_0\sqrt{1-\frac{v^2}{c^2}}}{1+\frac{v}{c}\cos\alpha} \tag{7.19}$$

(*relativistic Doppler effect formula*). Some special cases are worth considering. For $\alpha = 0$ we have the *radial Doppler effect* with $\omega = \omega_0\sqrt{\frac{1-\frac{v}{c}}{1+\frac{v}{c}}}$. To first order in v/c, it is

$$\omega = \omega_0\left(1-\frac{v}{c}\right) + \mathrm{O}\left(\frac{v^2}{c^2}\right)$$

and

$$\frac{\Delta\omega}{\omega_0} = -\frac{v}{c} + \mathrm{O}\left(\frac{v^2}{c^2}\right). \tag{7.20}$$

For $\alpha = \pi/2$ we have the *transverse Doppler effect* with

$$\omega = \omega_0\sqrt{1-\frac{v^2}{c^2}},$$

which is of order $\mathrm{O}(v^2/c^2)$. The transverse Doppler effect is purely relativistic and is due only to time dilation. It is usually covered by a first order longitudinal component and is, therefore, difficult to detect. Experiments performed using a rapidly rotating disk with a source at the centre and detectors at the rim, and using γ-rays and Mössbauer technology were able to detect this second order effect in the early 1960s [8, 9]. The formula of the pre-relativistic Doppler effect is

$$\omega_{prerelativistic} = \frac{\omega_0}{1+\frac{v}{c}\cos\alpha} \tag{7.21}$$

but in Special Relativity one has to take into account also the dilation for the period P of the wave, which yields

$$P = \frac{P_0}{\sqrt{1-\beta^2}} \tag{7.22}$$

(where P_0 is the period in the rest frame of the source), or $\omega = \omega_0\sqrt{1-\beta^2}$. Taking into account both effects one obtains Eq. (7.19). In the non-relativistic limit $\beta \to 0$ one recovers the pre-relativistic formula (7.21). Qualitatively we have that

$$v < 0 \Rightarrow \omega > \omega_0 \quad \text{BLUESHIFT}$$

$$v > 0 \Rightarrow \omega < \omega_0 \quad \text{REDSHIFT}$$

(remember the convention that $v > 0$ for motion of the source and the observer *away* from each other). In spectroscopy, the *redshift factor* is defined as

$$z \equiv \frac{\lambda_{observed} - \lambda_{emitted}}{\lambda_{emitted}}; \tag{7.23}$$

since $\lambda = \dfrac{2\pi c}{\omega}$ we have also

$$z = \frac{\omega_{emitted} - \omega_{observed}}{\omega_{observed}}. \tag{7.24}$$

In our case it is

$$z = \frac{\omega_{emitted} - \omega_{observed}}{\omega_{observed}} = \frac{\omega_{emitted} - \omega_{emitted}\frac{\sqrt{1-\frac{v^2}{c^2}}}{(1+\frac{v}{c}\cos\alpha)}}{\omega_{emitted}\frac{\sqrt{1-\frac{v^2}{c^2}}}{(1+\frac{v}{c}\cos\alpha)}}$$

$$= \frac{1+\frac{v}{c}\cos\alpha - \left[1+\mathrm{O}\left(\frac{v^2}{c^2}\right)\right]}{1+\mathrm{O}\left(\frac{v^2}{c^2}\right)} = \frac{v}{c}\cos\alpha + \mathrm{O}\left(\frac{v^2}{c^2}\right).$$

We have

- $z > 0$ for $v\cos\alpha > 0$ (source and observer are moving *away* from each other) and we have *redshift*: $\lambda_{observed} > \lambda_{emitted}$.
- $z < 0$ for $v\cos\alpha < 0$ (source and observer move *toward* each other) and we have *blueshift*: $\lambda_{observed} < \lambda_{emitted}$.

7.5 Aberration

The term *aberration* denotes the fact that the angle at which a moving object is seen depends on the relative velocity of the observer. A well known example of non-relativistic aberration occurs when the angle of falling rain depends on the velocity of the observer who is getting wet. If rain is falling vertically (in the absence of wind) for an inertial observer, a second observer moving with respect to the first one with constant velocity v will see the rain falling obliquely.

The phenomenon of relativistic aberration consists of the dependence of the direction of propagation of radiation from the relative velocity of two inertial frames. Consider an incoming light signal whose negative direction makes an angle α with the observer in an inertial frame S and an angle α' with a different observer in another inertial frame S' moving with speed v with respect to S in standard configuration. Assume that the light propagates in the (x, z) plane (Fig. 7.2). Apply the velocity transformation formula

$$u^{x'} = \frac{u^x - v}{1 - \frac{u^x v}{c^2}}, \quad u^{y'} = \frac{u^y}{\gamma\left(1 - \frac{u^x v}{c^2}\right)}, \quad u^{z'} = \frac{u^z}{\gamma\left(1 - \frac{u^x v}{c^2}\right)}$$

to the light signal[3] itself ($u^x = c$). In this case,

$$u^x = -c\,\cos\alpha\,, \qquad u^{x'} = -c\,\cos\alpha'$$

and, from the transformation property (2.31) of u^x, we obtain

$$-c\,\cos\alpha' = \frac{-c\,\cos\alpha - v}{1 - (-c\,\cos\alpha)\,\frac{v}{c^2}}$$

and

$$\cos\alpha' = \frac{\cos\alpha + \frac{v}{c}}{1 + \frac{v}{c}\cos\alpha}. \tag{7.25}$$

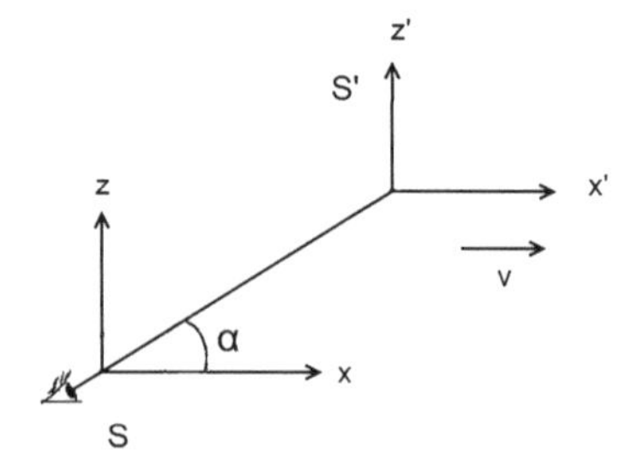

Fig. 7.2 A light signal arrives at the inertial observer in S, its negative direction of arrival making an angle α with the x-axis

[3] The velocity transformation formula applies to the situation in which $|u| \leq c$, although v is restricted by the condition $|v| < c$.

We can also use the transformation property (2.32) of u^y and

$$u^y = -c\sin\alpha, \qquad u^{y'} = -c\sin\alpha'$$

to obtain $-c\sin\alpha' = \dfrac{-c\sin\alpha}{\gamma\left[1-(-c\cos\alpha)\frac{v}{c^2}\right]}$ and

$$\sin\alpha' = \frac{\sin\alpha}{\gamma\left(1+\frac{v}{c}\cos\alpha\right)}. \tag{7.26}$$

(Alternatively, one can use Eq. (7.25) and $\sin\alpha = \sqrt{1-\cos^2\alpha}$ to obtain Eq. (7.26)). Equations (7.25) and (7.26) are known as *aberration formulae*. By using Eqs. (7.25) and (7.26) and a trigonometric identity one obtains

$$\tan\left(\frac{\alpha'}{2}\right) = \frac{\sin\alpha'}{1+\cos\alpha'} = \frac{\dfrac{\sin\alpha}{\gamma\left(1+\frac{v}{c}\cos\alpha\right)}}{1+\dfrac{\left(\cos\alpha+\frac{v}{c}\right)}{\left(1+\frac{v}{c}\cos\alpha\right)}}$$

$$= \frac{\sin\alpha}{\gamma\left[1+\frac{v}{c}\cos\alpha+\cos\alpha+\frac{v}{c}\right]}$$

$$= \frac{\sin\alpha}{\gamma\left(1+\frac{v}{c}\right)(1+\cos\alpha)} = \frac{\tan\left(\frac{\alpha}{2}\right)}{\gamma\left(1+\frac{v}{c}\right)}$$

$$= \frac{\sqrt{1-\frac{v^2}{c^2}}}{\left(1+\frac{v}{c}\right)}\tan\left(\frac{\alpha}{2}\right) = \sqrt{\frac{1-\frac{v}{c}}{1+\frac{v}{c}}}\tan\left(\frac{\alpha}{2}\right),$$

hence

$$\tan\left(\frac{\alpha'}{2}\right) = \sqrt{\frac{1-\frac{v}{c}}{1+\frac{v}{c}}}\tan\left(\frac{\alpha}{2}\right). \tag{7.27}$$

For rays propagating *outward* at angles α and α' one replaces c with $-c$.

Due to aberration, as the earth travels along its orbit, the apparent directions of the fixed stars trace small ellipses during the year (their major axes span approximately 40 arcseconds), an effect sufficiently large to have been discovered in the first half of the 18th century. Contrary to parallax, stellar aberration is independent of the star's distance.

The inverse laws of aberration can be obtained by invoking the Principle of Relativity and with the (now familiar) exchange of primed and unprimed quantities and $v \longleftrightarrow -v$,

$$\cos\alpha = \frac{\cos\alpha' - \frac{v}{c}}{1 - \frac{v}{c}\cos\alpha'}, \tag{7.28}$$

$$\sin\alpha = \frac{\sin\alpha'}{\gamma\left(1 - \frac{v}{c}\cos\alpha'\right)}, \tag{7.29}$$

$$\tan\left(\frac{\alpha}{2}\right) = \sqrt{\frac{1+\frac{v}{c}}{1-\frac{v}{c}}}\tan\left(\frac{\alpha'}{2}\right). \tag{7.30}$$

If, at this point, you think that relativistic aberration makes sense intuitively, think again. Relativistic aberration is counterintuitive: consider, for example, an object subtending a small angle α' in the frame S' (and, correspondingly, a small angle α in S). According to our low speed intuition when we approach an object its angular size increases and, when we recede from it, its angular size decreases. In Special Relativity the inverse law of aberration (7.29) gives, for small angles,

$$\alpha \simeq \frac{\alpha'}{\gamma\left(1 - v/c\right)}$$

and, expanding for small values of v/c and assuming that the object of angular size α is approached (i.e., $v < 0$), we obtain

$$\alpha \simeq \alpha'\left(1 - \frac{|v|}{c}\right) + \mathrm{O}\left(\frac{v^2}{c^2}\right). \tag{7.31}$$

Approaching the object at first *reduces* its angular size and, therefore, it appears to recede. As the object gets closer and closer the angles α and α' are no longer small, eventually passing through 90°, and the approximate formula (7.31) becomes invalid. When this happens, approaching an object increases the appearance of its angular size.

7.6 Relativistic Beaming

Consider a spherical surface emitting electromagnetic radiation isotropically in all directions in its rest frame. This isotropy will be broken when seen by an inertial observer O' moving with respect to the rest frame of the source, which is not surprising since the velocity of this observer introduces a preferred direction. The radiation will

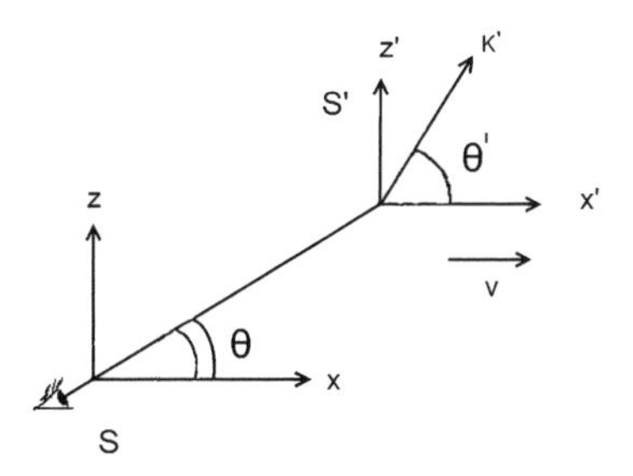

Fig. 7.3 Relativistic beaming

be concentrated along the direction of motion due to the transformation property of the photons' spatial momenta: this effect is called *relativistic beaming*.

Consider a photon propagating at an angle θ' with the x'-axis in the rest frame $\{ct', \mathbf{x}'\}$ of the source (Fig. 7.3).The aberration formula (7.25) calculated in the previous section gives

$$\cos\theta = \frac{\cos\theta' + \frac{v}{c}}{1 + \frac{v}{c}\cos\theta'} \tag{7.32}$$

(*relativistic beaming formula*). In general, $\cos\theta > \cos\theta'$ implies that $\theta < \theta'$ or, the forward radiation emitted in a cone is beamed in a narrower cone as seen by the observer O who sees the source moving. Consider the rays emitted in the forward hemisphere $\left(0 \le \theta' \le \pi/2\right)$ in the rest frame of the source. For the observer O these rays are concentrated in the cone defined by $0 \le \theta \le \theta_0$, where

$$\cos\theta_0 = \frac{\cos\frac{\pi}{2} + \frac{v}{c}}{1 + \frac{v}{c}\cos\frac{\pi}{2}} = \frac{v}{c}$$

and

$$\theta_0 = \cos^{-1}(\beta)\,. \tag{7.33}$$

As $v \longrightarrow c$, this cone will have a very small aperture.

A uniformly radiating extended source appears brighter if it is moving toward us than if it is moving in the opposite direction (*headlight* or *searchlight effect*). The *intensity* of the radiation, defined as the energy emitted per unit time, per unit of normal area, and per unit of solid angle,

$$I = \frac{dE}{dt dA d\Omega},$$

is higher in the forward direction than in the backward direction because:

- there is a change in the angular size of the emitting region. Rays are beamed in the forward direction and more rays fall on the unit of normal area than if the source were at rest.
- The Doppler shift increases the frequency and energy of the photons in the forward direction and decreases it in the backward direction (remember that the energy E of a photon is proportional to its frequency ν according to $E = h\nu$).
- There is an increase in the number of photons crossing the unit of normal area in the unit of time due to the combined motion of the observer toward the source and to time dilation.

Relativistic beaming is observed in the jets originating in active galactic nuclei and is usefully employed in synchrotron light imaging.

7.6.1 ⋆Synchrotron Light Imaging

As an example of applications of Special Relativity and of relativistic beaming in particular, we mention synchrotron light sources and their multiple uses in science and technology. One of the fundamental facts of electromagnetism is that accelerated charges emit electromagnetic radiation ("synchrotron radiation") and this property is particularly evident in both linear and circular particle accelerators. Beginning with the small accelerators of the post-World War II era, circular accelerators have grown to the giant size of accelerators used today for particle physics discoveries, such as the Large Hadron Collider (LHC) at CERN in Geneva. Light particles which can easily achieve large Lorentz factors, such as electrons and positrons, are subject to dramatic energy losses by synchrotron radiation in circular accelerators, which limits the usefulness of these devices for fundamental physics discoveries with electrons and positrons. However, if reaching very high electron energy is not the goal, the high Lorentz factors that can be achieved by man-made beams of light particles can still be put to good use. Relativistic electrons can easily be given large γ-factors and synchrotron radiation is focused in a narrow cone due to relativistic beaming. This cone can be made quite narrow: remember that its aperture α' as seen by an observer in the laboratory satisfies the relation $\sin\alpha' \approx \sin\alpha/\gamma$ (cf. Eq. (7.26)). Electrons with an energy of a few GeV have Lorentz factors $\gamma \approx 10^4$, which gives $\alpha' \approx 10^{-4}$ rad when $\alpha \sim \pi/2$ rad. Since there are advantages in having such a bright, concentrated beam of radiation for the imaging of small structures, relativistic beaming of synchrotron radiation emitted by accelerated electrons of moderate energy allows for the construction of dedicated facilities in which beams of synchrotron radiation at various wavelengths (infrared, ultraviolet, and X-ray, collectively called "synchrotron light") are available. Different applications require different wavelengths: in order to probe a certain structure one must use radiation with a wavelength comparable to the size of that structure, for example a bacterium of size $\sim 10^{-6}$ m is probed by visible light, a protein (size $\sim 10^{-8}$ m) by ultraviolet or soft X-rays, small molecules (size $\sim 10^{-9}$–10^{-10} m) by hard X-rays. Synchrotron radiation imaging has been applied to fundamental studies in biology, medicine, material science and condensed matter physics, and has many technological and industrial applications. Discoveries made with synchrotron light include the structure of certain DNA proteins, of neurotoxins, the 3-dimensional imaging of ancient insects encased in pre-historic amber, and the study of viruses.[4] Industrial applications include new drug design in the pharmaceutical industry, various X-ray fluorescence techniques to study trace contaminants in otherwise highly pure samples (for example silicon wafers used in the semiconductor industry, in which impurities whch alter their electronic properties must be kept strictly under control), and catalysts in chemical reactions for various purposes. The analysis of samples with synchrotron radiation is usually non-destructive and involves less or no preparation in comparison with more traditional techniques of analysis. Many synchrotron light facilities have been built around the world and new, higher frequency X-ray synchrotron sources are currently being designed.[5]

7.7 Visual Appearance of Extended Objects

When an extended object is passing by at relativistic speed, its appearance may be quite different from what it would look like at rest. The image may appear rotated, sheared, curved, and distorted. Consider, for example, a snapshot taken by a camera of an extended object passing by. Light rays reaching the plane of the camera at the same instant of time have started from different points of the object, which are further away from the camera, at different (earlier) times. Because the object is moving, its different points sending light rays to the camera were occupying different positions at the same time. As a result, the rays reaching the camera portray the points of the object when these point sources were at different positions. (In fact, as we know from the study of high school geometric optics, an observer perceives these points as being either on the rays reaching the camera or on their extensions.) A striking effect is that an object appears rotated with respect to its rest frame (*Terrell rotation*). These effects become noticeable when the object moves a significant distance during times comparable to the difference in travel times between its various points and the camera. Clearly, this effect occurs when the velocity of the object becomes relativistic and is due to the finiteness of the speed of light. In addition there are Lorentz contraction, aberration, and beaming. The color and luminosity of the object are also affected.

Let us consider two (over-)simplified examples to obtain a qualitative idea of the simplest effects at play. As the first example consider a thin rod which, in an inertial

---

[4] Even the virus of the 1918 Spanish influenza has been studied in 2004 using synchrotron light [10].

[5] See Ref. [11] for a list of synchrotron light facilities.

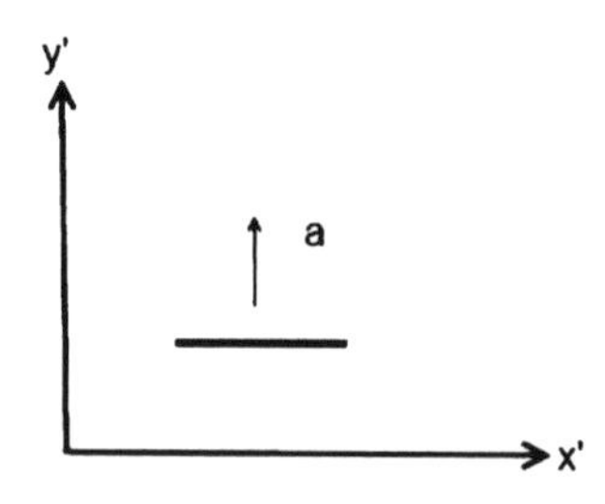

Fig. 7.4 A rod moving with constant acceleration a in the y-direction in the inertial frame S'

frame S', remains parallel to the x'-axis while it moves with constant acceleration a in the y' direction (Fig. 7.4) obeying the law[6]

$$y' = \frac{a}{2}t'^2. \tag{7.34}$$

For simplicity we assume that the rod has zero thickness: this assumption is, of course, unrealistic but corresponds to an observer not being able to distinguish details from a distance, which greatly simplifies the problem. In another inertial frame S related to S' by a Lorentz transformation (2.1)–(2.4) with velocity v, the equation of motion of the rod is

$$y = \frac{a}{2}\gamma^2\left(t - \frac{vx}{c^2}\right)^2. \tag{7.35}$$

At a constant time t the rod has the shape of a segment of a parabola. For example at $t = 0$ it is $y = \frac{a\gamma^2}{2}\frac{v^2}{c^4}x^2$ (see Fig. 7.5, drawn for the parameter values $a = 1\,\text{m/s}^2$ and $v = 0.99\,c$, in units in which $c = 1$). As time goes on, the shape of the rod seen in S changes because two coefficients in the equation of the parabola (7.35) are time-dependent.

As a second example consider a rod of negligible thickness which, in its rest frame $S' = \{ct', x', y', z'\}$, lies in the (x', y') plane, makes and angle θ_0 with the x'-axis, and has proper length l_0. The rod and its rest frame move with velocity v in standard configuration with respect to another inertial frame $S = \{ct, x, y, z\}$. In this frame, the rod makes an angle θ with the x-axis and has length l. Its $x-$component is Lorentz-contracted, $l^x = l^{x'}\sqrt{1-\beta^2} = l_0 \cos\theta_0\sqrt{1-\beta^2}$ while the y-component is not affected, $l^y = l^{y'} = l_0 \sin\theta_0$. As a result the length of the rod as measured in S is

[6] This acceleration can be maintained only for a short period of time, otherwise the velocity $u = u_0 + at$ would exceed the speed of light. This example is purely for illustration: a realistic situation must consider all the particles of the rod in motion with constant *proper* acceleration, on the lines of what discussed in Chap. 6 for uniformly accelerated point particles. One should then define uniformly accelerated reference frames, which involves some subtleties [12, 13]: we keep the discussion simple here. Nevertheless, the accelerated rod seems to be popular in the relativity literature [14–18].

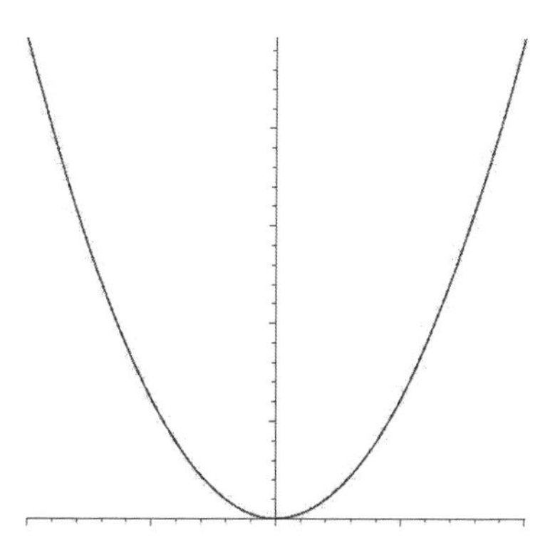

Fig. 7.5 The accelerating rod seen in the inertial frame S at $t = 0$

$$l = \sqrt{(l^x)^2 + (l^y)^2} = \sqrt{\left(l_0 \cos^2\theta_0 \sqrt{1-\beta^2}\right)^2 + (l_0 \sin\theta_0)^2}$$
$$= l_0 \sqrt{1 - \beta^2 \cos^2\theta_0}.$$

The angle θ which the rod makes with the x-axis is given by

$$\tan\theta = \frac{l^y}{l^x} = \gamma \tan\theta_0,$$

so the rod appears rotated with respect to its rest frame.

The description of the rods in the previous two examples is intentionally oversimplified by neglecting one of their dimensions. If all three dimensions of an object are taken into account and a camera takes instantaneous pictures in sharp focus of this object going by at constant speed v, the object will not just appear Lorentz-contracted: its image will be sheared, rotated, and acquire curvature even if it is has a straight contour, such as a cube. The process of imaging an extended object with light rays starting from different points of the object at different times is rather complex and very detailed and will not be discussed here. It began being studied only five decades after the introduction of Special Relativity and has been the subject of many articles, gaining more popularity as the power of modern computers progressively allowed better and better simulations of objects moving at relativistic speeds [19–37]. Recently, computer simulations of relativistic optics have achieved a high level of sophistication which allows one to build intuition in Special Relativity [38–41].[7]

[7] Due to the popularity of space travel in science fiction and, consequently, in examples in the pedagogical and popular literature on relativity, also the appearance of the sky to an observer traveling at relativistic speed was studied [42, 43].

7.8 Conclusion

The 4-dimensional world view has uncovered another deep relation between the frequency (or energy) and the wave vector (or momentum) of massless particles. Moreover, the appearance of objects (including their shape, apparent orientation, angular size, color, and luminosity) turns out to be quite different in different inertial frames moving at high speeds relative to each other. Shape, orientation, angular size, and color are relative to the state of motion of the observer, another good reason to name the theory "Relativity".

Problems

7.1 Derive the inverse laws of aberration (7.28)–(7.30) without invoking the Principle of Relativity as done in the text.

7.2 Is there a particular angle of view at which the Doppler shift vanishes *exactly*? If so, compute it
(a) for small values of v/c;
(b) for observer and source approaching each other at speed $0.9\,c$.

7.3 Aberration occurs in pre-relativistic physics. On a windless day, rain falls vertically on an inertial observer O. Consider a raindrop which, in the frame $\{x, y, z\}$ of O falls vertically in the (x, y) plane according to

$$x = x_0,$$
$$y = y_0 - ut,$$
$$z = 0,$$

with x_0, y_0, and u positive constants (the droplet has reached terminal velocity and is in uniform rectilinear motion in this frame).
At what angle θ' with the horizontal will another inertial observer O' (moving in standard configuration with velocity v with respect to O) see the rain coming or, what is the trajectory of a raindrop in the $\left\{x', y', z'\right\}$ frame of O'? Discuss the situations $v < u$, $v = u$, $v > u$, and the limit $\frac{v}{u} \longrightarrow \infty$. Can O' see horizontal rain (an English expression used in foul weather)?

7.4 Discuss the relativistic version of the previous exercise: compute the angle θ' which the incoming rain makes with the horizontal as seen by the moving inertial observer O' and compare it with the non-relativistic case. At what speed v between the two inertial frames is the angle θ' equal to 45°? What happens in the limit $v \longrightarrow c$? Derive your answer
(a) using the Lorentz transformation;
(b) using the law of transformation of velocities.
(c) Compare the two results.

7.5 In the $S' = \{ct', x', y', z'\}$ inertial frame a rod of zero thickness lies in the (x', y') plane, is parallel to the x'-axis, and covers the segment $(0, l_0)$. The rod moves in this plane, always remaining parallel to the x'-axis, in the direction of the positive y'-axis with constant velocity u. Describe the appearance and the motion of the rod in the $S = \{ct, x, y, z\}$ inertial frame with respect to which S' is moving with constant velocity v in standard configuration. Is your result consistent with the relativistic transformation of velocities?

References

1. H. Fizeau, Comptes Rendus **33**, 349 (1851)
2. A.A. Michelson, E.W. Morley, Am. J. Sci. **31**, 377 (1886)
3. P. Zeeman, Proc. Kon. Acad. Van Weten. **17**, 445 (1914)
4. J. Laub, Ann. Phys. **328**, 738 (1907)
5. M. von Laue Ann, Physik **23**, 989 (1907)
6. C. Møller, *The Theory of Relativity* (Oxford University Press, Oxford, 1952)
7. D.R. Frankl, Am. J. Phys. **52**, 374 (1984)
8. H.J. Hay, J.P. Schiffer, T.E. Cranshaw, P.A. Egelstaff, Phys. Rev. Lett. **4**, 165 (1960)
9. D.C. Champeney, G.R. Isaak, A.M. Khan, Nature **198**, 1186 (1963)
10. J. Stevens, A.L. Corper, C.F. Basler, J.K. Taubenberger, P. Palese, I.A. Wilson, Science **303**, 1866 (2004)
11. Light sources website http://www.lightsources.org/cms/
12. E.A. Desloge, R.J. Philpott, Am. J. Phys. **55**, 252 (1987)
13. E.A. Desloge, Am. J. Phys. **57**, 598 (1989)
14. G. Cavalleri, G. Spinelli, Nuovo Cimento B **66**, 11 (1970)
15. K. Nordtvedt, Am. J. Phys. **43**, 256 (1975)
16. Ø. Grøn, Am. J. Phys. **45**, 65 (1977)
17. H. Nikolić, Am. J. Phys. **67**, 1007 (1999)
18. J.B. Hartle, *Gravity: An Introduction to Einstein's General Relativity* (Addison-Wesley, San Francisco, 2003)
19. J. Terrell, Phys. Rev. **116**, 1041 (1959)
20. R. Penrose, Proc. Camb. Phil. Soc. **55**, 137 (1959)
21. V.F. Weisskopf, Phys. Today **13**(9), 24 (1960)
22. T.M. Helliwell, *Introduction to Special Relativity* (Allyn and Bacon, Boston, 1960)
23. G.D. Scott, M.R. Viner, Am. J. Phys. **33**, 534 (1965)
24. G. Gamow, *Mr. Tompkins in Paperback* (Cambridge University Press, Cambridge, 1965)
25. A. Komar, Am. J. Phys. **33**, 1024 (1965)
26. G.D. Scott, M.J. van Driel, Am. J. Phys. **38**, 971 (1970)
27. R. Bandhari, Am. J. Phys. **38**, 1200 (1970)
28. M.L. Boas, R.C. Calhoun, O. Horan, Am. J. Phys. **39**, 782 (1971)
29. P.M. Mathews, M. Lakshmanan, Nuovo Cimento B **12**, 168 (1972)
30. R. Bandhari, Am. J. Phys. **46**, 760 (1971)
31. M. McKinley, Am. J. Phys. **47**, 309 (1979)
32. M. McKinley, P. Doherty, Am. J. Phys. **47**, 602 (1979)
33. A. Peres, Am. J. Phys. **55**, 516 (1987)
34. E. Sheldon, Am. J. Phys. **56**, 199 (1988)
35. J. Terrell, Am. J. Phys. **57**, 9 (1988)
36. K.G. Suffern, Am. J. Phys. **56**, 729 (1988)
37. W. Rindler, *Introduction to Special Relativity*, 2nd edn. (Clarendon Press, Oxford, 1991)
38. E.F. Taylor, Am. J. Phys. **58**, 889 (1990)

Chapter 8
Measurements in Minkowski Spacetime

Truth is what stands the test of experience.
—Albert Einstein

8.1 Introduction

In Special Relativity an observer is characterized by a timelike 4-velocity u^μ. Quantities such as lengths, time intervals, and special-relativistic energies are not Lorentz-invariant and different observers (even different *inertial* observers) will measure different values for these quantities—it seems as if they will measure *components* of vectors or tensors in their own rest frame. From both the physical and mathematical points of view this is not satisfactory because we want the physical observables to be quantities defined in a coordinate-independent way, not coordinate components (after all, a given observer can change coordinates without making Lorentz boosts). This need motivates the present chapter. The strategy to solve this problem consists of characterizing the observer by means of its 4-velocity u^μ and of projecting vectors and tensors onto this 4-vector, thereby obtaining covariant quantities.

8.2 Energy of a Particle Measured by an Observer

Consider a particle of mass m and 4-momentum $p^\mu = \left(\frac{E_p}{c}, \gamma \, m\mathbf{v}_p \right)$ and an observer with 4-velocity $u^\mu = (\gamma_o \, c, \gamma_o \, \mathbf{v}_o)$, where the subscripts p and o stand for "particle" and "observer", respectively. The particle energy measured by the observer is

$$E_p = -p_\mu u^\mu , \tag{8.1}$$

V. Faraoni, *Special Relativity*, Undergraduate Lecture Notes in Physics,
DOI: 10.1007/978-3-319-01107-3_8, © Springer International Publishing Switzerland 2013

i.e., (minus) the projection of the particle's 4-momentum onto the 4-velocity ("time direction") of the observer, or the time component of the particle 4-momentum according to the time of this observer. To make sense of this expression note that $p_\mu u^\mu \equiv \eta_{\mu\nu}\, p^\mu u^\nu = -p^0 u^0 + \mathbf{p}\cdot\mathbf{u}$. Using $p^\mu = \left(\frac{E_p}{c}, \gamma_p\, m\, \mathbf{v}_p\right)$ and $u^\mu = (\gamma_o\, c, \gamma_o\, \mathbf{v}_o)$ one finds that

$$p_\mu u^\mu = -\frac{E_p}{c}\,\gamma_o\, c + \gamma_p\, m\mathbf{v}_p \cdot \gamma_o \mathbf{v}_o. \tag{8.2}$$

In the rest frame of the observer $\mathbf{v}_o = 0$ and $\gamma_o = 1$, hence

$$p_\mu u^\mu = -E_p. \tag{8.3}$$

If, furthermore, the particle is at rest with respect to the observer we have

$$E = -p_\mu u^\mu = \gamma\, mc^2 = mc^2. \tag{8.4}$$

Since u^μ is the "time direction of the observer" (i.e., $u^\mu \propto \delta^{0\mu}$ in its rest frame), projecting p^μ onto u^μ corresponds to experiencing the particle's 4-momentum from the point of view of that observer. If one adopts the quantum-mechanical relations for photons $E = \hbar\omega$ and $\mathbf{p} = \hbar\mathbf{k}$, then $p^\mu = \hbar k^\mu$ and $\hbar\omega = E = -p_\mu u^\mu = -\hbar k_\mu u^\mu$ so $\omega = -k_\mu u^\mu$. This is not a derivation because Special Relativity is a classical theory, but it shows that $E = -p_\mu u^\mu$ and $\omega = -k_\mu u^\mu$ are consistent with the quantum theory of light, to which Einstein himself greatly contributed with his 1905 explanation of the photoelectric effect.

8.3 Frequency Measured by an Observer

Consider a light source at rest in some inertial frame and another (possibly accelerated) observer moving with respect to the first one along a worldline $x^\mu(\tau)$ with 4-tangent $u^\mu = dx^\mu/d\tau$. Let ω denote the angular frequency of light. What angular frequency $\omega(\tau)$ will the non-stationary (possibly accelerated) observer measure?

The angular frequency is related to the time component k^0 of the wave 4-vector by $k^0 = \omega/c$, so

$$\omega_0 = -k^\mu u_\mu. \tag{8.5}$$

Proof We have $k^\mu = \left(k^0, k^1, 0, 0\right) = \left(\frac{\omega}{c}, \frac{\omega}{c}, 0, 0\right)$ (if, for simplicity, the ray propagates along the x-axis; otherwise, we can always rotate the spatial axes to make this happen) and $u^\mu = \left(u^0, u^1, 0, 0\right) = (\gamma_o c, \gamma_o v_o, 0, 0)$, then

$$k^\mu u_\mu = \eta_{\mu\nu} k^\mu u^\nu = \eta_{00} k^0 u^0 + \eta_{11} k^1 u^1$$
$$= -\frac{\omega}{c} u^0 + \frac{\omega}{c} u^1 = \frac{\omega}{c}\left(-u^0 + u^1\right).$$

In the rest frame of the observer it is $\mathbf{v}_o = 0$, $\gamma_o = 1$, $u^1 = 0$, and $u^0 = c$. Hence, it is $k^\mu u_\mu = -\omega$. This quantity is a 4-scalar and does not depend on the coordinate frame used but it does, of course, depend on the observer u^μ chosen. □

The Doppler effect discussed in the previous chapter relates the frequency emitted by the source in its rest frame with that measured by any inertial observer.

8.4 A More Systematic Treatment of Measurement

Special Relativity in formulated in terms of inertial observers, a serious limitation of this theory that is overcome only in General Relativity. However, accelerations and forces do exist and it is sometimes necessary to consider accelerated observers and measurements in accelerated frames (for example, a frame connected with the rotating earth in the analysis of the GPS system).

An observer traces a worldline $x^\mu = x^\mu(\tau)$ with 4-tangent $u^\mu = dx^\mu/d\tau$ satisfying $u_\mu u^\mu = -c^2$. This observer carries a laboratory along its worldline, equipped with clocks, rulers, and possibly more sophisticated equipment. Measurements are referred to axes in this laboratory, with their spatial directions defined by unit vectors $\{\mathbf{e}_1, \mathbf{e}_2, \mathbf{e}_3\}$ plus the time direction (a *tetrad*). For this reason, we associate to the laboratory four orthogonal (in the 4-dimensional sense) unit 4-vectors

$$e^\mu_{(0)}, \quad e^\mu_{(1)}, \quad e^\mu_{(2)}, \quad e^\mu_{(3)}, \tag{8.6}$$

of which $e^\mu_{(0)}$ describes the time direction and $e^\mu_{(i)}$ $(i = 1, 2, 3)$ describe perpendicular (in the three-dimensional sense) spatial directions in the 3-space orthogonal (in the 4-dimensional sense) to $e^\mu_{(0)}$ (Fig. 8.1).

The three spatial vectors can be thought of as unit vectors pointing along the directions of the axes of three gyroscopes carried by the observer. The time direction $e^\mu_{(0)} = u^\mu/c$ coincides with the 4-velocity of the observer and laboratory and

$$e^\mu_{(1)} e^{(1)}_\mu = e^\mu_{(2)} e^{(2)}_\mu = e^\mu_{(3)} e^{(3)}_\mu = 1, \tag{8.7}$$

$$e^\mu_{(0)} e^{(0)}_\mu = -1, \tag{8.8}$$

$$e^\mu_{(0)} e^{(i)}_\mu = 0, \qquad e^\mu_{(i)} e^{(j)}_\mu = \delta^j_i \qquad (i, j = 1, 2, 3). \tag{8.9}$$

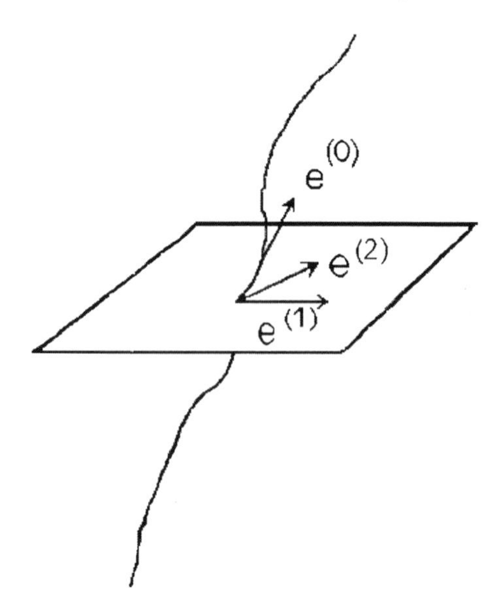

Fig. 8.1 The unit vector $e^{\mu}_{(0)} = u^{\mu}/c$ coincides with the time direction of the observer (the direction of the 4-tangent to its worldline), while $e^{\mu}_{(i)}$ are three orthogonal unit vectors in the 3-space of this observer. One spatial direction is suppressed

If the observer is inertial its worldline is not accelerated, which in Cartesian coordinates is expressed by the property

$$a^{\nu} \doteq u^{\mu} \partial_{\mu} u^{\nu} \doteq 0; \tag{8.10}$$

otherwise $a^{\mu} = \left(f^{\mu} - \dot{m}u^{\mu}\right)/m \neq 0$ for an accelerated observer.

When an observer measures 4-vectors or 4-tensors in the laboratory, it is *components* projected along the four axes that are measured. For example the measured energy of a particle with 4-momentum p^{μ} is

$$E = -p^{\mu} e^{(0)}_{\mu} = -p_{\mu} \frac{u^{\mu}}{c} \tag{8.11}$$

and the 3-momentum of the particle in the x-direction is

$$p^{\mu} e^{(1)}_{\mu}.$$

The 4-momentum of the particle can be decomposed along the tetrad

$$\left\{e^{\mu}_{(0)}, e^{\mu}_{(1)}, e^{\mu}_{(2)}, e^{\mu}_{(3)}\right\} \equiv \left\{e^{\mu}_{(\alpha)}\right\},$$

where the index (α) in brackets is a *tetrad index* (it will always appear in brackets):

$$p^{\mu} = p^{(\alpha)} e^{\mu}_{(\alpha)}, \tag{8.12}$$

and where $p^{(\alpha)}$ are the components of the 4-momentum in the chosen tetrad (*tetrad components*).

We can introduce the tetrad tensor $\eta_{(\alpha\beta)} = \text{diag}\,(-1, 1, 1, 1)$ which operates only on the tetrad indices, its inverse $\eta^{(\alpha\beta)} = \text{diag}\,(-1, 1, 1, 1)$ and, for any 4-vector $A^\mu = A^{(\alpha)}\, e^\mu_{(\alpha)}$ decomposed along the tetrad $\left\{e^\mu_{(\alpha)}\right\}$, the tetrad components are $A^{(\alpha)}$ and

$$A_{(\alpha)} \equiv \eta_{(\alpha\beta)} A^{(\beta)}. \tag{8.13}$$

The tetrad components of a tensor are obtained by projecting it on the tetrad basis, for example the tetrad components of a 4-vector A^μ are

$$A^{(\mu)} = e^{(\mu)}{}_\alpha A^\alpha \tag{8.14}$$

and those of a 2-tensor $B^{\mu\nu}$ are

$$B^{(\mu)(\nu)} = e^{(\mu)}{}_\rho e^{(\nu)}{}_\sigma B^{\rho\sigma}. \tag{8.15}$$

Tetrad indices can be raised and lowered with $\eta^{(\alpha\beta)}$ and $\eta_{(\alpha\beta)}$. For the 4-momentum of a particle, $p^{(0)} = -p^\mu e^{(0)}_\mu \equiv E/c$ is the energy of the particle, while $p^{(1)} = p^\mu e^{(1)}_\mu$, $p^{(2)} = p^\mu e^{(2)}_\mu$, and $p^{(3)} = p^\mu e^{(3)}_\mu$.

Example 8.1 Consider a particle at rest in an inertial frame and an observer moving with velocity v with respect to that frame in standard configuration. The observer sees the particle moving with velocity $-v$ in his/her own frame. In the rest frame of the particle, the particle 4-momentum is simply $p^\mu = (mc, 0, 0, 0)$ and the observer has 4-velocity

$$u^\mu = c\, e^\mu_{(0)} = (\gamma\, c, \gamma\, v, 0, 0)\,. \tag{8.16}$$

The energy of the particle measured by the observer is

$$\begin{aligned} E &= -p^\mu u_\mu = -p^\mu e^{(0)}_\mu c = -\eta_{\mu\nu}\, p^\mu e^\nu_{(0)} c \\ &= -c\left(\eta_{00}\, p^0 e^0_{(0)} + \eta_{11}\, p^1 e^1_{(0)}\right) = -\left(-\gamma\, mc^2\right) = \gamma\, mc^2 \end{aligned}$$

so $E = \gamma\, mc^2$ in the frame of the observer, a result that we already know.

8.5 The 3+1 Splitting

There is a coordinate-invariant way of slicing Minkowski spacetime into the 3-dimensional space and the time "seen" by an arbitrary (timelike) observer with 4-velocity v^μ. One writes the Minkowski metric as

$$g_{\mu\nu} = \underbrace{-\frac{v_\mu v_\nu}{c^2}}_{\text{“time part”}} + \underbrace{h_{\mu\nu}}_{\text{“space part”}} \tag{8.17}$$

by defining

$$h_{\mu\nu} \equiv g_{\mu\nu} + \frac{v_\mu v_\nu}{c^2}. \tag{8.18}$$

$h_{\mu\nu}$ has the meaning of 3-dimensional Riemannian metric on the 3-dimensional space of the observer characterized by its 4-velocity v^μ. In fact $h_{\mu\nu}$ is *orthogonal to* v^μ:

$$h_{\mu\nu} v^\mu = h_{\mu\nu} v^\nu = 0 \tag{8.19}$$

and is symmetric.

Proof We have

$$\begin{aligned}
h_{\mu\nu} v^\mu &\equiv \left(g_{\mu\nu} + \frac{v_\mu v_\nu}{c^2}\right) v^\mu = g_{\mu\nu} v^\mu + \left(\frac{v_\mu v^\mu}{c^2}\right) v_\nu = v_\nu - v_\nu = 0, \\
h_{\mu\nu} v^\nu &\equiv \left(g_{\mu\nu} + \frac{v_\mu v_\nu}{c^2}\right) v^\nu = g_{\mu\nu} v^\nu + \left(\frac{v_\nu v^\nu}{c^2}\right) v_\mu = v_\mu - v_\mu = 0, \\
h_{\nu\mu} &\equiv g_{\nu\mu} + \frac{v_\nu v_\mu}{c^2} = g_{\mu\nu} + \frac{v_\mu v_\nu}{c^2} \equiv h_{\mu\nu}.
\end{aligned}$$

□

The mixed tensor $h_\mu{}^\nu$ is a projection operator (or *projector*) onto the 3-dimensional space orthogonal to the time direction v^α of the observer. Any 4-vector A^μ can be decomposed into a part parallel to v^μ and a part orthogonal to it (Fig. 8.2):

$$A_\parallel^\mu \equiv -\left(A^\alpha v_\alpha\right) \frac{v^\mu}{c^2} \tag{8.20}$$

(obviously lying in the direction identified by v^μ) and

$$A_\perp^\mu \equiv h^\mu{}_\nu A^\nu. \tag{8.21}$$

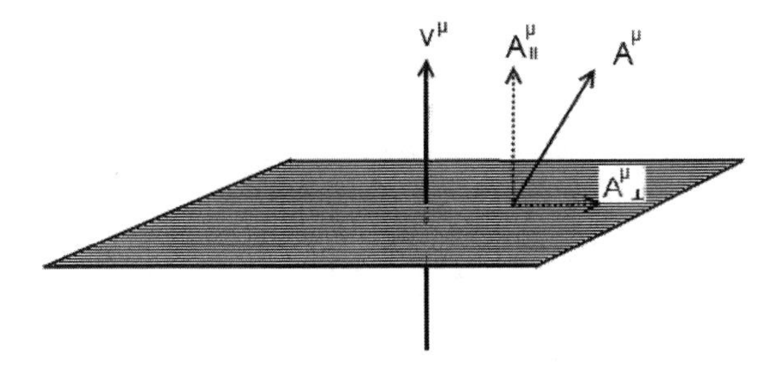

Fig. 8.2 The splitting of a 4-vector A^μ into a part parallel and a part orthogonal (in the 4-dimensional sense) to the 4-velocity v^μ of an observer

Let us check that $A^\mu_\perp$ is orthogonal to v^μ: we have

$$A^\mu_\perp v_\mu = \left(h^\mu{}_\nu A^\nu\right) v_\mu = \left(h^\mu{}_\nu v_\mu\right) A^\nu = 0.$$

and

$$\begin{aligned} A^\mu_\perp + A^\mu_\parallel &= h^\mu{}_\nu A^\nu - \left(A^\alpha v_\alpha\right) \frac{v^\mu}{c^2} \\ &\equiv \left(\delta^\mu{}_\nu + \frac{v^\mu v_\nu}{c^2}\right) A^\nu - \left(A^\alpha v_\alpha\right) \frac{v^\mu}{c^2} \\ &= A^\mu + v^\mu \left(\frac{v_\nu A^\nu}{c^2}\right) - \frac{\left(A^\alpha v_\alpha\right) v^\mu}{c^2} = A^\mu, \end{aligned}$$

hence

$$A^\mu = A^\mu_\parallel + A^\mu_\perp. \tag{8.22}$$

In the slow-motion limit $v/c \to 0$ it is

$$h_{\mu\nu} \equiv g_{\mu\nu} + \frac{v_\mu v_\nu}{c^2} \approx g_{\mu\nu} + \delta_{0\mu}\delta_{0\nu} \doteq \text{diag}\,(0, 1, 1, 1)$$

in Cartesian coordinates, so $h_{\mu\nu}$ reduces to the purely spatial Euclidean metric of 3-dimensional space. In this limit $v^\mu = \dfrac{dx^\mu_{(observer)}}{d\tau_{(observer)}} \approx (c, 0, 0, 0)$.

The projection onto v^μ of a derivative of a function f ("time derivative") becomes

$$v^\mu \partial_\mu f = c\,\frac{df}{d\tau} \approx c\,\frac{df}{dt} \tag{8.23}$$

and the "spatial gradient" $h_\mu{}^\alpha\, \partial_\alpha f$ of f becomes

$$h_\mu{}^\alpha\,\partial_\alpha f = \left(g^\alpha_\mu + \frac{v_\mu v^\alpha}{c^2}\right)\partial_\alpha f = \delta^\alpha_\mu\,\partial_\alpha f + \frac{v_\mu}{c}\frac{df}{d\tau}$$
$$\approx \delta^\alpha_\mu\,\partial_\alpha f - \delta^\mu_0\,\frac{df}{dt} = \delta^i_\mu\,\partial_i f \qquad (i = 1, 2, 3)$$

because

$$\left|\frac{\partial f}{c\,\partial t}\right| \ll \left|\frac{\partial f}{\partial x^i}\right| \tag{8.24}$$

in this limit.

8.6 Conclusion

We can now end this more formal but conceptually important chapter discussing measurements in Special Relativity. It is now clear that physical quantities measured in simple experiments are not just components of tensors but, rather, quantities which are invariant under coordinate transformations. They do, of course, depend on the choice of observer who is characterized by his or her 4-velocity v^μ, a geometric object. We have also given a formal description of the 3-space "seen" by an (inertial or accelerated) observer. Often, but not always, matter is characterized by a special family of observers, those for which the matter distribution is at rest. This is the case, for example, of perfect fluids studied in the next chapter.

Problems

8.1 Show that, if the 4-momentum p^μ of a particle of mass m is decomposed as

$$p^\mu = \frac{1}{c}\left(E v^\mu + \sqrt{E^2c^2 - m^2c^4}\,e^\mu\right),$$

where $v^\nu v_\nu = -c^2$ and e^μ is a spacelike unit vector orthogonal to v^μ, then the normalization $p_\mu p^\mu = -m^2c^2$ is satisfied. Provide a physical interpretation of the quantities v^μ, e^μ, and E.

8.2 Let u^μ be the 4-velocity of an observer. Using the 3 + 1 splitting, the gradient of a scalar function ϕ can be decomposed as

$$\nabla_\mu\phi = -\dot{\phi}\left(\frac{u_\mu}{c^2}\right) + \tilde{\nabla}_\mu\phi.$$

Find covariant expressions for $\dot{\phi}$ and $\tilde{\nabla}\phi$.

8.3 Consider a uniformly accelerated particle with worldline

$$t(\tau) = \frac{c}{a} \sinh\left(\frac{a\tau}{c}\right),$$
$$x(\tau) = \frac{c^2}{a} \cosh\left(\frac{a\tau}{c}\right),$$
$$y = 0,$$
$$z = 0,$$

in Cartesian coordinates. Show explicitly the decomposition $g_{\mu\nu} = -\frac{u_\mu u_\nu}{c^2} + h_{\mu\nu}$ for the Rindler observer comoving with the particle. What are u_μ and $h_{\mu\nu}$? Check explicitly that $h_{\mu\nu}u^\nu = 0$. What is the energy of this uniformly accelerated particle according to a stationary observer? Let the particle emit a light signal travelling along the x-axis toward a stationary observer located at $\mathbf{x} = \left(\frac{c^2}{a}, 0, 0\right)$. Compute the frequency received by this stationary observer and discuss the limits $\tau \longrightarrow \pm\infty$.

8.4 Consider the Minkowski metric in Cartesian coordinates x^μ and the tetrad $\left\{\mathrm{d}x^\mu\right\}$. The components of the metric tensor can be written as

$$g_{\mu\nu} = -c^2 t_\mu t_\nu + x_\mu x_\nu + y_\mu y_\nu + z_\mu z_\nu$$

where $t_\mu \equiv \partial t/\partial x^\mu$, etc. Define a *null tetrad* $\left\{e^{\mu'}{}_{(\alpha)}\right\} = \left\{k_\mu, l_\mu, m_\mu, \bar{m}_\mu\right\}$ by

$$k_\mu = \frac{x_\mu + ct_\mu}{\sqrt{2}},$$
$$l_\mu = \frac{x_\mu - ct_\mu}{\sqrt{2}},$$
$$m_\mu = \frac{y_\mu + \mathrm{i}z_\mu}{\sqrt{2}},$$
$$\bar{m}_\mu = \frac{y_\mu - \mathrm{i}z_\mu}{\sqrt{2}}.$$

Verify that these four tetrad vectors are null and compute their products with all the other vectors of the null tetrad. Check that the metric can be written as

$$g_{\mu\nu} = k_\mu l_\nu + k_\nu l_\mu + m_\mu \bar{m}_\nu + m_\nu \bar{m}_\mu.$$

Chapter 9
Matter in Minkowski Spacetime

Equations are more important to me, because politics is for the present, but an equation is something for eternity.
—Albert Einstein

9.1 Introduction

We have studied the Lorentzian geometry of Minkowski spacetime, the arena in which things happen, and the kinematics and dynamics of massive and massless particles. It is now time to explore the laws ruling various forms of mass-energy and their physics in the 4-dimensional world. A fundamental role for any form of matter is played by its stress-energy tensor, which is introduced below. After general considerations we discuss the fluid dynamics of perfect fluids and then we study the scalar field and the electromagnetic field.

9.2 The Energy-Momentum Tensor

It is important to remember that Special Relativity does not describe gravity, which is completely ignored in this theory. Therefore, all the matter distributions considered in the following do not gravitate, i.e., they are considered to be *test matter*: they do not generate an appreciable gravitational field. This is a reasonable approximation in many situations. After all, when studying electromagnetism or nuclear or particle physics, we ignore gravity completely.

We already know how to describe single massive and massless particles and distributions of point particles in Special Relativity. We now want to consider *continuous distributions of matter* and *fields*, systems with an infinite number of degrees of

V. Faraoni, *Special Relativity*, Undergraduate Lecture Notes in Physics,

DOI: 10.1007/978-3-319-01107-3_9,

freedom, as opposed to systems of point particles which only have a finite number of degrees of freedom.[1]

A continuous matter distribution (including a field) in Special Relativity is described by a symmetric tensor $T_{\mu\nu}$ called *stress-energy tensor* or *energy-momentum tensor*

$$T_{\mu\nu} = T_{\nu\mu}. \tag{9.1}$$

This tensor will describe continuous distributions of energy such as fluids, continuous media, or fields which we expect to possess energy density, momentum density, and internal stresses. Before we proceed, a geometric consideration is in order. A timelike 4-vector A^μ admits a *unique* frame in which all its spatial components A^i are zero ($i = 1, 2, 3$). A spacelike 4-vector admits *no* frame in which all its spatial components are zero. By contrast, a symmetric tensor $A^{\mu\nu}$ admits one, none, or many frames in which the time-space components A^{0i} ($i = 1, 2, 3$) are zero. For timelike matter described by a symmetric 2-tensor $T^{\mu\nu}$ it is *assumed* that there exists a frame in which all the components T^{0i} vanish (as we will see, these components describe the momentum density of this matter). Consider now an observer with 4-velocity v^μ (in short, "an observer v^μ"):

- the *energy density* ρc^2 of the matter distribution is obtained by projecting $T_{\mu\nu}$ onto v^μ twice:

$$\rho c^2 \equiv T_{\mu\nu} \frac{v^\mu}{c} \frac{v^\nu}{c}; \tag{9.2}$$

  this is the energy per unit volume as measured by this observer. Remembering the equivalence of mass and energy, a mass density ρ corresponds to an energy density ρc^2. "Reasonable" matter is believed to have non-negative energy density ρc^2.
- Let s^μ be a unit 4-vector orthogonal to the 4-velocity of the observer (a purely spatial vector in the rest frame of this observer), $s_\mu v^\mu = 0$. Then s^μ identifies a spatial direction with respect to this observer. The *momentum flux density in the s^μ direction* (times c) is the 4-vector

$$-T_{\mu\nu} \frac{v^\mu}{c} s^\nu. \tag{9.3}$$

  Up to the factor c, this is the momentum flowing per unit time and per unit of area normal to the direction s^μ with respect to the observer. In general, by letting the vector s^μ vary spanning all the three spatial directions orthogonal to v^μ, one obtains the 4-vector

$$J_\mu \equiv -T_{\mu\nu} \frac{v^\nu}{c} \tag{9.4}$$

  (the *energy 4-current density* times c).

[1] Remember that rigid bodies, which are also described by a finite number of degrees of freedom, correspond to an infinite sound speed and do not exist in Special Relativity. The same conclusion applies to perfectly incompressible fluids.

- Let $s^{\mu}_{(1)}$ and $s^{\mu}_{(2)}$ be two spatial vectors orthogonal to v^{μ}; then

$$T_{\mu\nu} s^{\mu}_{(1)} s^{\nu}_{(2)} \tag{9.5}$$

is the $s^{\mu}_{(1)}$–$s^{\nu}_{(2)}$ component of the stress-energy tensor of the matter distribution. In other words, if we are in the frame of the observer v^{μ} and we consider only the spatial indices, T_{ij} $(i, j = 1, 2, 3)$ is a 3-dimensional tensor identified with the stress tensor of the material (or field).

9.3 Covariant Conservation

In general, for any continuous matter distribution or field in Minkowski space *in Cartesian coordinates*, mass-energy conservation is described by

$$\partial^{\nu} T_{\mu\nu} \doteq 0 \qquad \Longleftrightarrow \qquad \partial^{\mu} T_{\mu\nu} \doteq 0 \tag{9.6}$$

(the proof of the equivalence between the two equations is left as an exercise), where $T_{\mu\nu}$ is the symmetric energy momentum tensor describing that form of matter (for visualization think of this matter as a perfect fluid).

Consider a family of inertial observers with parallel 4-velocities v^{μ} defined throughout spacetime. Since they are inertial (non-accelerated) observers it is $\partial_{\mu} v^{\nu} \doteq 0$ in these coordinates. In fact,

$$0 = \frac{d^2 x^{\mu}}{d\tau^2} = \frac{dv^{\mu}}{d\tau} = \frac{\partial v^{\mu}}{\partial x^{\alpha}}\, v^{\alpha}$$

with $v^{\mu} \neq 0$ implies $\partial v^{\mu}/\partial x^{\alpha} = 0$. Now define the *energy 4-current density* "seen" by these observers as

$$J_{\mu} \equiv -T_{\mu\nu}\, \frac{v^{\nu}}{c}. \tag{9.7}$$

Then Eq. (9.6) implies that

$$c\, \partial^{\mu} J_{\mu} = -\partial^{\mu} \left(T_{\mu\nu} v^{\nu} \right) = -\left(\partial^{\mu} T_{\mu\nu} \right) v^{\nu} - T_{\mu\nu} \partial^{\mu} v^{\nu} \doteq 0,$$

therefore,

$$\partial^{\mu} J_{\mu} \doteq 0 \tag{9.8}$$

(*4-current conservation*). This is a *local* conservation equation which can be translated into a *global* conservation statement as follows. Consider two instants of time t_1 and t_2 and all 3-dimensional physical space at these times (Fig. 9.1).

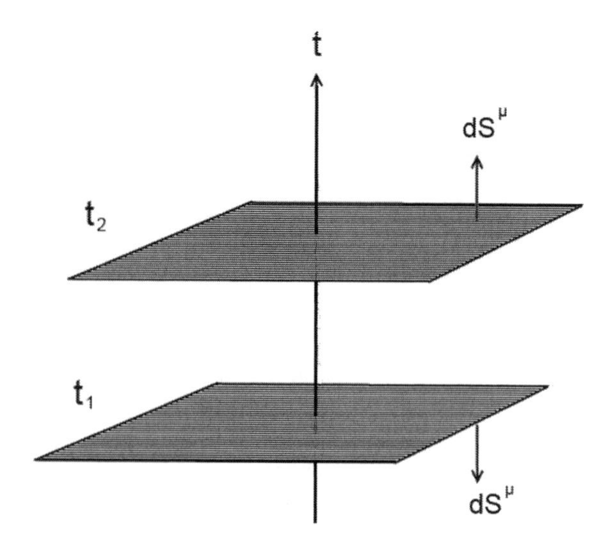

Fig. 9.1 Two hypersurfaces of constant time $t = t_1$ and $t = t_2 > t_1$. The unit normal n^μ is parallel to the surface element dS^μ and points outward (in the direction of time at t_2 and in the opposite direction at t_1)

Apply Gauss' theorem to the 4-dimensional simply connected volume Ω bounded by the surfaces $t = t_1, t = t_2$ and by spatial infinity,

$$\int_\Omega \mathrm{d}^4x\, \partial_\mu J^\mu = \int_{\partial\Omega} \mathrm{d}S^\mu J_\mu, \tag{9.9}$$

where $\partial\Omega$ is the 3-dimensional boundary of the 4-dimensional region Ω. Any physically meaningful distribution of energy is localized, in the sense that $T_{\mu\nu}$ goes to zero at spatial infinity,[2] therefore the contributions to the integral $\int_{\partial\Omega} \mathrm{d}S^\mu J_\mu$ come only from the time hypersurfaces[3] $t = t_1$ and $t = t_2$. Since the unit normal n^μ to the hypersurface $\partial\Omega$ (with $n^\mu \mathrm{d}S \equiv \mathrm{d}S^\mu$) is pointing outward, it is opposite to the time direction at the surface $t = t_1$, or $n^\mu\Big|_{t_1} = -v^\mu/c\Big|_{t_1}$, $n^\mu\Big|_{t_2} = v^\mu/c\Big|_{t_2}$, and

$$\int_{t=t_1} \mathrm{d}S^\mu J_\mu = -\int_{t=t_1} \mathrm{d}^3\mathbf{x}\, \frac{v^\mu}{c}\, J_\mu,$$
$$\int_{t=t_2} \mathrm{d}S^\mu J_\mu = \int_{t=t_2} \mathrm{d}^3\mathbf{x}\, \frac{v^\mu}{c}\, J_\mu.$$

[2] An exception occurs in the study of cosmology but this situation belongs to General, not to Special, Relativity.

[3] Alternatively, one can consider a distribution of energy exactly confined to a finite region, and take the volume Ω such that its "side" lies outside of the matter distribution, where $T_{\mu\nu} = 0$.

Moreover, it is

$$\frac{v^\mu}{c} J_\mu = \frac{v^\mu}{c}\left(-T_{\mu\nu}\frac{v^\nu}{c}\right) = -T_{\mu\nu}\frac{v^\mu v^\nu}{c^2} = -\rho c^2 \tag{9.10}$$

so Eqs. (9.8) and (9.9) yield $\int_{\partial\Omega} \mathrm{d}S^\mu J_\mu = 0$ which, in turn, gives

$$\int_{t=t_1} \mathrm{d}^3\mathbf{x}\, \rho = \int_{t=t_2} \mathrm{d}^3\mathbf{x}\, \rho \tag{9.11}$$

or, the total energy contained in space at any instant of time t_1 is equal to the total energy in space at any later time t_2 (*conservation of mass-energy*). This argument applies to arbitrary timelike observers v^μ.

The covariant conservation equation $\partial^\mu T_{\mu\nu} = 0$ has also more local consequences. If we integrate it over a *finite*, simply connected, closed 4-volume Ω with finite 3-dimensional boundary $S \equiv \partial\Omega$ (Fig. 9.2), it is

$$0 = \int_\Omega \mathrm{d}^4x\, \partial^\mu J_\mu = \int_{\partial\Omega} \mathrm{d}S^\mu J_\mu. \tag{9.12}$$

Let n^μ be the outer unit normal to the 3-surface $\partial\Omega$, then

$$\int_S \mathrm{d}^3\mathbf{x}\, n^\mu J_\mu = 0, \tag{9.13}$$

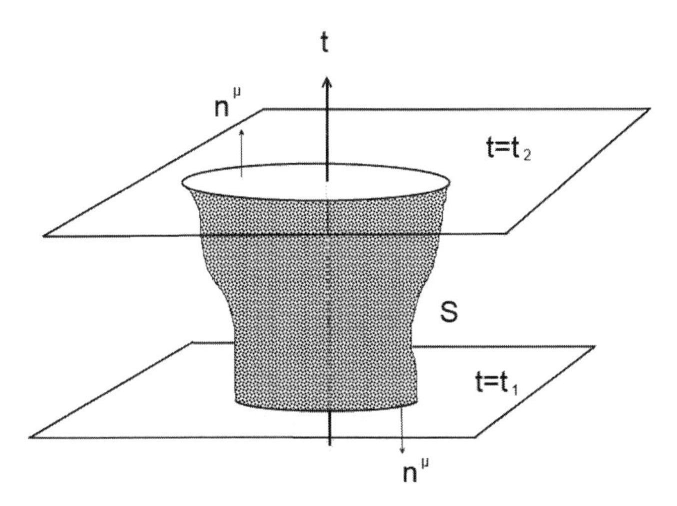

Fig. 9.2 A spacetime region limited by a worldtube and its intersections with surfaces of constant time t_1 and t_2

which tells us that the energy change of the matter distribution between top and bottom of S equals the time-integrated energy flux into the volume (i.e., the contribution from the "sides" of S)—this is conservation of energy for the 4-volume Ω.

An important fact is that

the covariant conservation equation $\partial^\nu T_{\mu\nu} \doteq 0$ contains the equations of motion of the matter described by $T_{\mu\nu}$.

This statement will be explained in applications to specific forms of matter such as perfect fluids and scalar and electromagnetic fields in the following sections.

Let Ω be a 3-dimensional volume in physical space, then

$$P^\mu = \frac{1}{c}\int_\Omega \mathrm{d}^3\mathbf{x}\, J^\mu = \frac{1}{c}\int_\Omega \mathrm{d}^3\mathbf{x}\, T^{0\mu} \tag{9.14}$$

is the 4-momentum of the mass-energy in Ω if

$$E = \int_\Omega \mathrm{d}^3\mathbf{x}\, \rho c^2, \qquad P^i = \frac{1}{c}\int_\Omega \mathrm{d}^3\mathbf{x}\, T^{0i} \qquad (i = 1, 2, 3).$$

Then

$$P^\mu = \left(\frac{E}{c}, \mathbf{P}\right) = \frac{1}{c}\int_\Omega \mathrm{d}^3\mathbf{x}\, J^\mu \tag{9.15}$$

is a 4-vector. This vector will be timelike or null under certain conditions, called *energy conditions*. There are various energy conditions which can be *assumed* to be satisfied by what we believe is "reasonable" matter; they essentially exclude the possibility of stresses which are too large compared with the energy density of matter (for example, they exclude a fluid which has pressure but no energy density) and of spacelike energy flows.

9.4 *Energy Conditions

Here we summarize the point-wise and averaged energy conditions of relativity.[4] The energy conditions satisfied by a specific form of matter are formulated in terms of the stress-energy tensor $T_{\mu\nu}$ that describes it. In many situations it is meaningful to describe matter as a perfect fluid, which is characterized by the particular form (9.30) of the stress-energy tensor and has energy density ρc^2 and pressure P. Here we use such a fluid to illustrate the physical meaning of the energy conditions. Violations

---

[4] There are slight differences in the definitions of some of these energy conditions adopted by different authors in the literature (see, e.g., Refs. [1–4]). Here we report the most common definitions.

of the various energy conditions are usually associated with exotic forms of matter (such as those which are suspected to occur in wormholes or in cosmology) or with quantum fields.

- The *weak energy condition* is satisfied if

$$T_{\mu\nu} t^{\mu} t^{\nu} \geq 0 \qquad \text{for all timelike 4-vectors } t^{\mu}. \tag{9.16}$$

For the fluid (9.30) this condition is equivalent to

$$\rho \geq 0 \quad \text{and} \quad \rho + \frac{P}{c^2} \geq 0. \tag{9.17}$$

- The *dominant energy condition* is satisfied if the weak energy condition is satisfied and, in addition, $T^{\mu\nu} t_{\mu}$ is a null or timelike 4-vector (i.e., $T_{\mu\nu} T^{\nu}{}_{\gamma} t^{\mu} t^{\gamma} \leq 0$) for any timelike 4-vector t^{μ}. For the perfect fluid (9.30) this condition means that

$$\rho \geq \frac{|P|}{c^2}, \tag{9.18}$$

or that the speed at which energy flows in this form of matter (as experienced by any material observer) does not exceed the speed of light.
- The *null energy condition* consists of

$$T_{\mu\nu} l^{\mu} l^{\nu} \geq 0 \qquad \text{for all null 4-vectors } l^{\mu}. \tag{9.19}$$

For the perfect fluid (9.30) this requirement means

$$\rho + \frac{P}{c^2} \geq 0. \tag{9.20}$$

- The *null dominant energy condition* consists of

$$T_{\mu\nu} l^{\mu} l^{\nu} \geq 0 \quad \text{and} \quad T^{\mu\nu} l_{\nu} \text{ is null or timelike for any null 4-vector } l^{\mu}. \tag{9.21}$$

This condition is the same as the dominant energy condition but here l^{μ} is a null instead of timelike 4-vector. For the perfect fluid (9.30) this condition means

$$\rho \geq \frac{|P|}{c^2} \quad \text{or} \quad \rho = -\frac{P}{c^2}. \tag{9.22}$$

- The *strong energy condition* consists of

$$\left(T_{\mu\nu} - \frac{1}{2} T g_{\mu\nu}\right) t^{\mu} t^{\nu} \geq 0 \quad \text{for any timelike 4-vector } t^{\mu}. \tag{9.23}$$

For the perfect fluid (9.30) this requirement is equivalent to

$$\rho + \frac{P}{c^2} \geq 0 \quad \text{and} \quad \rho + \frac{3P}{c^2} \geq 0. \tag{9.24}$$

The conditions above are point-wise energy conditions, which are imposed at each spacetime point. There are also energy conditions averaged over causal spacetime curves:

- The *averaged weak energy condition* consists of

$$\int_{-\infty}^{+\infty} \mathrm{d}\tau \, T_{\mu\nu} u^{\mu} u^{\nu} \geq 0 \tag{9.25}$$

for all timelike non-accelerated curves (*geodesics*) with 4-tangent u^{μ} and proper time τ.
- The *averaged null energy condition* consists of

$$\int_{-\infty}^{+\infty} \mathrm{d}\lambda \, T_{\mu\nu} l^{\mu} l^{\nu} \geq 0 \tag{9.26}$$

for all null non-accelerated curves with 4-tangent l^{μ} and affine parameter λ.

On short time scales, quantum systems are expected to violate all of the energy conditions, including positivity of the energy density.[5]

9.5 Angular Momentum

Consider an *isolated* distribution of mass-energy in a Cartesian inertial frame and let O be the origin event. Let V be a simply connected closed volume of 3-dimensional space at time t_1 and let Ω be the worldtube determined by the surfaces $t = t_1$, $t = t_2 > t_1$, and by the "side" described by the spacetime evolution of the boundary ∂V of V (Fig. 9.2). The *angular momentum tensor* of the distribution is defined as

$$L^{\mu\nu} \equiv \frac{1}{c} \int_V \mathrm{d}^3\mathbf{x} \left(x^{\mu} T^{\nu 0} - x^{\nu} T^{\mu 0} \right), \tag{9.27}$$

where x^{μ} is the position vector of the volume element $\mathrm{d}^3\mathbf{x} \equiv \mathrm{d}V$ in the mass-energy distribution with respect to a fixed origin of the coordinates. Let us consider an arbitrary constant tensor $Y_{\mu\nu}$ and apply the Gauss theorem to the 4-vector $(x^{\mu} T^{\nu\alpha} - x^{\nu} T^{\mu\alpha}) \, Y_{\mu\nu}$ on the 4-volume Ω, obtaining

---

[5] See, e.g., Ref. [5]. However, while a negative energy density is permitted for a quantum system over a certain interval of time, it is conjectured that later on the system more than compensates with a positive energy density (*quantum interest conjecture* [6]), thus respecting some of the averaged energy conditions.

$$\int_{\Omega} \mathrm{d}^4 x\, \partial_\alpha \left[\left(x^\mu T^{\nu\alpha} - x^\nu T^{\mu\alpha} \right) Y_{\mu\nu} \right] = \int_{\partial\Omega} \mathrm{d}^3 \mathbf{x}\, n_\alpha \left(x^\mu T^{\nu\alpha} - x^\nu T^{\mu\alpha} \right) Y_{\mu\nu}.$$

We have

$$\begin{aligned}
&\partial_\alpha \left[\left(x^\mu T^{\nu\alpha} - x^\nu T^{\mu\alpha} \right) Y_{\mu\nu} \right] \\
&= \left(\partial_\alpha x^\mu \right) T^{\nu\alpha} Y_{\mu\nu} + x^\mu \left(\partial_\alpha T^{\nu\alpha} \right) Y_{\mu\nu} + x^\mu T^{\nu\alpha} \partial_\alpha Y_{\mu\nu} \\
&\quad - \left(\partial_\alpha x^\nu \right) T^{\mu\alpha} Y_{\mu\nu} - x^\nu \left(\partial_\alpha T^{\mu\alpha} \right) Y_{\mu\nu} - x^\nu T^{\mu\alpha} \partial_\alpha Y_{\mu\nu} \\
&= \underbrace{\delta^\mu_\alpha T^{\nu\alpha} Y_{\mu\nu}}_{T^{\nu\mu} Y_{\mu\nu}} + x^\mu \left(\partial_\alpha T^{\nu\alpha} \right) Y_{\mu\nu} - \underbrace{\delta^\nu_\alpha T^{\mu\alpha} Y_{\mu\nu}}_{T^{\mu\nu} Y_{\mu\nu}} - x^\nu \left(\partial_\alpha T^{\mu\alpha} \right) Y_{\mu\nu} \\
&= \left[x^\mu \left(\partial_\alpha T^{\nu\alpha} \right) - x^\nu \left(\partial_\alpha T^{\mu\alpha} \right) \right] Y_{\mu\nu} \doteq 0,
\end{aligned}$$

where the last two quantities in brackets in the last line are equal to zero because of covariant conservation. Therefore,

$$\begin{aligned}
0 &= \int_{\partial\Omega} \mathrm{d}^3 \mathbf{x}\, n_\alpha \left(x^\mu T^{\nu\alpha} - x^\nu T^{\mu\alpha} \right) Y_{\mu\nu} \\
&= Y_{\mu\nu} \left(\int_{t=t_1} + \int_{t=t_2} + \int_{\text{``side''}} \right) \mathrm{d}^3 \mathbf{x}\; n_\alpha \left(x^\mu T^{\nu\alpha} - x^\nu T^{\mu\alpha} \right),
\end{aligned}$$

where the integral on the "side" is zero because $T_{\mu\nu}$ vanishes there. Since the outward unit normal n^μ points in the positive t-direction at t_2 and in the negative t-direction at $t = t_1$, we have

$$Y_{\mu\nu} \int_{t=t_1} \mathrm{d}^3 \mathbf{x} \left(x^\mu T^{\nu 0} - x^\nu T^{\mu 0} \right) = Y_{\mu\nu} \int_{t=t_2} \mathrm{d}^3 \mathbf{x} \left(x^\mu T^{\nu 0} - x^\nu T^{\mu 0} \right) \tag{9.28}$$

and

$$L^{\mu\nu}(t_2) = L^{\mu\nu}(t_1), \tag{9.29}$$

i.e.,

the angular momentum of an isolated distribution of mass-energy is conserved.

9.6 Perfect Fluids

A *perfect fluid* is a continuous distribution of matter with an energy-momentum tensor of the form

$$T_{\mu\nu} = \left(P + \rho c^2\right) \frac{u_\mu u_\nu}{c^2} + P\, g_{\mu\nu}, \tag{9.30}$$

where u^μ is a unit timelike vector $\left(u_\mu u^\mu = -c^2\right)$ representing the *4-velocity* of the fluid, ρc^2 is the *energy density*, and P is the fluid *pressure*. The name "energy density" given to ρc^2 is consistent with the previous definition of this quantity as $\rho c^2 = T_{\mu\nu} v^\mu v^\nu / c^2$ provided that we choose an *observer comoving with the fluid* "particles" (or volume elements), i.e., $v^\mu = u^\mu$. Then it is

$$\begin{aligned} T_{\mu\nu} u^\mu u^\nu &= \left[\left(P + \rho c^2\right) \frac{u_\mu u_\nu}{c^2} + P g_{\mu\nu}\right] u^\mu u^\nu \\ &= \left(P + \rho c^2\right) \left(\frac{u_\mu u^\mu}{c}\right) \left(\frac{u_\nu u^\nu}{c}\right) + P \underbrace{g_{\mu\nu} u^\mu u^\nu}_{-c^2} \\ &= P c^2 + \rho c^4 - P c^2 = \rho c^4, \end{aligned}$$

and

$$\begin{aligned} T_{\mu\nu} u^\nu &= \left[\left(P + \rho c^2\right) \frac{u_\mu u_\nu}{c^2} + P g_{\mu\nu}\right] u^\nu \\ &= -\left(P + \rho c^2\right) u_\mu + P u_\mu \\ &= -\rho c^2 u_\mu = -c^2 \times (\text{energy flux density}). \end{aligned}$$

This matter distribution is a "perfect" fluid because only the 4-velocity, pressure, and density appear in its description and no heat conduction or viscosity terms are present. In a frame comoving with the fluid,[6] in Cartesian coordinates, it is

$$T_{\mu\nu} \doteq \left(P + \rho c^2\right) \frac{u_\mu u_\nu}{c^2} + P \eta_{\mu\nu}, \quad u^\mu \doteq c\, \delta^{0\mu}\ , \quad u_\mu \doteq -\, c\, \delta_{0\mu} \tag{9.31}$$

and, therefore,

$$\left(T_{\mu\nu}\right) \doteq \begin{pmatrix} \rho c^2 & 0 & 0 & 0 \\ 0 & P & 0 & 0 \\ 0 & 0 & P & 0 \\ 0 & 0 & 0 & P \end{pmatrix}. \tag{9.32}$$

From this form of $T_{\mu\nu}$ it is clear that the fluid is spatially isotropic—there is a single pressure P, a triply degenerate eigenvalue of T_{ij}, and there are no tangential pressures. The 3-dimensional stress tensor T_{ij} of the perfect fluid is diagonal in the comoving frame of this fluid. The trace $T^\mu{}_\mu = g_{\mu\nu} T^{\mu\nu}$ is

[6] This *comoving frame* is the rest frame of the volume elements of the fluid.

$$T = -\rho c^2 + 3P \tag{9.33}$$

(this quantity is a world scalar and assumes the same value in all coordinate systems, although here we used Cartesian coordinates to evaluate it). Therefore, remembering that $P = \rho c^2/3$ is the equation of state of a bath of electromagnetic radiation in thermal equilibrium in a cavity,

the radiation equation of state $P = \rho c^2/3$ corresponds to vanishing trace T.

The stress-energy tensor of a perfect fluid is locally isotropic in the frame comoving with the fluid (the longitudinal stresses are all equal to P in this frame) and there are no tangential stresses T_{ij} $(i, j = 1, 2, 3$ with $i \neq j)$ due to the absence of viscosity.

9.6.1 Equation of Motion of the Perfect Fluid

The equation of motion of a fluid in the absence of external forces and in Cartesian coordinates is

$$\partial^\nu T_{\mu\nu} \doteq 0 \tag{9.34}$$

(*covariant conservation equation*). This name is given to Eq. (9.34) because it contains the conservation of mass-energy and momentum. To see this fact we need to project Eq. (9.34) onto time and 3-space.

First write down $T_{\mu\nu}$ explicitly:

$$\partial^\nu \left[\left(P + \rho c^2\right) \frac{u_\mu u_\nu}{c^2} + P\eta_{\mu\nu} \right] = 0$$

or

$$\left[\partial^\nu \left(P + \rho c^2\right)\right] \frac{u_\mu u_\nu}{c^2} + \left(P + \rho c^2\right) \frac{u^\nu \partial_\nu u_\mu}{c^2} + \left(P + \rho c^2\right) \frac{u_\mu \partial^\nu u_\nu}{c^2} + \underbrace{\left(\partial^\nu P\right) \eta_{\mu\nu}}_{\partial_\mu P} = 0. \tag{9.35}$$

Then project this equation onto u^μ:

$$\frac{u^\mu u_\mu}{c^2} \left[u^\nu \partial_\nu \left(P + \rho c^2\right)\right] + \left(P + \rho c^2\right) \underbrace{\frac{u^\mu u^\nu}{c^2} \partial_\nu u_\mu}_{= \frac{u^\mu a_\mu}{c^2} = 0}$$

$$+ \left(P + \rho c^2\right) \frac{u^\mu u_\mu}{c^2} \partial^\nu u_\nu + u^\mu \partial_\mu P = 0$$

or

$$-u^{\nu}\partial_{\nu}P - u^{\nu}\partial_{\nu}\left(\rho c^2\right) - \left(P + \rho c^2\right)\partial^{\nu}u_{\nu} + u^{\mu}\partial_{\mu}P = 0.$$

Now use $u^{\nu}\partial_{\nu}\rho = \dfrac{dx^{\nu}}{d\tau}\dfrac{\partial\rho}{\partial x^{\nu}} \equiv \dfrac{d\rho}{d\tau}$, where τ is the proper time of the comoving observer (or of fluid particles), to obtain

$$\frac{d\rho}{d\tau}c^2 + \left(P + \rho c^2\right)\partial^{\nu}u_{\nu} \doteq 0. \tag{9.36}$$

In the non-relativistic limit $P \ll \rho c^2$, $\tau \approx t$ (Newtonian time) it is

$$\partial^{\nu}u_{\nu} = \partial_{\nu}u^{\nu} = \left(\partial_{ct}u^0\right) + \nabla\cdot\mathbf{v} \approx \nabla\cdot\mathbf{v}. \tag{9.37}$$

In this limit Eq. (9.37) reduces to

$$\frac{d\rho}{dt} + \rho\nabla\cdot\mathbf{v} \approx 0 \qquad \text{as } \frac{\mathrm{v}}{\mathrm{c}} \longrightarrow 0. \tag{9.38}$$

Equation (9.38) is the continuity equation of non-relativistic fluid mechanics expressing the local conservation of mass. Therefore, Eq. (9.36) generalizes the conservation equation but:

- ρc^2 is the energy density, not the mass density.
- Instead of ρ, the combination $\left(\rho + \dfrac{P}{c^2}\right)$ appears in the second term on the left hand side of Eq. (9.36), which tells us that pressure and, in general, stresses, are a form of energy density in Special Relativity and will contribute to inertia and to the energy balance of a physical system.

Now project Eq. (9.35) onto the 3-space orthogonal to u^{α} using $h^{\mu}{}_{\alpha}$:

$$\begin{aligned} &h^{\mu}{}_{\alpha}\frac{u_{\mu}u^{\nu}}{c^2}\partial_{\nu}\left(P + \rho c^2\right) + \left(P + \rho c^2\right)h^{\mu}{}_{\alpha}\frac{u^{\nu}}{c^2}\partial_{\nu}u_{\mu} \\ &+ \left(P + \rho c^2\right)h^{\mu}{}_{\alpha}\frac{u_{\mu}\partial_{\nu}u^{\nu}}{c^2} + h^{\mu}{}_{\alpha}\partial_{\mu}P = 0, \end{aligned}$$

but

$$h^{\mu}{}_{\alpha}u^{\nu}\partial_{\nu}u_{\mu} \doteq h^{\mu}{}_{\alpha}a_{\mu} = \left(\delta^{\mu}_{\alpha} + \frac{u^{\mu}u_{\alpha}}{c^2}\right)a_{\mu} = a_{\alpha} + \left(u^{\mu}a_{\mu}\right)\frac{u_{\alpha}}{c^2} = a_{\alpha}$$

(in other words, a^{μ} is purely spatial and projecting it onto 3-space has no effect, $a^{\mu}_{\perp} = a^{\mu}$). Then it is

$$\underbrace{h^{\mu}_{\alpha}\partial_{\mu}P}_{\substack{\text{spatial}\\ \text{gradient}\\ \text{of } P}} + \left(P + \rho c^2\right) a_{\alpha} = 0. \tag{9.39}$$

This equation generalizes the Euler equation $\rho\,\mathbf{a} = -\nabla P$ of non-relativistic fluid mechanics (Newton's second law for a fluid). In fact in the slow-motion limit $\frac{P}{c^2} \ll \rho, \quad \frac{v}{c} \longrightarrow 0$ it reduces to

$$\nabla P + \rho\,\mathbf{a} = 0 \tag{9.40}$$

or

$$\nabla P + \rho\,\frac{d\mathbf{v}}{dt} = 0.$$

Since $\mathbf{v} = \mathbf{v}\,(t, \mathbf{x})$ we have $\frac{d\mathbf{v}}{dt} = \frac{\partial \mathbf{v}}{\partial t} + \frac{\partial \mathbf{v}}{\partial \mathbf{x}} \cdot \frac{d\mathbf{x}}{dt} = \frac{\partial \mathbf{v}}{\partial t} + \mathbf{v} \cdot \nabla\mathbf{v}$, therefore,

$$\underbrace{\rho\left(\frac{\partial \mathbf{v}}{\partial t} + \mathbf{v} \cdot \nabla\mathbf{v}\right)}_{\substack{\rho\frac{d\mathbf{v}}{dt}\\ \text{mass per unit}\\ \text{volume}\times\\ \text{acceleration}}} = \underbrace{-\nabla P}_{\substack{\text{force}\\ \text{per unit}\\ \text{volume}}}. \tag{9.41}$$

Equation (9.39) tells us again that pressure is a form of energy density and contributes to inertia together with ρc^2 in this special-relativistic version of $\mathbf{F} = d\mathbf{p}/dt$.

A perfect fluid is not completely specified by the form (9.30) of the stress-energy tensor but one must also assign an *equation of state*, i.e., a thermodynamic relation $f\,(\rho, P, T) = 0$ between pressure P, energy density ρc^2, and temperature T. Fluids with different equations of state differ wildly in their physical characteristic and behaviours. If the equation of state is of the simple form

$$P = P\,(\rho), \tag{9.42}$$

it is called a *barotropic equation of state*. A barotropic equation of state may take the even simpler form

$$P = w\,\rho c^2, \tag{9.43}$$

where $w =$ constant is called *equation of state parameter*. Barotropic equations of state which are widely used in relativity include:

- $P = 0$ (*dust*), which describes a fluid of non-interacting particles with no pressure and zero pressure gradient. There is no force acting on the fluid particles, which have no proper motions and move in unison. Admittedly, this kind of rarified fluid is a rather artificial model in ordinary situations but it does have its uses in relativity and cosmology.
- $P = \rho c^2/3$ (*radiation*). This is the equation of state of a gas of incoherent photons in a blackbody cavity in thermal equilibrium and is derived in statistical mechanics. It can be described as the equation of state of a random distribution of photons or massless particles in its rest frame (the only one in which it is random).

More exotic equations of state are also used. For example the *stiff equation of state* $P = \rho c^2$ is believed to be appropriate for the description of matter at nuclear densities in the core of neutron stars, and quantum vacuum is described by the equation of state $P = -\rho c^2$.

9.7 The Scalar Field

Many fields are considered in physics: you are probably familiar with the electromagnetic field represented by the vectors **E** and **B** (or, alternatively, by the scalar and vector potentials Φ and **A**). There are also fields representing the nuclear interactions, particles with spin (spinor fields), the various particles of the particle zoo, the gravitational field, the Higgs boson, etc. Restricting ourselves to classical fields, the simplest one is the *scalar field*. Although no classical scalar fields are known to exist in nature, quantum scalar fields are found to represent the Higgs boson, certain mesons in quantum mechanics, and fundamental and effective scalar fields are encountered in particle physics, condensed matter, cosmology, string theories, and extended theories of gravity [7–10]. Classically, a scalar field is a real function[7] of the coordinates $\phi\left(x^{\alpha}\right)$ that transforms as a scalar ($\phi' = \phi$) under coordinate transformations $x^{\mu} \longrightarrow x^{\mu'}$. Many scalar fields relevant for particle physics and cosmology obey the *Klein-Gordon equation*

$$\Box\phi - \frac{m^2c^2}{\hbar^2}\phi = 0, \tag{9.44}$$

where m is the *mass* of the field (which could possibly be zero), $\hbar$ is the reduced Planck constant, and $\Box\phi \doteq \partial^{\mu}\partial_{\mu}\phi$ in Cartesian coordinates. This linear equation is well-known in field theory. The stress-energy tensor of the scalar field is

---

[7] Also complex scalar fields are used in particle physics but they are not discussed here for simplicity.

$$T_{\mu\nu} = \partial_\mu \phi \, \partial_\nu \phi - \frac{1}{2} g_{\mu\nu} \left(\partial^\alpha \phi \partial_\alpha \phi \right) - \frac{m^2 c^2 \phi^2}{2\hbar^2} g_{\mu\nu}. \qquad (9.45)$$

This energy-momentum tensor is manifestly symmetric, $T_{\mu\nu} = T_{\nu\mu}$. Note that $T_{\mu\nu}$ is quadratic in the first order derivatives of ϕ with respect to time and space and it contains only first order derivatives in

$$\partial_\mu \phi \partial_\nu \phi - \frac{1}{2} g_{\mu\nu} \left(\partial^\alpha \phi \, \partial_\alpha \phi \right)$$

("kinetic" terms), and contains a term $-\frac{m^2 c^2 \phi^2}{2\hbar^2} g_{\mu\nu}$ quadratic in ϕ ("potential" term). To be more precise, consider the vector

$$v_\mu \equiv \partial_\mu \phi; \qquad (9.46)$$

if $v^\mu v_\mu \neq 0$ we can define the vector

$$u_\mu \equiv c \frac{\partial_\mu \phi}{\sqrt{|\partial^\alpha \phi \partial_\alpha \phi|}}. \qquad (9.47)$$

Let us assume that $\nabla_\mu \phi$ is timelike; then $u_\mu u^\mu = -c^2$ and by projecting $T_{\mu\nu}$ twice onto u_μ we obtain the energy density

$$\rho_\phi c^2 = T_{\mu\nu} \frac{u^\mu u^\nu}{c^2} = \partial_\mu \phi \partial_\nu \phi \frac{\partial^\mu \phi \partial^\nu \phi}{|\partial^\alpha \phi \partial_\alpha \phi|} - \frac{1}{2} \frac{\left(\partial^\mu \phi \partial_\mu \phi \right)^2}{|\partial^\alpha \phi \partial_\alpha \phi|} - \frac{m^2 c^2 \phi^2}{2\hbar^2} \frac{\partial^\mu \phi \partial_\mu \phi}{|\partial^\alpha \phi \partial_\alpha \phi|}$$

$$= -\frac{1}{2} \partial^\mu \phi \partial_\mu \phi + \frac{m^2 c^2 \phi^2}{2\hbar^2}.$$

The first term can be seen as a "kinetic" energy density of ϕ and the second term as a "potential" energy density. A point particle analogy would be a particle moving in one dimension with unit mass, position $\phi(t)$, kinetic energy $\frac{1}{2} \dot{\phi}^2$, and harmonic potential $\frac{m^2 c^2}{2\hbar^2} \phi^2$. In fact, if the field ϕ depends only on the time t (as in the case of cosmology in General Relativity), $\phi = \phi(t)$, then it is

$$\rho_\phi c^2 = \frac{\dot{\phi}^2}{2} + \frac{m^2 c^2 \phi^2}{2\hbar^2},$$

which is reminiscent of the energy of a particle with unit mass and position ϕ subject to a harmonic force.

The covariant conservation equation (9.6) contains the equation of motion for ϕ, which is the Klein-Gordon equation (9.44). In fact we have, in Cartesian coordinates,

$$
\begin{aligned}
0 \doteq \partial^\nu T_{\mu\nu} \doteq \partial^\nu & \left[\partial_\mu \phi \partial_\nu \phi - \frac{1}{2} \eta_{\mu\nu} \left(\partial^\alpha \phi \partial_\alpha \phi \right) - \frac{m^2 c^2 \phi^2}{2\hbar^2} \eta_{\mu\nu} \right] \\
&= \partial^\nu \left(\partial_\mu \phi \partial_\nu \phi \right) - \frac{1}{2} \eta_{\mu\nu} \partial^\nu \left(\partial^\alpha \phi \partial_\alpha \phi \right) - \frac{m^2 c^2}{\hbar^2} \phi \left(\partial^\nu \phi \right) \eta_{\mu\nu} \\
&= \left(\partial^\nu \partial_\mu \phi \right) \left(\partial_\nu \phi \right) + \left(\partial_\mu \phi \right) \left(\partial^\nu \partial_\nu \phi \right) - \frac{1}{2} \eta_{\mu\nu} \left(\partial^\nu \partial_\alpha \phi \right) \left(\partial^\alpha \phi \right) \\
&\quad - \frac{1}{2} \eta_{\mu\nu} \left(\partial_\alpha \phi \right) \left(\partial^\nu \partial_\alpha \phi \right) - \frac{m^2 c^2}{\hbar^2} \phi \, \partial_\mu \phi \\
&= \left(\partial^\nu \partial_\mu \phi \right) \left(\partial_\nu \phi \right) + \left(\partial_\mu \phi \right) \Box \phi - \left(\partial_\mu \partial_\alpha \phi \right) \left(\partial^\alpha \phi \right) - \frac{m^2 c^2}{\hbar^2} \phi \, \partial_\mu \phi \\
&= \left(\partial_\mu \phi \right) \left(\Box \phi - \frac{m^2 c^2}{\hbar^2} \phi \right)
\end{aligned}
$$

so, if $\phi \neq$ const. (i.e., if $\partial_\mu \phi \neq 0$) Eq. (9.6) implies

$$
\Box \phi - \frac{m^2 c^2}{\hbar^2} \phi = 0.
$$

The Klein-Gordon equation can be generalized to a non-linear equation. By writing the term $-\dfrac{m^2 c^2 \phi^2}{2\hbar^2} \eta_{\mu\nu}$ as $-V(\phi)\,\eta_{\mu\nu}$ and regarding $V(\phi)$ as a potential energy density, it is possible to conceive of functions $V(\phi)$ that are not quadratic, and these are indeed motivated by high energy physics and condensed matter physics. In this case the stress-energy tensor of the scalar field is generalized to

$$
T_{\mu\nu} = \partial_\mu \phi \, \partial_\nu \phi - \frac{1}{2} g_{\mu\nu} \left(\partial^\alpha \phi \, \partial_\alpha \phi \right) - V(\phi) \, g_{\mu\nu} \tag{9.48}
$$

and the covariant conservation equation $\partial^\nu T_{\mu\nu} \doteq 0$ then yields the equation of motion

$$
\Box \phi - \frac{dV}{d\phi} = 0 \tag{9.49}
$$

where, as usual, $\Box \doteq \partial^\nu \partial_\nu \doteq \eta^{\mu\nu} \partial_\mu \partial_\nu$ in Cartesian coordinates.

9.8 The Electromagnetic Field

In Special Relativity the spatial vector fields **E** (electric field) and **B** (magnetic field) of Maxwell's theory are combined into a 2-index antisymmetric tensor $F_{\mu\nu}$ (*Maxwell tensor*). This unified description is necessary because **E** and **B**, separately, do not transform as 4-vectors but $F_{\mu\nu}$ transforms as a true tensor. To realize this property, consider an inertial observer O who sees an electric charge at rest. For this observer there is a Coulomb electric field but no magnetic field, $\mathbf{E} \neq 0$ and $\mathbf{B} = 0$. Another inertial observer O' moving with constant velocity $\mathbf{v}$ with respect to O in a generic direction[8] will see a charge moving with velocity $-\mathbf{v}$ (a current) and will measure a magnetic field $\mathbf{B} \neq 0$. Therefore, electric and magnetic fields are not invariant but their components are mixed by a Lorentz transformation, in the same way that space and time are mixed in a change of inertial frame. The Maxwell tensor $F_{\mu\nu}$ combines both electric and magnetic fields and is antisymmetric,

$$F_{\mu\nu} = -F_{\nu\mu}, \tag{9.50}$$

therefore, it has six independent components (three components for **E** and three for **B**). In fact, since electric and magnetic field mix under Lorentz transformations and must be described by the same tensorial quantity, the simplest such object is a 2-index tensor. But a 2-index tensor, in general, has sixteen independent components (ten if the tensor is symmetric), therefore it must be antisymmetric to reduce to six independent components in all frames and avoid introducing extra unphysical degrees of freedom into Maxwell's theory.

These properties are sufficient to define the $F_{\mu\nu}$ tensor which, in Cartesian coordinates, assumes the form

$$(F_{\mu\nu}) \doteq \begin{pmatrix} 0 & -E_x & -E_y & -E_z \\ E_x & 0 & B_z & -B_y \\ E_y & -B_z & 0 & B_x \\ E_z & B_y & -B_x & 0 \end{pmatrix}. \tag{9.51}$$

Because the components of $F_{\mu\nu}$ are mixed up by a Lorentz transformation, in general electric and magnetic fields get converted into each other in a change of inertial frame. Electric and magnetic fields are not 4-vectors and, like space and time, they do not have a separate identity. They depend on the state of motion of the observer who measures them because they are both components of a tensor which transforms with the usual tensor law. This fact is another manifestation of the 4-dimensional nature of the world. One can verify, using Eq. (9.51), that the 3-dimensional electric and magnetic fields measured by an observer with timelike 4-velocity u^{μ} are given by

[8] For a special direction, that is if the velocity **v** is parallel to a static electric field, there is no magnetic field in the moving frame.

$$E_\mu = F_{\mu\nu} \frac{u^\nu}{c}, \tag{9.52}$$

$$B_\mu = -\frac{1}{2} \varepsilon_{\mu\nu}{}^{\rho\sigma} F_{\rho\sigma} \frac{u^\nu}{c}. \tag{9.53}$$

It is clear from these formulas that the electric and magnetic fields are not absolute entities but depend on the observer who is measuring them (who is characterized by the 4-velocity u^μ) and on his or her state of motion.

The Maxwell equations expressed in terms of $F_{\mu\nu}$ take the form

$$\partial^\mu F_{\mu\nu} \doteq -4\pi j_\nu, \tag{9.54}$$

$$\partial_{[\alpha} F_{\beta\gamma]} = 0, \tag{9.55}$$

in Cartesian coordinates, where j^μ is the *4-current density*. The last equation can be written also as

$$\partial_\alpha {}^*F^{\beta\gamma} = 0, \tag{9.56}$$

where ${}^*F^{\mu\nu} = \frac{1}{2} \varepsilon^{\mu\nu\alpha\beta} F_{\alpha\beta}$ is the dual of $F_{\mu\nu}$. For a charge distribution with charge density ρ, velocity $\mathbf{v}$, and current density $\mathbf{j} = \rho\mathbf{v}$, the 4-current density has components

$$j^\mu = (c\rho, \mathbf{j}). \tag{9.57}$$

The charge density ρ in any frame can be related to the charge density ρ_0 in the rest frame of this charge distribution, in which $\mathbf{v} = 0$ and $\mathbf{j} = 0$, by $\rho = \gamma\, \rho_0$, which follows from the relation between 3-dimensional coordinate volume and proper volume (1.32). In the rest frame it is

$$j^\mu = (c\rho_0, \mathbf{0}).$$

The 4-current density j^μ is conserved,

$$\partial^\mu j_\mu \doteq 0, \tag{9.58}$$

which has the meaning of local conservation of electric charge. In fact, the antisymmetry of $F_{\mu\nu}$ implies that $\partial^\mu \partial^\nu F_{\mu\nu} \doteq 0$, hence[9]

[9] There is a subtle point here: we use the fact that $\partial^\mu \partial^\nu$ is symmetric but, in general coordinates, we must use covariant derivatives $\nabla^\mu \nabla^\nu$ of tensors (see Chap. 10) to ensure that we have true tensor equations, and these covariant derivatives are not symmetric when acting on a tensor. However, the difference is immaterial here because $\nabla^\mu \nabla^\nu F_{\mu\nu}$ is a scalar and, if it vanishes in a coordinate system, then it vanishes in any other coordinate system. Therefore, it is legitimate to show that it vanishes in Cartesian coordinates in which $\nabla^\mu \nabla^\nu F_{\mu\nu} \doteq \partial^\mu \partial^\nu F_{\mu\nu}$.

$$\partial^{\mu}\left(\partial^{\nu}F_{\nu\mu}\right) \doteq -4\pi\,\partial^{\mu}j_{\mu} \doteq 0.$$

In 3-dimensional notation the conservation equation (9.58) is written as

$$\frac{\partial\rho}{\partial t} + \nabla\cdot\mathbf{j} = 0, \tag{9.59}$$

the familiar continuity equation of electromagnetism expressing local conservation of electric charge. It is rather straightforward to verify, by listing all the components of Eqs. (9.54) and (9.55), that the 3-dimensional Maxwell equations (1.12)–(1.15) are reproduced.

A charge moving in an electromagnetic field is subject to the Lorentz force (1.16). In four dimensions this law is generalized as follows: let u^{μ} be the 4-velocity of the particle with mass m and charge q. Then its equation of motion is

$$m\,a^{\mu} = \frac{q}{c}\,F^{\mu}{}_{\alpha}\,u^{\alpha} \quad \text{(Lorentz force)} \tag{9.60}$$

where $a^{\mu} \doteq u^{\alpha}\partial_{\alpha}u^{\mu}$ in Cartesian coordinates. The stress-energy tensor of the electromagnetic field which reproduces the 4-dimensional Maxwell equations is

$$T_{\mu\nu} = \frac{1}{4\pi}\left(F_{\mu\alpha}F_{\nu}{}^{\alpha} - \frac{1}{4}g_{\mu\nu}\,F_{\alpha\beta}\,F^{\alpha\beta}\right) \tag{9.61}$$

and is symmetric $\left(T_{\mu\nu} = T_{\nu\mu}\right)$ and traceless ($T^{\mu}{}_{\mu} = 0$). The time–time component is

$$T_{00} = \rho c^2 = \frac{c^2}{4\pi}\left(E^2 + B^2\right) \tag{9.62}$$

while the time–space components form the 3-vector

$$\left(T^{01}, T^{02}, T^{03}\right) = \frac{c}{4\pi}\,\mathbf{E}\times\mathbf{B}, \tag{9.63}$$

which is the Poynting vector describing the 3-momentum density of the electromagnetic field.

If $j^{\mu} = 0$, then $T_{\mu\nu}$ is covariantly conserved, $\partial^{\nu}T_{\mu\nu} \doteq 0$, expressing conservation of energy for the Maxwell field. If $j^{\mu} \neq 0$, this is no longer true because the field and the charge and current distributions exchange energy and momentum. In this case, one should write down a total stress-energy tensor composed of (9.61) and of a $T_{\mu\nu}$ for the charges and currents. Then it is the *total* $T_{\mu\nu}$ which is covariantly conserved.

The fact that $\partial_{[\mu}F_{\nu\rho]} = 0$ implies that there exists a 4-vector field A^{μ} (called *4-vector potential* or simply *4-potential*) such that

$$F_{\mu\nu} = \partial_\mu A_\nu - \partial_\nu A_\mu \tag{9.64}$$

(this combination is manifestly antisymmetric). The 4-vector potential contains as components the usual 3-dimensional scalar and vector potentials Φ and $\mathbf{A}$:

$$A^\mu = (\Phi, \mathbf{A}). \tag{9.65}$$

Then the Maxwell equation (9.55) is just an identity in the 4-potential A^μ. Instead of reasoning on the Maxwell tensor $F_{\mu\nu}$ we can reason on A^μ. In Cartesian coordinates the Maxwell equation (9.54) in terms of A^μ becomes

$$\partial^\mu \left(\partial_\mu A_\nu - \partial_\nu A_\mu\right) \doteq -4\pi j_\nu. \tag{9.66}$$

The theory has *gauge freedom*, that is, one can equivalently consider multiple vector potentials A^μ which reproduce the same Maxwell tensor $F_{\mu\nu}$ and the same physics. We say that the 4-potential A^μ is a *gauge-dependent* quantity, while the field $F_{\mu\nu}$ is *gauge-independent*. More precisely, we can add to A^μ the gradient of a scalar function f (*gauge function*),

$$A_\mu \longrightarrow A'_\mu = A_\mu + \partial_\mu f \quad (\textit{gauge transformation}) \tag{9.67}$$

and this change leaves $F_{\mu\nu}$, and therefore the Maxwell equations expressed in terms of it, unchanged:

$$\begin{aligned} F'_{\mu\nu} &= \partial_\mu A'_\nu - \partial_\nu A'_\mu = \partial_\mu (A_\nu + \partial_\nu f) - \partial_\nu \left(A_\mu + \partial_\mu f\right) \\ &= \partial_\nu A_\nu - \partial_\nu A_\mu + \partial_\mu \partial_\nu f - \partial_\nu \partial_\mu f \\ &= \partial_\mu A_\nu - \partial_\nu A_\mu = F_{\mu\nu}. \end{aligned}$$

In 3-dimensional language the gauge transformation (9.67) is equivalent to

$$\Phi \longrightarrow \Phi' = \Phi - \frac{1}{c}\frac{\partial f}{\partial t}, \tag{9.68}$$

$$\mathbf{A} \longrightarrow \mathbf{A}' = \mathbf{A} + \nabla f. \tag{9.69}$$

We can use this gauge freedom to simplify the Maxwell equations (9.66) in terms of A^μ. Choose the gauge function f in such a way that[10]

[10] Here the symbol $\doteq$ denotes the fact that an equality is only valid in a particular gauge. However, a gauge choice is not the same as a choice of coordinates.

$$\partial^\mu A'_\mu \doteq 0 \qquad \text{(Lorentz gauge).} \tag{9.70}$$

This condition can be achieved as follows: Eq. (9.67) gives that

$$\partial^\mu A'_\mu \doteq 0 \quad \Longleftrightarrow \quad \partial^\mu A_\mu + \partial^\mu \partial_\mu f \doteq 0,$$

hence, it is sufficient to choose f as a solution of the inhomogeneous wave equation

$$\Box f \doteq -\partial^\mu A_\mu, \tag{9.71}$$

which is always possible.[11] In 3-dimensional notation the Lorentz gauge (9.70) reads

$$\frac{1}{c}\frac{\partial \Phi'}{\partial t} + \nabla \cdot \mathbf{A}' = 0. \tag{9.72}$$

In the Lorentz gauge, the Maxwell equations for A^μ simplify to

$$\Box A_\nu = -4\pi j_\nu \qquad \text{(in the Lorentz gauge)}, \tag{9.73}$$

the inhomogeneous wave equation. Outside of the distributions of charges and currents the 4-potential satisfies the homogeneous wave equation $\Box A_\nu = 0$, the solutions of which are Fourier superpositions of monochromatic plane waves

$$A_\mu = C_\mu \, \mathrm{e}^{\mathrm{i}S(x^\alpha)}, \tag{9.74}$$

where the constant amplitude C^μ is the *polarization vector* and $S = S(x^\alpha)$ is the *phase* of the wave. The phase S obeys the equations

$$\partial^\mu \partial_\mu S \doteq 0, \tag{9.75}$$

$$\partial_\mu S \, \partial^\mu S = 0, \tag{9.76}$$

$$C_\mu \partial^\mu S = 0, \tag{9.77}$$

i.e., the phase S satisfies the homogeneous wave equation and its gradient $\partial_\mu S$ is a null 4-vector orthogonal to the amplitude 4-vector C_μ. Equation (9.77) expresses the transversality of electromagnetic waves.

Proof Insert Eq. (9.74) into the wave equation $\Box A_\mu = 0$. This yields

---

[11] The solution of Eq. (9.71) is not unique. The Lorentz gauge is preserved by adding to f any function h such that $\Box h = 0$, which leaves some further gauge freedom while remaining in the Lorentz gauge.

$$\partial^{\nu}\partial_{\nu}\left(C_{\mu}\,\mathrm{e}^{\mathrm{i}S}\right) = C_{\mu}\partial^{\nu}\left(\mathrm{i}\mathrm{e}^{\mathrm{i}S}\partial_{\nu}S\right) = C_{\mu}\mathrm{i}\left(\mathrm{i}\mathrm{e}^{\mathrm{i}S}\,\partial^{\nu}S\,\partial_{\nu}S + \mathrm{e}^{\mathrm{i}S}\partial^{\nu}\partial_{\nu}S\right) \doteq 0$$

and

$$\partial^{\mu}\partial_{\mu}S + \mathrm{i}\,\partial_{\mu}S\,\partial^{\mu}S \doteq 0.$$

Taking the real part of this equation one obtains $\partial^{\mu}\partial_{\mu}S \doteq 0$, while taking the imaginary part gives $(\partial_{\mu}S)(\partial^{\mu}S) = 0$. Now, A_{μ} must satisfy also the Lorentz gauge (9.70) or

$$0 = \partial^{\mu}\left(C_{\mu}\,\mathrm{e}^{\mathrm{i}S}\right) = C_{\mu}\mathrm{i}\,\mathrm{e}^{\mathrm{i}S}\,\partial^{\mu}S,$$

which implies the transversality condition $C^{\mu}\partial_{\mu}S = 0$, concluding the proof. □

The *angular frequency* of a monochromatic wave measured by an observer with 4-velocity v^{μ} is

$$\omega = -v^{\mu}\partial_{\mu}S = -v^{\mu}k_{\mu}, \tag{9.78}$$

where $k_{\mu} = \partial_{\mu}S$ is the *wave 4-vector* which satisfies

$$k^{\mu}k_{\mu} = 0, \tag{9.79}$$

$$\partial_{\mu}k^{\mu} \doteq 0, \tag{9.80}$$

$$C_{\mu}k^{\mu} = 0, \tag{9.81}$$

and also

$$k^{\mu}\partial_{\mu}k_{\nu} \doteq 0 \tag{9.82}$$

("geodesic equation" in Cartesian coordinates). Equation (9.82) expresses the fact that k^{μ} is tangent to a null "geodesic", i.e., a non-accelerated curve of extremal length according to the metric $\eta_{\mu\nu}$.

For plane waves, the phase has the form

$$S = k_{\mu}x^{\mu} \tag{9.83}$$

with k^{μ} constant:

$$A_{\mu} = C_{\mu}\,\mathrm{e}^{\mathrm{i}k_{\alpha}x^{\alpha}} \qquad \text{(plane waves)} \tag{9.84}$$

and the general solution of the wave equation is a Fourier superposition of plane waves

$$A_{\mu}(x^{\alpha}) = \frac{1}{(2\pi)^2}\int \mathrm{d}^4k\,\hat{A}_{\mu}(k^{\alpha})\,\mathrm{e}^{\mathrm{i}k_{\nu}x^{\nu}}. \tag{9.85}$$

The Lorentz gauge (9.70) is most convenient to study radiation problems. Other gauges are possible and are used: for example the *Coulomb gauge*

$$\nabla \cdot \mathbf{A} = 0 \tag{9.86}$$

is widely used in static situations. The Coulomb gauge is ruined by a Lorentz boost and, to be restored, it needs a gauge transformation to be performed in the new inertial frame after the Lorentz transformation. By contrast, the Lorentz gauge (9.70) is Lorentz-invariant.

Since the Maxwell field is described by the 4-vector A^{μ}, it couples to vectorial sources and a (time-varying) dipole moment, and electromagnetic radiation is dipole to lowest order.[12]

We would now be ready to begin studying electrodynamics problems such as the generation of electromagnetic fields by moving charges, the scattering of charged particles, their interaction with given electromagnetic fields, and their backreaction on them. We leave these problems for more advanced textbooks [11–13], contenting ourselves with providing some basic notions here.

9.9 Conclusion

From the point of view of Special Relativity, different branches of physics deal with different forms of mass-energy living in Minkowski spacetime. The relevant physical quantities and the equations that they obey must be formulated in a covariant way in the 4-dimensional world. Determining the appropriate energy-momentum tensor is a crucial step in this direction.

We have discussed perfect fluids, the simplest field (the scalar field), and the Maxwell field. Ultimately, quantum matter is also described by relativistic theories. Although most atomic and molecular physics encountered in standard textbooks does not require relativity or only requires small relativistic corrections to non-relativistic quantum mechanics,[13] fields do require a relativistic formulation. Dirac provided the first relativistic theory of the quantum electron and the relativistic quantum theory of the electromagnetic field (quantum electrodynamics or, in short, QED) followed. The nuclear interactions require a relativistic description, and so does any theory of higher energy physics. Gravity is discussed relativistically in Einstein's theory of General Relativity but we do not yet have a satisfactory theory of gravity which is both relativistic and quantum-mechanical.

---

[12] By the same token, in (linearized) General Relativity the gravitational field is described by a 2-tensor $h_{\mu\nu}$ which couples to (time-varying) quadrupole-type sources, and gravitational radiation is quadrupole to lowest order.

[13] There are, however, exceptions, for example highly conducting materials in which electrons move fast, such as graphene, which is studied intensively.

Problems

9.1 Prove the equivalence of the two forms of Eq. (9.6).

9.2 Check that, using the $3+1$ splitting of Chap. 8, the stress-energy tensor of a perfect fluid (9.30) can be written as

$$T_{\mu\nu} = \rho u_\mu u_\nu + P h_{\mu\nu}.$$

9.3 Using the $3+1$ splitting of Minkowski spacetime, compute $T_{\mu\nu}u^\mu u^\nu$ and $T_{\mu\nu}h^{\mu\nu}$ for a perfect fluid characterized by the stress-energy tensor (9.30).

9.4 Compute the invariants $T_{\mu\nu}T^{\mu\nu}$ and $T_{\mu\alpha}T_{\nu\beta}T^{\beta\mu}T^{\alpha\nu}$ for a perfect fluid described by a stress-energy tensor of the form (9.30).

9.5 *From the form (9.30) of the energy-momentum tensor of a perfect fluid, obtain that the weak energy condition implies $\rho \geq 0$ and $\rho + P/c^2 \geq 0$.

9.6 A certain imperfect fluid is described by the energy-momentum tensor

$$T_{\mu\nu} = \left(P + \rho c^2\right)\frac{u_\mu u_\nu}{c^2} + P g_{\mu\nu} + q_\mu \frac{u_\nu}{c} + q_\nu \frac{u_\mu}{c},$$

where q^μ is a spacelike 4-vector describing a diffusive energy (heat) current[14] and orthogonal to u^μ (i.e., $q^\mu u_\mu = 0$).[15]
(a) Check explicitly that $\rho c^2 = T_{\mu\nu}u^\mu u^\nu$.
(b) Check explicitly that $P = h^{\mu\nu}T_{\mu\nu}/3$.
(c) Compute the trace $T \equiv g^{\mu\nu}T_{\mu\nu}$.

9.7 Compute the invariant $T_{\mu\nu}T^{\mu\nu}$ for an imperfect fluid characterized by the energy-momentum tensor

$$T_{\mu\nu} = \left(P + \rho c^2\right)\frac{u_\mu u_\nu}{c^2} + P g_{\mu\nu} + q_\mu \frac{u_\nu}{c} + q_\nu \frac{u_\mu}{c},$$

where $q_\mu u^\mu = 0$.

9.8 An imperfect fluid is described by the stress-energy tensor

$$T_{\mu\nu} = \rho u_\mu u_\nu + P h_{\mu\nu} + q_\mu \frac{u_\nu}{c} + q_\nu \frac{u_\mu}{c} + \Pi_{\mu\nu},$$

where u^μ is the fluid 4-velocity, ρc^2 is the energy density, P is the pressure, $h_{\mu\nu}$ is defined by the $3+1$ splitting with respect to u^α, q^μ is the *heat current density*, and $\Pi_{\mu\nu}$ is the *anisotropic stress tensor*. q^μ and $\Pi^{\mu\nu}$ are purely spatial ($q^\mu u_\mu = 0$, $\Pi^{\mu\nu}u_\mu = \Pi^{\mu\nu}u_\nu = 0$) and $\Pi^{\mu\nu}$ is trace-free ($\Pi^\alpha{}_\alpha = 0$). Verify

---

[14] No anisotropic stresses are present in this exercise although they are allowed, in general, for an imperfect fluid.

[15] This heat current density is spacelike, therefore, it cannot describe a realistic flow. This stress-energy tensor is just a toy model which ignores the fact that instantaneous diffusion cannot exist in relativity, but is useful because of its simplicity.

that

$$\rho c^2 = T_{\alpha\beta} \frac{u^\alpha u^\beta}{c^2},$$

$$P = \frac{1}{3} h^{\alpha\beta} T_{\alpha\beta},$$

$$q_\alpha = -h_\alpha{}^\beta \frac{u^\sigma}{c} T_{\sigma\beta},$$

$$\Pi_{\alpha\beta} = T_{\langle\alpha\beta\rangle},$$

where $\langle\ldots\rangle$ denotes the spatial part of a tensor followed by trace removal, for example the spatial part of $T_{\alpha\beta}$ is

$$\tilde{T}_{\alpha\beta} = h_\alpha{}^\sigma h_\beta{}^\rho T_{\sigma\rho}$$

and

$$T_{<\alpha\beta>} = \tilde{T}_{\alpha\beta} - \frac{1}{3} \tilde{T}_i^i .$$

9.9 Consider two non-interacting perfect fluids with different 4-velocity fields u^μ and $\tilde{u}^\mu$, respectively. Each of the two fluids is represented by a stress-energy tensor of the form

$$T^{(i)}_{\alpha\beta} = \rho^{(i)} u^{(i)}_\alpha u^{(i)}_\beta + P^{(i)} h^{(i)}_{\alpha\beta}$$

$(i = 1, 2)$ in its rest (or comoving) frame. The total stress-energy tensor is

$$T_{\alpha\beta} = \rho_1 u_\alpha u_\beta + P_1 h_{\alpha\beta} + \rho_2 \tilde{u}_\alpha \tilde{u}_\beta + P_2 \tilde{h}_{\alpha\beta}$$

using two different 3 + 1 splittings with respect to u^μ and $\tilde{u}^\mu$. The velocity field of the second fluid may be written as a Lorentz boost of that of the first fluid,[16] according to Eqs. (6.23)–(6.26) by introducing a spacelike 4-vector v^μ and

$$\gamma \equiv \left(1 - v^\alpha v_\alpha / c^2\right)^{-1/2}.$$

Show that $T_{\mu\nu}$ can be written as the stress-energy tensor of an *imperfect* fluid

$$T_{\alpha\beta} = \bar{\rho}\, u_\alpha u_\beta + \bar{P} h_{\alpha\beta} + q_\alpha u_\beta + q_\beta u_\alpha + \Pi_{\alpha\beta}$$

and find expressions for the effective quantities $\bar{\rho} c^2$ (energy density), $\bar{P}$ (isotropic pressure), q_μ (energy flux density), and $\Pi_{\mu\nu}$ (anisotropic stress ten-

[16] It is said that the second fluid is *tilted* with respect to the first one.

sor) of this imperfect fluid in terms of ρ_1, ρ_2, P_1, P_2, v^μ, γ, and $h_{\mu\nu}$. As a special case of this exercise, conclude that *a* perfect fluid looks like an imperfect one when seen from an inertial frame which is Lorentz-boosted with respect to the comoving frame of the fluid.

9.10 Consider a massive scalar field ϕ satisfying the Klein-Gordon equation (9.44). Since this equation is linear one can Fourier-expand its solution in terms of plane waves,

$$\phi(t, \mathbf{x}) = \frac{1}{(2\pi)^{3/2}} \int d^3\mathbf{k}\, \hat{\phi}(\mathbf{k})\, e^{i\,\mathbf{k}\cdot\mathbf{x}}\, e^{-i\,\omega t} = \frac{1}{(2\pi)^{3/2}} \int d^3\mathbf{k}\ \hat{\phi}(\mathbf{k})\, e^{i\,k_\alpha x^\alpha}.$$

What is the normalization of the wave 4-vector k^μ for each plane wave $\hat{\phi}(\mathbf{k})\, e^{i\,k_\alpha x^\alpha}$? Turn this relation into an expression of the form $\omega = \omega(\mathbf{k})$ (*dispersion relation*).

9.11 A solution of the massless Klein-Gordon equation $\partial_\mu \partial^\mu \phi \doteq 0$ in Cartesian coordinates is the plane monochromatic wave $\phi(x^\alpha) = \phi_0 \cos\left(k_\mu x^\mu\right)$, where ϕ_0 is a constant and $k^\mu = \left(\frac{\omega}{c}, \mathbf{k}\right)$ is a constant 4-vector. Compute the energy-momentum tensor of this scalar field and its trace and provide a physical interpretation of its components.

9.12 * Does a scalar field described by the energy-momentum tensor (9.48) satisfy any of the point-wise energy conditions?

9.13 Assume that the gradient $\nabla_\mu \phi$ of a scalar field ϕ is always timelike and show that its energy-momentum tensor can be written in the form of a perfect fluid stress-energy tensor (9.30). Compute the effective energy density and pressure. *Hint:* assume that the 4-velocity u^μ of this effective perfect fluid is proportional to $\nabla^\mu \phi$, which makes sense because this gradient is timelike. Use units in which $c = 1$.

9.14 If, for a 2-index tensor $A_{\mu\nu}$, there exist a number λ and a 4-vector v^μ such that $A_{\mu\nu} v^\nu = \lambda v_\mu$, we say that λ is an *eigenvalue* and v^μ is the associated *eigenvector* of $A_{\mu\nu}$. Find eigenvalues and eigenvectors of the stress-energy tensor (9.30) of a perfect fluid. Discuss the case of a perfect fluid with equation of state $P = -\rho c^2$.

9.15 Verify Eqs. (9.52) and (9.53).

9.16 In the inertial[17] frame S an electric field $\mathbf{E} = (E_0, 0, 0)$ is parallel to the x-axis and there is no magnetic field (the Maxwell field is "purely electric"). Compute the Maxwell tensor in an inertial frame S' moving in standard configuration with velocity v with respect to to S. Is the Maxwell field in S' purely electric? Is the electric field parallel to the x'-axis?

9.17 (a) Compute the invariants $F_{\mu\nu} F^{\mu\nu}$ and $\varepsilon_{\mu\nu\rho\sigma} F^{\mu\nu} F^{\rho\sigma}$ of the Maxwell tensor in terms of the 3-dimensional electric and magnetic fields.
(b) Find the Maxwell tensor for a plane monochromatic wave described by the 4-potential (9.84).

[17] Cf. Refs. [13, 14].

(c) Compute the invariants $F_{\mu\nu}F^{\mu\nu}$ and $\varepsilon_{\mu\nu\rho\sigma}F^{\mu\nu}F^{\rho\sigma}$ for the plane monochromatic wave (9.84). What is the physical meaning of your result?

9.18 Which one of the following partial differential equations, the wave equation

$$\nabla^2 u - \frac{1}{c^2}\frac{\partial^2 u}{\partial t^2} = 0,$$

or the heat/diffusion equation

$$\frac{\partial u}{\partial t} = a\, \nabla^2 u$$

(where a is a constant), is relativistic?

References

1. S.W. Hawking, G.F.R. Ellis, *The Large Scale Structure of Spacetime* (Cambridge University Press, Cambridge, 1973)
2. R.M. Wald, *General Relativity* (Chicago University Press, Chicago, 1984)
3. S.M. Carroll, *Spacetime and Geometry, An Introduction to General Relativity* (Addison-Wesley, San Francisco, 2004)
4. R. d'Inverno, *Introducing Einstein's Relativity* (Clarendon Press, Oxford, 2002)
5. L.H. Ford, T.A. Roman, Phys. Rev. D **64**, 024023 (2001)
6. L.H. Ford, T.A. Roman, Phys. Rev. D **60**, 104018 (1999)
7. A. Zee, *Quantum Field Theory in a Nutshell*, 2nd edn. (Princeton University Press, Princeton, 2010)
8. M. Green, J. Schwarz, E. Witten, *Superstring Theory* (Cambridge University Press, Cambridge, 1988)
9. B. Zwiebach, *A First Course in String Theory* (Cambridge University Press, Cambridge, 2004)
10. S. Capozziello, V. Faraoni, *Beyond Einstein Gravity* (Springer, New York, 2011)
11. J.D. Jackson, *Classical Electrodynamics* (Wiley, New York, 1999)
12. L.D. Landau, E. Lifschitz, *The Classical Theory of Fields* (Pergamon Press, Oxford, 1989)
13. W. Rindler, *Introduction to Special Relativity*, 2nd edn. (Clarendon Press, Oxford, 1991)
14. W. Rindler, J. Denur, Am. J. Phys. **56**, 795 (1988)

Chapter 10
*Special Relativity in Arbitrary Coordinates

The important thing is not to stop questioning.
—Albert Einstein

10.1 Introduction

Thus far we have limited ourselves to studying Special Relativity using Cartesian coordinates to cover Minkowski spacetime and you are probably sick of the phrase "the equation is ... in Cartesian coordinates" and are wondering why we are not discussing equations in general coordinate systems as we did in Chap. 4. There is no a priori reason to adopt Cartesian coordinates, only they allow greater mathematical simplicity tailored to the beginner. It is time to re-examine Special Relativity using arbitrary coordinates $\{x^\mu\}$. They involve a little more mathematics and this is the reason why they were put off until now. After generalizing the geometrical formulation of Minkowski spacetime to arbitrary coordinate systems, we will revisit the laws of motion of particles and the physics of various forms of mass-energy in this spacetime. We will conclude by briefly introducing some of the basic ideas of General Relativity.

In an arbitrary coordinate system the components of the Minkowski metric are not $\eta_{\mu\nu} = \text{diag}(-1, 1, 1, 1)$ but general components $g_{\mu\nu} = g_{\nu\mu}$. The Minkowski metric tensor, however, has an existence independent of its components in a particular coordinate system. In general, the metric $g_{\mu\nu}$ will not be diagonal in an arbitrary coordinate system. If $x^\mu \longrightarrow x^{\mu'}(x^\alpha)$ is a transformation from Cartesian to arbitrary coordinates, the new metric components will be

$$g_{\mu'\nu'} = \frac{\partial x^\alpha}{\partial x^{\mu'}} \frac{\partial x^\beta}{\partial x^{\nu'}} g_{\alpha\beta}.$$

V. Faraoni, *Special Relativity*, Undergraduate Lecture Notes in Physics,
DOI: 10.1007/978-3-319-01107-3_10, © Springer International Publishing Switzerland 2013

In order to proceed we need to generalize the notion of ordinary and partial derivative and introduce *covariant differentiation*. We have already seen that, if $\phi\,(x^\alpha)$ is a scalar function of the coordinates, $\partial_\mu\phi$ is a covariant 4-vector but $\partial_\mu\partial_\nu\phi$ is *not* a covariant 4-tensor. In general, if A^μ is a 4-vector, $\partial_\mu A^\nu$ is *not* a tensor of type $\begin{pmatrix}1\\1\end{pmatrix}$ and, more generally, the partial derivative of a tensor is not a tensor. This fact can be appreciated by computing the transformation law for $\partial_\mu A^\nu$. We have

$$\begin{aligned}
\partial_{\mu'} A^{\nu'} &= \frac{\partial A^{\nu'}}{\partial x^{\mu'}} = \frac{\partial}{\partial x^{\mu'}}\left(\frac{\partial x^{\nu'}}{\partial x^\alpha} A^\alpha\right) \\
&= \frac{\partial x^\beta}{\partial x^{\mu'}}\frac{\partial}{\partial x^\beta}\left(\frac{\partial x^{\nu'}}{\partial x^\alpha} A^\alpha\right) \\
&= \frac{\partial^2 x^{\nu'}}{\partial x^\alpha \partial \beta} A^\alpha \frac{\partial x^\beta}{\partial x^{\mu'}} + \frac{\partial x^{\nu'}}{\partial x^\alpha}\frac{\partial A^\alpha}{\partial x^\beta}\frac{\partial x^\beta}{\partial x^{\mu'}} \\
&= \frac{\partial x^{\nu'}}{\partial x^\alpha}\frac{\partial x^\beta}{\partial x^{\mu'}}\partial_\beta A^\alpha + \frac{\partial x^\beta}{\partial x^{\mu'}}\frac{\partial^2 x^{\nu'}}{\partial x^\alpha\,\partial\beta} A^\alpha .
\end{aligned}$$

Because of the second term in the last line, $\partial_\mu A^\nu$ does not transform as a tensor of type $\begin{pmatrix}1\\1\end{pmatrix}$, except for linear coordinate transformations. We need to generalize the concept of derivative of a tensor so that an object such as $\nabla_\mu A^\nu$ (where ∇_μ will be the new derivative operator) transforms as a true tensor.

If we consider only Cartesian coordinates and linear coordinate transformations, such as the Lorentz transformation in standard configuration, then $\frac{\partial^2 x^{\nu'}}{\partial x^\alpha \partial \beta} = 0$ and $\partial_\mu A^\nu$ does transform as a tensor under these special coordinate changes. However, these conditions are too restrictive.

10.2 The Covariant Derivative

One can introduce the covariant derivative operator in Minkowski spacetime (or, in general, on a manifold M describing the curved spacetime of General Relativity and endowed with a metric tensor $g_{\mu\nu}$) by using an axiomatic approach.

Let $\tau\begin{pmatrix}k\\l\end{pmatrix}$ denote the space of smooth tensor fields of type $\begin{pmatrix}k\\l\end{pmatrix}$ on M, let $T(M) = \tau\begin{pmatrix}1\\0\end{pmatrix}$ be the tangent space to M (this is the space of all vectors in spacetime seen as operators on functions), $\tau\begin{pmatrix}k\\l\end{pmatrix}$ be the space of tensors of rank $\begin{pmatrix}k\\ \\l\end{pmatrix}$, and $\mathscr{F}=\tau\begin{pmatrix}0\\0\end{pmatrix}$ be the space of smooth functions on an open region $O\subseteq M$.

Definition 10.1 A *covariant derivative operator* is a map

$$\nabla\colon\quad \tau\begin{pmatrix}k\\l\end{pmatrix}\longrightarrow\tau\begin{pmatrix}k\\l+1\end{pmatrix}$$

which takes each differentiable tensor field of type $\begin{pmatrix}k\\ \\l\end{pmatrix}$ into a tensor field of type $\begin{pmatrix}k\\ \\l+1\end{pmatrix}$ and satisfies the following properties:

1. Linearity:
   $\forall A^{\mu_1\dots\mu_k}{}_{\nu_1\dots\nu_l}\,,\ B^{\mu_1\dots\mu_k}{}_{\nu_1\dots\nu_l}\in\tau\begin{pmatrix}k\\l\end{pmatrix},\ \forall\alpha,\beta\in\mathbb{R},$

$$\nabla_\rho\left(\alpha A^{\mu_1\dots\mu_k}{}_{\nu_1\dots\nu_l}+\beta B^{\mu_1\dots\mu_k}{}_{\nu_1\dots\nu_l}\right)=\alpha\,\nabla_\rho A^{\mu_1\dots\mu_k}{}_{\nu_1\dots\nu_l}+\beta\,\nabla_\rho B^{\mu_1\dots\mu_k}{}_{\nu_1\dots\nu_l}\,.$$

2. Leibnitz property
   $\forall A^{\mu_1\dots\mu_k}{}_{\nu_1\dots\nu_l}\in\tau\begin{pmatrix}k\\l\end{pmatrix},\quad \forall B^{\mu_1\dots\mu_{k'}}{}_{\nu_1\dots\nu_{l'}}\in\tau\begin{pmatrix}k'\\l'\end{pmatrix},$

$$\nabla_\rho\left(A^{\mu_1\dots\mu_k}{}_{\nu_1\dots\nu_l}B^{\mu_1\dots\mu_{k'}}{}_{\nu_1\dots\nu_{l'}}\right)=\left(\nabla_\rho A^{\mu_1\dots\mu_k}{}_{\nu_1\dots\nu_l}\right)B^{\mu_1\dots\mu_{k'}}{}_{\nu_1\dots\nu_{l'}}+A^{\mu_1\dots\mu_k}{}_{\nu_1\dots\nu_l}\nabla_\rho B^{\mu_1\dots\mu_{k'}}{}_{\nu_1\dots\nu_{l'}}$$

3. Commutation with contraction:
   $\forall A^{\mu_1\dots\mu_k}{}_{\nu_1\dots\nu_l}\in\tau\begin{pmatrix}k\\l\end{pmatrix},$

$$\nabla_\sigma\left(A^{\mu_1\dots\rho\dots\mu_k}{}_{\nu_1\dots\rho\dots\nu_l}\right)=\nabla_\sigma A^{\mu_1\dots\rho\dots\mu_k}{}_{\nu_1\dots\rho\dots\nu_l}$$

4. Consistency with the notion of directional derivative on a scalar:
   $\forall f\in\mathscr{F},\quad \forall \mathsf{X}\in T(M),$

$$\mathsf{X}(f)=X^\mu\nabla_\mu f$$

5. Torsion-free property:
   $\forall f\in\mathscr{F},$

$$\nabla_\mu\nabla_\nu f=\nabla_\nu\nabla_\mu f$$

   (second covariant derivatives of ascalar commute).

In general, infinitely many covariant derivative operators can be defined but, if a metric tensor $g_{\mu\nu}$ is present, it determines a preferred covariant derivative operator. This operator is the one that satisfies the property

$$\nabla_\alpha g_{\mu\nu} = 0. \tag{10.1}$$

It can be shown that an operator with the property (10.1) is unique. The operator ∇_α is called *covariant derivative operator* and applying it on a tensor $T^{\mu_1\mu_2\ldots\mu_k}{}_{\nu_1\nu_2\ldots\nu_l}$ results in the operation called *covariant differentiation* and produces the *covariant derivative* $\nabla_\alpha T^{\mu_1\mu_2\ldots\mu_k}{}_{\nu_1\nu_2\ldots\nu_l}$ of that tensor. Of course, once a covariant differentiation operator ∇_α is defined, one can take successive covariant derivatives $\nabla_\alpha\nabla_\beta\ldots T^{\mu_1\mu_2\ldots\mu_k}{}_{\nu_1\nu_2\ldots\nu_l}$ of a tensor field $T^{\mu_1\mu_2\ldots\mu_k}{}_{\nu_1\nu_2\ldots\nu_l}$.

10.2.1 Computing Covariant Derivatives

We are interested in practical rules to compute covariant derivatives, which we list here without proof (for a proof see Refs. [1, 2]). In order to covariantly differentiate vectors and tensors, one must first introduce the *connection coefficients*

$$\Gamma^\mu_{\alpha\beta} \equiv \frac{1}{2}\, g^{\mu\sigma}\left(g_{\sigma\alpha,\beta} + g_{\sigma\beta,\alpha} - g_{\alpha\beta,\sigma}\right), \tag{10.2}$$

where a comma denotes partial differentiation, i.e., $f_{,\alpha} \equiv \frac{\partial f}{\partial x^\alpha}$, etc. By construction, we have the symmetry

$$\Gamma^\mu_{\alpha\beta} = \Gamma^\mu_{\beta\alpha}. \tag{10.3}$$

The $\Gamma^\mu_{\alpha\beta}$ are not tensors since they do not transform as such under coordinate transformations. Their transformation property under a coordinate change $x^\mu \longrightarrow x^{\mu'}$ is instead

$$\Gamma^{\alpha'}_{\mu'\nu'} = \frac{\partial x^{\alpha'}}{\partial x^\delta}\frac{\partial x^\rho}{\partial x^{\mu'}}\frac{\partial x^\sigma}{\partial x^{\nu'}}\Gamma^\delta_{\rho\sigma} + \frac{\partial x^{\alpha'}}{\partial x^\delta}\frac{\partial^2 x^\delta}{\partial x^{\mu'}\partial x^{\nu'}} \tag{10.4}$$

(the derivation of this transformation law is left as an exercise).

The *covariant derivative of a contravariant vector field* A^μ is

$$\nabla_\mu A^\alpha = \partial_\mu A^\alpha + \Gamma^\alpha_{\mu\nu} A^\nu, \tag{10.5}$$

i.e., the ordinary derivative plus an additional term constructed with the connection coefficients and the vector field itself. Note that in this additional term the contravariant index α of the 4-vector is "stolen" by the Γ-symbol. The positive sign with which the second term on the right hand side enters the expression of $\nabla_\mu A^\alpha$ is strictly associated with the contravariant nature of the index α in A^α. By contrast, the *covariant derivative of a covariant vector field* B_μ is

$$\nabla_\mu B_\alpha = \partial_\mu B_\alpha - \Gamma^\nu_{\mu\alpha} B_\nu \tag{10.6}$$

(note the negative sign of the second term on the right hand side). Using Eq. (10.4) it is straightforward to verify that $\nabla_\mu A^\alpha$ transforms as a tensor of type $\begin{pmatrix}1\\1\end{pmatrix}$ and $\nabla_\mu B_\alpha$ transforms as a tensor of type $\begin{pmatrix}0\\2\end{pmatrix}$.

The rules for computing covariant derivatives extend to all covariant and contravariant indices of a tensor field $T^{\alpha_1 \dots \alpha_l}{}_{\beta_1 \dots \beta_m}$:

$$\begin{aligned}
\nabla_\mu T^{\alpha_1 \dots \alpha_l}{}_{\beta_1 \dots \beta_m} &= \partial_\mu T^{\alpha_1 \dots \alpha_l}{}_{\beta_1 \dots \beta_m} \\
&\left.\begin{aligned}
&+\Gamma^{\alpha_1}_{\mu\sigma} T^{\sigma\alpha_2 \dots \alpha_l}{}_{\beta_1 \dots \beta_m} \\
&+\Gamma^{\alpha_2}_{\mu\sigma} T^{\alpha_1\sigma\alpha_2 \dots \alpha_l}{}_{\beta_1 \dots \beta_m} + \dots \\
&+\Gamma^{\alpha_l}_{\mu\sigma} T^{\alpha_1\alpha_2 \dots \sigma}{}_{\beta_1 \dots \beta_m}
\end{aligned}\right]\ \text{l terms for the l contravariant indices} \\
&\left.\begin{aligned}
&-\Gamma^{\sigma}_{\mu\beta_1} T^{\alpha_1 \dots \alpha_l}{}_{\sigma\beta_2 \dots \beta_m} \\
&-\Gamma^{\sigma}_{\mu\beta_2} T^{\alpha_1 \dots \alpha_l}{}_{\beta_1\sigma\beta_2 \dots \beta_m} - \dots \\
&-\Gamma^{\sigma}_{\mu\beta_m} T^{\alpha_1 \dots \alpha_l}{}_{\beta_1\beta_2 \dots \sigma}.
\end{aligned}\right]\ \text{m terms for the m covariant indices}
\end{aligned} \tag{10.7}$$

Again, using Eqs. (10.4) and (10.7), it is straightforward to check that the covariant derivatives $\nabla_\mu T^{\alpha_1 \dots \alpha_l}{}_{\beta_1 \dots \beta_m}$ transform as true tensors under coordinate transformations.

In Cartesian coordinates in Minkowski spacetime it is $g_{\mu\nu} \doteq \text{diag}\,(-1, 1, 1, 1)$ and the connection coefficients vanish identically,

$$\Gamma^{\mu}_{\alpha\beta} \doteq \frac{1}{2}\,\eta^{\mu\sigma}\left(\eta_{\sigma\alpha,\beta} + \eta_{\sigma\beta,\alpha} - \eta_{\alpha\beta,\sigma}\right) \doteq 0 \tag{10.8}$$

hence, in Cartesian coordinates, the covariant derivative reduces to an ordinary derivative, $\nabla_\mu \doteq \partial_\mu$, with obvious formal simplifications and this is the reason why we have mostly used Cartesian coordinates until this chapter.

10.3 Spacetime Curves and Covariant Derivative

Let $x^\mu(\lambda)$ be a spacetime curve parametrized by a parameter λ and with 4-tangent $u^\mu = dx^\mu/d\lambda$. We are interested in timelike and null curves, which are the 4-trajectories of physical particles. For timelike curves we adopt the parameter $\lambda = \tau$, where τ is the proper time along the curve[1] and the tangent to the curve is the 4-velocity

$$u^\mu = \frac{dx^\mu}{d\tau}, \qquad u_\mu u^\mu = -c^2. \tag{10.9}$$

The *derivative of a tensor field* $T^{\mu_1 \dots \mu_k}{}_{\nu_1 \dots \nu_l}$ *along the curve* is defined as

$$\frac{DT^{\mu_1 \dots \mu_k}{}_{\nu_1 \dots \nu_l}}{D\lambda} \equiv u^\alpha \nabla_\alpha T^{\mu_1 \dots \mu_k}{}_{\nu_1 \dots \nu_l}, \tag{10.10}$$

i.e., as the projection of the covariant derivative of the tensor along the direction of the curve (the direction of its tangent).

10.3.1 Geodesics

Definition 10.2 A *geodesic curve* is a curve of zero acceleration $a^\mu = 0$, or

$$u^\beta \nabla_\beta u^\mu = 0. \tag{10.11}$$

This definition requires that the variation of the tangent u^μ in the direction of the curve vanishes. Since

[1] For timelike curves the parameter λ has the dimensions of a time while for null curves it has the dimension of a length.

$$\frac{Du^\mu}{D\lambda} \equiv u^\beta \nabla_\beta u^\mu = u^\beta \left(\partial_\beta u^\mu + \Gamma^\mu_{\beta\alpha} u^\alpha\right)$$
$$= \frac{dx^\beta}{d\lambda}\frac{\partial u^\mu}{\partial x^\beta} + \Gamma^\mu_{\beta\alpha} u^\alpha u^\beta$$
$$= \frac{du^\mu}{d\lambda} + \Gamma^\mu_{\beta\alpha} u^\alpha u^\beta,$$

a geodesic curve obeys the *geodesic equation*

$$\frac{Du^\mu}{D\lambda} = u^\beta \nabla_\beta u^\mu = 0 \quad \Longleftrightarrow \quad \frac{d^2 x^\mu}{d\lambda^2} + \Gamma^\mu_{\alpha\beta}\frac{dx^\alpha}{d\lambda}\frac{dx^\beta}{d\lambda} = 0. \tag{10.12}$$

In Cartesian coordinates the connection coefficients vanish identically, $\Gamma^\mu_{\beta\alpha} \doteq 0$, and the geodesic equation reduces to

$$\frac{du^\mu}{d\lambda} = \frac{d^2 x^\mu}{d\lambda^2} = 0,$$

which has the linear solution $x^\mu(\lambda) = A^\mu \lambda + B^\mu$, where A^μ and B^μ are constants determined by the initial conditions $x^\mu(\lambda_0)$ and $u^\mu(\lambda_0)$ at an initial value λ_0 of the parameter. Therefore,

the geodesics of Minkowski spacetime are straight lines.

A property of geodesics is that of being curves that extremize the length between two points. It is well known that the minimum length between two points in $\mathbb{R}^n$ with Euclidean metric is obtained by following the straight line joining them. This property holds true also in Minkowski spacetime and can be obtained (as in the Euclidean case) using the calculus of variations [3–5].

In actual fact the form of the geodesic equation can be more general than (10.11). A geodesic can be defined more generally as a curve such that its tangent is transported parallel to itself when moving along the geodesic, that is

$$\frac{Du^\mu}{D\lambda} \equiv u^\nu \nabla_\nu u^\mu = \alpha\, u^\mu, \tag{10.13}$$

where α is a constant, or

$$\frac{d^2 x^\mu}{d\lambda^2} + \Gamma^\mu_{\rho\sigma}\frac{dx^\rho}{d\lambda}\frac{dx^\sigma}{d\lambda} = \alpha\,\frac{dx^\mu}{d\lambda}. \tag{10.14}$$

This equation expresses the fact that the change of the tangent vector u^μ in the direction of the curve is parallel to u^μ itself, i.e., this vector is transported parallel to itself when moving along the curve.

As for all curves, many parametrizations can be chosen for the same geodesic. A particular class of parameters, called *affine parameters*, is such that Eq. (10.13) reduces to the simpler form (10.11), which is called *affinely parametrized* form. Non-affinely parametrized geodesics are still curves of extremal length, only their parameters are non-affine.[2]

Definition 10.3 A vector u^μ is a *geodesic vector* if it is tangent to a geodesic, i.e., if it satisfies Eq. (10.11) or Eq. (10.13).

10.4 Physics in Minkowski Spacetime Revisited

Let us revisit now the physics in Minkowski spacetime discussed in Chaps. 5–9 using arbitrary coordinate systems instead of restricting ourselves to Cartesian coordinates. The rule to generalize Cartesian coordinates is to replace partial derivatives of tensors with covariant derivatives, the "delta-to-nabla" rule $\partial_\mu \longrightarrow \nabla_\mu$. As we already know, covariant and partial derivatives coincide for a scalar function in arbitrary coordinate systems, $\partial_\mu f = \nabla_\mu f$, so we only need to worry about 4-vectors and 4-tensors with two or more indices. The d'Alembertian of a scalar function f is defined as

$$\Box f \equiv g^{\mu\nu} \nabla_\mu \nabla_\nu f. \tag{10.15}$$

This quantity consists of $\partial^\mu \partial_\mu f$ plus a second term:

$$g^{\mu\nu} \nabla_\mu \nabla_\nu f = g^{\mu\nu} \nabla_\mu \partial_\nu f = \partial^\mu \partial_\mu f - g^{\mu\nu} \Gamma^\alpha_{\mu\nu} \partial_\alpha f.$$

A useful property is

$$\Box f = \frac{1}{\sqrt{-g}} \partial_\mu \left(\sqrt{-g}\, \partial^\mu f \right), \tag{10.16}$$

the proof of which is left as an exercise.

Example 10.1 Compute the d'Alembertian $\Box\phi$ of a scalar field ϕ in spherical coordinates $\{ct, r, \theta, \varphi\}$.
The metric in spherical coordinates is $g_{\mu\nu} = \operatorname{diag}\left(-1, 1, r^2, r^2 \sin^2\theta\right)$ and the inverse metric and metric determinant are $g^{\mu\nu} = \operatorname{diag}\left(-1, 1, \frac{1}{r^2}, \frac{1}{r^2 \sin^2\theta}\right)$ and $g = -r^4 \sin^2\theta$. The d'Alembertian of ϕ is

[2] The curve itself is a geometric object defined independently of the particular parametrization used (it is an equivalence class of parametrizations, with the equivalence relation obtained by identifying parametric representations which have the same image).

$$\Box\phi = \frac{1}{\sqrt{-g}}\,\partial_\mu\left(\sqrt{-g}\,\partial^\mu\phi\right) = \frac{1}{r^2\sin\theta}\partial_\mu\left(r^2\sin\theta\, g^{\mu\nu}\partial_\nu\phi\right)$$

$$= \frac{1}{r^2\sin\theta}\left\{\partial_{ct}\left[r^2\sin\theta\,(-\partial_{ct}\phi)\right] + \partial_r\left(r^2\sin\theta\,\partial_r\phi\right) + \partial_\theta\left(\frac{r^2\sin\theta}{r^2}\,\partial_\theta\phi\right)\right.$$

$$\left. + \partial_\varphi\left(\frac{r^2\sin\theta}{r^2\sin^2\theta}\,\partial_\varphi\phi\right)\right\} = -\frac{1}{c^2}\frac{\partial^2\phi}{\partial t^2} + \nabla^2_{(r,\theta,\varphi)}\phi,$$

where

$$\nabla^2_{(r,\theta,\varphi)} = \frac{1}{r^2}\frac{\partial}{\partial r}\left(r^2\frac{\partial\phi}{\partial r}\right) + \frac{1}{r^2\sin\theta}\frac{\partial}{\partial\theta}\left(\sin\theta\,\frac{\partial}{\partial\theta}\right) + \frac{1}{r^2\sin^2\theta}\frac{\partial^2}{\partial\varphi^2}$$

is the well-known 3-dimensional Laplacian in spherical coordinates.

10.4.1 Mechanics

We have already seen that, in general coordinates, the definition of 4-velocity $u^\mu \equiv dx^\mu/d\tau$ remains valid but the 4-acceleration is now

$$a^\mu \equiv \frac{Du^\mu}{D\tau} = u^\nu\nabla_\nu u^\mu. \tag{10.17}$$

The 4-momentum of a particle of mass m is again $p^\mu = mu^\mu$, while the 4-force becomes

$$f^\mu = \frac{Dp^\mu}{D\tau} = u^\nu\nabla_\nu p^\mu \tag{10.18}$$

and the formula

$$f_\mu u^\mu = -\dot{m}c^2 \tag{10.19}$$

remains valid because $f_\mu u^\mu$ is a scalar and has the same value in all coordinate systems, therefore it can be computed in a special one (Cartesian coordinates) without loss of generality. In particular, $f_\mu u^\mu = 0$ for particles of constant mass.

10.4.2 Optics

Again, the 4-tangent to a null ray is defined as $u^\mu \equiv dx^\mu/d\lambda$ where λ is a parameter and $u_\mu u^\mu = 0$. The null ray is a geodesic which can be affinely parametrized. If λ is an affine parameter then

$$\frac{Du^\mu}{D\lambda} = u^\nu \nabla_\nu u^\mu = 0. \tag{10.20}$$

The wave 4-vector k^μ is introduced as done earlier and is normalized to $k_\mu k^\mu = 0$.

10.4.3 General Matter Distributions

The main change with respect to Cartesian coordinates is the covariant conservation equation for the stress-energy tensor of a matter distribution, which now reads

$$\nabla^\nu T_{\mu\nu} = 0 \tag{10.21}$$

and the conservation of the 4-current density of energy becomes

$$\nabla_\mu J^\mu = 0. \tag{10.22}$$

10.4.4 Perfect Fluids

For a perfect fluid the equations already seen remain unchanged, except that the general covariant conservation law (10.21) projected along the 4-velocity u^μ of the fluid now gives

$$\frac{d(\rho c^2)}{d\tau} + \left(P + \rho c^2\right) \nabla_\nu u^\nu = 0. \tag{10.23}$$

10.4.5 The Maxwell Field

The Maxwell equations become

$$\nabla_\mu F^{\mu\nu} = -4\pi j^\nu, \tag{10.24}$$

$$\nabla_{[\alpha} F_{\beta\gamma]} = 0, \tag{10.25}$$

where the last equation coincides with the form

$$\partial_{[\alpha} F_{\beta\gamma]} = 0$$

which we already know. Because of the particular combination of derivatives appearing in this equation the Γ-terms in the covariant derivatives cancel out. The expression of the stress-energy tensor of the Maxwell field remains the same that we have already seen. The expression of the Maxwell tensor in terms of the 4-potential coincides with the familiar one:

$$F_{\mu\nu} = \nabla_\mu A_\nu - \nabla_\nu A_\mu = \partial_\mu A_\nu - \partial_\nu A_\mu \tag{10.26}$$

because

$$\nabla_\mu A_\nu - \nabla_\nu A_\mu = \partial_\mu A_\nu - \Gamma^\alpha_{\mu\nu} A_\alpha - \left(\partial_\nu A_\mu - \Gamma^\alpha_{\nu\mu} A_\alpha\right)$$

$$= \partial_\mu A_\nu - \partial_\nu A_\mu$$

due to the symmetry $\Gamma^\alpha_{\nu\mu} = \Gamma^\alpha_{\mu\nu}$ of the connection coefficients.

A gauge transformation can now be written as

$$A_\mu \longrightarrow A'_\mu = A_\mu + \nabla_\mu f, \tag{10.27}$$

which coincides with the one already known since $\nabla_\mu f = \partial_\mu f$. The Lorentz gauge (9.70) is written

$$\nabla_\mu A^\mu = 0. \tag{10.28}$$

The Maxwell equations for the 4-potential A^μ in the Lorentz gauge are still

$$\Box A_\mu = -4\pi j_\mu \tag{10.29}$$

provided that the correct expression $g^{\mu\nu}\nabla_\mu\nabla_\nu$ of the d'Alembertian $\Box$ is adopted. The solutions of these equations are of the plane wave form $A_\mu = C_\mu\, \mathrm{e}^{\mathrm{i}S(x^\alpha)}$ with

$$\Box S = 0,$$

$$\nabla_\mu S\, \nabla^\mu S = 0,$$

$$C^\mu \nabla_\mu S = 0.$$

Then the wave 4-vector is $k_\mu = \nabla_\mu S = \partial_\mu S$ and we have

$$k_\mu k^\mu = 0,$$

$$\nabla_\mu k^\mu = 0,$$

$$C_\mu k^\mu = 0,$$

and

$$k^\nu \nabla_\nu k^\mu = 0.$$

10.5 Conclusions

We have concluded our exposition of the basics of Special Relativity and we are now acquainted also with some of the formalism which is necessary to introduce General Relativity, the generalization which includes gravity explicitly. Locally, i.e., over small regions of space and for short time intervals, General Relativity reduces to Special Relativity. The intrinsic limitation of Special Relativity of being formulated in terms of inertial observers has been overcome only partially by our occasional discussion of accelerated observers and worldlines. The restriction to inertial observers has not been eliminated completely—this is only done in General Relativity. Considering accelerations opens the door to new possibilities and, indeed, Einstein's path to General Relativity was to consider the equivalence between gravity and acceleration. That is, the effects of gravity can be eliminated locally by adopting a freely falling reference frame. Since in the absence of other forces all test particles are subject to the same universal acceleration, independent of their masses and chemical compositions (an experimental finding dating back to Galilei and called the Equivalence Principle), Einstein conceived the idea of describing gravity as a property of spacetime instead of a force. A basic assumption of General Relativity is that the gravity generated by energy distributions curves spacetime and all test particles feel this curved geometry. This geometric description is only possible because of the universality of free fall and is peculiar to gravity.

If gravity can be locally eliminated by acceleration and acceleration is equivalent to gravity, at least locally, acceleration curves the spacetime geometry as can be argued by the following argument due to Ehrenfest and originally known as the *Ehrenfest paradox*. Consider a disk of radius r in uniform rotation. A meter stick at rest relative to it measures the length of its circumference as $C_0 = 2\pi r$. However, an observer on the rotation axis, with respect to whom the disk is rotating with uniform angular velocity ω, measures the tangential velocity $v = \omega r$ and a circumference

$$C = 2\pi\, r/\gamma, \tag{10.30}$$

where $\gamma = 1/\sqrt{1-\frac{v^2}{c^2}}$ due to Lorentz contraction, while the transversal (radial) direction is not Lorentz-contracted.[3] For this observer the ratio of the circumference to the radius is

$$\frac{C}{r} = 2\pi\sqrt{1-\frac{\omega^2 r^2}{c^2}} < 2\pi, \tag{10.31}$$

as it happens in non-Euclidean geometry. This observation suggests that acceleration curves spacetime. Einstein apparently took inspiration from this example in formu-

[3] One can obtain Eq. (10.30) also by considering a frame on which the disk is rolling. Then this equation is the result of time dilation for the time taken by the disk (deformed to an ellipse) to perform a complete roll [6].

lating the Equivalence Principle which underlines General Relativity and states the local equivalence of gravity and acceleration. General Relativity is not based on the rigid Minkowski spacetime endowed with the Minkowski metric which we have used so far, but rather on a curved spacetime with a Lorentzian metric $g_{\mu\nu}(x^\alpha)$ which depends on the spacetime position. The details of the curvature of spacetime describe the particular gravitational field associated with a mass-energy distribution residing in that spacetime. But this is another story ...

Problems

10.1. Derive the transformation property (10.4) of the connection coefficients.

10.2. Prove that

$$\partial_\sigma g_{\alpha\beta} = \Gamma^\rho_{\alpha\sigma} g_{\rho\beta} + \Gamma^\rho_{\beta\sigma} g_{\alpha\rho}.$$

10.3. Given the geodesic equation in the general form (10.14), find an affine parameter $\bar{\lambda}$ such that it reduces to the form (10.11). Write explicitly the functional relation between λ and $\bar{\lambda}$. Show that, if λ is an affine parameter, a reparametrization $\lambda \longrightarrow \lambda'$ will produce an affine parameter λ' only if $\lambda' = \alpha\,\lambda + \beta$, where α and β are constants.

10.4. The Euclidean 3-dimensional metric induces a Riemannian metric $h_{\mu\nu}$ on the 2-dimensional surface of a sphere of fixed radius R, given by

$$\mathrm{dl}^2_{(2)} = h_{\mu\nu}\mathrm{dx}^\mu \mathrm{dx}^\nu = R^2 \left(\mathrm{d}\theta^2 + \sin^2\theta\,\mathrm{d}\varphi^2\right).$$

Write explicitly the metric $h_{\mu\nu}$, its inverse $h^{\mu\nu}$ and determinant h, and compute all the connection coefficients $\Gamma^\mu_{\rho\sigma}$ of this metric. Write the geodesic equation on the sphere and show that the equator and the meridians are geodesic curves.

10.5. Find the connection coefficients of the Euclidean 3-dimensional metric
(a) in cylindrical coordinates $\{r, \varphi, z\}$;
(b) in spherical coordinates $\{r, \theta, \varphi\}$.

10.6. Find the components of the stress-energy tensor of a perfect fluid in Minkowski spacetime
(a) in cylindrical coordinates $\{ct, r, \varphi, z\}$;
(b) in spherical coordinates $\{ct, r, \theta, \varphi\}$.

10.7. Is there any point in using the vanishing of the differential invariant $\nabla^\mu \nabla^\nu T_{\mu\nu}$ of a stress-energy tensor $T_{\mu\nu}$ to characterize a form of matter, in the same way that, for example, the vanishing of the trace T of a perfect fluid energy-momentum tensor is used to characterize the radiation equation of state?

10.8. Prove that particles of a dust ($T_{\mu\nu} = \rho u_\mu u_\nu$) follow geodesics.

10.9. Let[4] $T_{\mu\nu}$ be the stress-energy tensor of a perfect fluid with the form (9.30). Introduce the number density of particles n and the 4-current density $n^\mu = nu^\mu$ (where u^μ is the fluid 4-velocity, then $n = \frac{1}{c}\sqrt{|n^\mu n_\mu|}$, check!).
(a) Show that $T_{\mu\nu}$ can be written in the form

$$T_{\mu\nu} = \frac{f'}{n} n_\mu n_\nu + \left(nf' - f\right) c^2 g_{\mu\nu},$$

where $\rho = f(n)$ is a regular function of n and $P = \left(nf' - f\right) c^2$, a prime denoting differentiation of f with respect to n.
(b)* In this description, find $f(n)$ for a fluid with barotropic equation of state $P = w\rho c^2$ and constant w. Specify your result for a dust fluid ($w = 0$), a radiation fluid ($w = -1/3$), and quantum vacuum ($w = -1$). What does the null energy condition correspond to in this description?
(c) Show that covariant conservation of $T_{\mu\nu}$ implies conservation of the number of particles, $\nabla^\mu n_\mu = 0$.

10.10. A *null dust* is described by the stress-energy tensor

$$T_{\mu\nu} = \rho u_\mu u_\nu \,, \qquad \rho \geq 0, \quad u^\mu u_\mu = 0,$$

where the 4-vector u^μ is null instead of timelike. This energy-momentum tensor is interpreted as describing the superposition of waves with random phases and polarizations but propagating in the same direction.[5]
(a) Show that one can set the function ρ equal to unity without loss of generality.
(b) Compute $T^\mu{}_\mu$ and $T^\alpha{}_\mu T^{\mu\beta}$.
(c) Prove that l^μ is a geodesic vector.

References

1. R.M. Wald, *General Relativity* (Chicago University Press, Chicago, 1984)
2. S.M. Carroll, *Spacetime and Geometry: An Introduction to General Relativity* (Addison-Wesley, San Francisco, 2004)
3. B.L. Moiseiwitsch, *Variational Principles* (Interscience, London, 1966)
4. H.J. Weber, G.B. Arfken, *Essential Mathematical Methods for Physicists* (Elsevier, Amsterdam, 2004)
5. H. Goldstein, *Classical Mechanics*, 2nd edn. (Addison-Wesley, Reading, 1972)
6. K. Vøyenli, Am. J. Phys. **45**, 876 (1977)
7. R.V. Buny, S.D.H. Hsu, Phys. Lett. B **632**, 543 (2006)

---

[4] Cf. Ref. [7].

[5] By contrast a perfect fluid with radiation equation of state $P = \rho c^2/3$ describes waves with random phases, polarizations, *and* propagation directions.

Appendix

Physical Constants

The values of the constants are taken from Ref. [1] and from the Particle Data Group [2]. Digits in parentheses denote the 1-σ uncertainty in the previous two digits.

Speed of light in vacuo $\quad c = 2.99792458 \times 10^{8}\,\mathrm{m/s}$

gravitational constant $\quad G = 6.67259(85) \times 10^{-11}\,\mathrm{N\,m^{2}\,kg^{-2}}$

Planck constant $\quad h = 6.6260755(40) \times 10^{-34}\,\mathrm{J\,s}$

reduced Planck constant $\quad \hbar \equiv \dfrac{h}{2\pi} = 1.05457266(63) \times 10^{-34}\,\mathrm{J\,s}$

Boltzmann constant $\quad K_{B} = 1.380658(12) \times 10^{-23}\,\mathrm{J/K}$

Stefan-Boltzmann constant $\quad \sigma = 5.67051(19) \times 10^{-8}\,\mathrm{J\,m^{-2}\,s^{-1}\,K^{-4}}$

electron mass $\quad m_{e} = 9.1093897(54) \times 10^{-31}\,\mathrm{kg} \simeq 511.0\,\mathrm{keV}$

proton mass $\quad m_{p} = 1.6726231(10) \times 10^{-27}\,\mathrm{kg} \simeq 938.3\,\mathrm{MeV}$

atomic mass unit $\quad 1\,\mathrm{a.m.u.} = 1.6605402(10) \times 10^{-27}\,\mathrm{kg} \simeq 931.5\,\mathrm{MeV}$

fine structure constant $\quad \alpha = 7.29735308(33) \times 10^{-3} \approx \dfrac{1}{137}$

Compton wavelength of the electron $\quad \lambda_{c} = 2.426 \times 10^{-12}\,\mathrm{m}.$

Conversion Factors

Armstrong: $\quad 1\,\mathrm{\AA} = 10^{-8}\,\mathrm{cm} = 10^{-10}\,\mathrm{m}$

Fermi: $\quad 1\,\mathrm{fm} = 10^{-13}\,\mathrm{cm} = 10^{-15}\,\mathrm{m}$

V. Faraoni, *Special Relativity*, Undergraduate Lecture Notes in Physics,
DOI: 10.1007/978-3-319-01107-3,

arcsecond: $1'' = 4.8481 \times 10^{-6}$ rad

$1\,\text{erg} = 10^{-7}\,\text{J}$

$1\,\text{eV} = 1.602177 \times 10^{-19}\,\text{J} = 1.602177 \times 10^{-12}\,\text{erg}$

Astronomical Unit $1\,\text{A.U.} = 1.496 \times 10^{11}\,\text{m}$

light year: $1\,l\,y = 9.46073 \times 10^{15}\,\text{m}$

parsec: $1\,\text{pc} = 30.85678 \times 10^{15}\,\text{m} = 3.26161\,y$

1 ton of TNT $= 4.2 \times 10^{9}$ J.

Astronomical Quantities

average acceleration of gravity on earth $g = 9.806\,\text{m/s}^2$

mass of the sun $M_\odot = 1.989 \times 10^{30}\,\text{kg}$

mass of the earth $M_e = 5.978 \times 10^{24}\,\text{kg}$

average radius of the earth $R_e = 6.370 \times 10^{6}\,\text{m}$

radius of the sun $R_\odot = 6.96 \times 10^{8}\,\text{m}$

Solutions to Selected Problems

If you are out to describe the truth, leave elegance to the tailor.
—Albert Einstein

Problems of Chapter 1

1.1 In standard configuration, a Galilei transformation is given by

$$\begin{aligned} x' &= x - vt, \\ y' &= y, \\ z' &= z, \\ t' &= t. \end{aligned}$$

Therefore, we have

$$\begin{aligned} \Delta x' &\equiv x'_2 - x'_1 = (x_2 - vt) - (x_1 - vt) = \Delta x, \\ \Delta y' &\equiv y'_2 - y'_1 = y_2 - y_1 \equiv \Delta y, \\ \Delta z' &\equiv z'_2 - z'_1 = z_2 - z_1 \equiv \Delta z, \end{aligned}$$

and

$$\left(l'_{(3)}\right)^2 = (\Delta x')^2 + (\Delta y')^2 + (\Delta z')^2 = (\Delta x)^2 + (\Delta y)^2 + (\Delta z)^2 = \left(l_{(3)}\right)^2 .$$

V. Faraoni, *Special Relativity*, Undergraduate Lecture Notes in Physics,
DOI: 10.1007/978-3-319-01107-3, © Springer International Publishing Switzerland 2013

1.2 Differentiate the equations defining parabolic coordinates

$$
\begin{aligned}
x &= x^1 x^2 \cos(x^3), \\
y &= x^1 x^2 \sin(x^3), \\
z &= \frac{1}{2}\left[(x^1)^2 - (x^2)^2\right],
\end{aligned}
$$

to obtain

$$
\begin{aligned}
\mathrm{d}x &= \frac{\partial x}{\partial x^1}\mathrm{d}x^1 + \frac{\partial x}{\partial x^2}\mathrm{d}x^2 + \frac{\partial x}{\partial x^3}\mathrm{d}x^3 = x^2\cos(x^3)\mathrm{d}x^1 + x^1\cos(x^3)\mathrm{d}x^2 \\
&\quad - x^1x^2\sin(x^3)\mathrm{d}x^3, \\
\mathrm{d}y &= \frac{\partial y}{\partial x^1}\mathrm{d}x^1 + \frac{\partial y}{\partial x^2}\mathrm{d}x^2 + \frac{\partial y}{\partial x^3}\mathrm{d}x^3 = x^2\sin(x^3)\mathrm{d}x^1 + x^1\sin(x^3)\mathrm{d}x^2 \\
&\quad + x^1x^2\cos(x^3)\mathrm{d}x^3, \\
\mathrm{d}z &= \frac{\partial z}{\partial x^1}\mathrm{d}x^1 + \frac{\partial z}{\partial x^2}\mathrm{d}x^2 + \frac{\partial z}{\partial x^3}\mathrm{d}x^3 = x^1\mathrm{d}x^1 - x^2\mathrm{d}x^2.
\end{aligned}
$$

The 3-dimensional line element (squared) is

$$
\begin{aligned}
\mathrm{d}l_{(3)}^2 &= \cos^2(x^3)\left(x^2\mathrm{d}x^1 + x^1\mathrm{d}x^2\right)^2 + (x^1x^2)^2\sin^2(x^3)(\mathrm{d}x^3)^2 \\
&\quad - 2x^1x^2\cos(x^3)\sin(x^3)\left(x^2\mathrm{d}x^1 + x^1\mathrm{d}x^2\right)\mathrm{d}x^3 \\
&\quad + \sin^2(x^3)\left(x^2\mathrm{d}x^1 + x^1\mathrm{d}x^2\right)^2 + (x^1x^2)^2\cos^2(x^3)(\mathrm{d}x^3)^2 \\
&\quad + 2x^1x^2\sin(x^3)\cos(x^3)\left(x^1\mathrm{d}x^2 + x^2\mathrm{d}x^1\right)\mathrm{d}x^3 + \left(x^1\mathrm{d}x^1 - x^2\mathrm{d}x^2\right)^2 \\
&= \left(x^2\mathrm{d}x^1 + x^1\mathrm{d}x^2\right)^2\left[\sin^2(x^3) + \cos^2(x^3)\right] \\
&\quad + \left(x^1\mathrm{d}x^1 - x^2\mathrm{d}x^2\right)^2 + (x^1x^2)^2(\mathrm{d}x^3)^2\left[\sin^2(x^3) + \cos^2(x^3)\right] \\
&= \left(x^2\mathrm{d}x^1 + x^1\mathrm{d}x^2\right)^2 + \left(x^1\mathrm{d}x^1 - x^2\mathrm{d}x^2\right)^2 + (x^1x^2)^2(\mathrm{d}x^3)^2 \\
&= \left[(x^1)^2 + (x^2)^2\right](\mathrm{d}x^1)^2 + \left[(x^1)^2 + (x^2)^2\right](\mathrm{d}x^2)^2 + 2x^1x^2\mathrm{d}x^1\mathrm{d}x^2 \\
&\quad - 2x^1x^2\mathrm{d}x^1\mathrm{d}x^2 + (x^1x^2)^2(\mathrm{d}x^3)^2.
\end{aligned}
$$

Therefore, the line element (squared) in parabolic coordinates is

$$
\mathrm{d}l_{(3)}^2 = \left[(x^1)^2 + (x^2)^2\right]\left[(\mathrm{d}x^1)^2 + (\mathrm{d}x^2)^2\right] + (x^1x^2)^2(\mathrm{d}x^3)^2.
$$

1.3 Proceed as in the previous exercise. The answer is

$$dl^2_{(3)} = \left(\sinh^2\chi\cos^2\theta + \cosh^2\chi\sin^2\theta\right)\left(d\chi^2 + d\theta^2\right) + \sinh^2\chi\sin^2\theta\, d\varphi^2.$$

1.7 In these units it is $[L] = [T] = \left[M^{-1}\right]$.

1.8 It is

$$R^{-1} = \begin{pmatrix} \cos\theta & -\sin\theta & 0 \\ \sin\theta & \cos\theta & 0 \\ 0 & 0 & 1 \end{pmatrix}$$

and $\text{Det}(R) = 1$.

Problems of Chapter 2

2.2 The eigenvalues are $\lambda_1 = 1$ (doubly degenerate) and

$$\lambda_{2,3} = \gamma\left(1 \pm \frac{v}{c}\right) = \sqrt{\frac{1 \pm v/c}{1 \mp v/c}}$$

(non-degenerate). The eigendirections associated with the eigenvalue $\lambda_1 = 1$ are the y- and z-axes. The eigendirections associated with the other two eigenvalues form the light cone. It is intuitive that all these directions are left invariant by a Lorentz transformation in standard configuration.

2.8 The angular size of the moon is $\theta = 2r/d$, where r is its radius and d is the earth-moon distance. Therefore, the angular velocity of the spot is

$$\dot{\theta} = \frac{2r}{d\Delta t} = 2 \cdot \frac{\left(1.737 \times 10^6\,\text{m}\right)}{\left(3.844 \times 10^8\,\text{m}\right)(0.010\,\text{s})} = 9.04 \times 10^{-1}\,\frac{\text{rad}}{\text{s}}.$$

The apparent linear velocity of the laser spot on the surface of the moon is

$$v = \dot{\theta}\, d = \frac{2r}{\Delta t} = \frac{2\left(1.737 \times 10^6\,\text{m}\right)}{0.010\,\text{s}} = 3.474 \times 10^8\,\frac{\text{m}}{\text{s}} > c.$$

The fact that this apparent velocity is larger than c does not constitute a problem since this is not the velocity of propagation of a particle or signal: it is merely a projection, an optical illusion.

2.9 The dispersion relation is

$$\omega(k) = \sqrt{c^2k^2 + \omega_p^2} = ck\sqrt{1 + \frac{\omega_p^2}{c^2k^2}}$$

and the phase velocity is

$$v_p \equiv \frac{\omega}{k} = c\sqrt{1 + \frac{\omega_p^2}{c^2k^2}} > c.$$

The phase velocity is always larger than c and asymptotes to it as $k \to +\infty$ (see Fig. A.1).
The group velocity is

$$v_g \equiv \frac{d\omega}{dk} = \frac{c^2k}{\sqrt{c^2k^2 + \omega_p^2}} = \frac{c}{\sqrt{1 + \frac{\omega_p^2}{c^2k^2}}} < c.$$

The group velocity v_g is always smaller than c and asymptotes to it as $k \to +\infty$ (see Fig. A.2).
The geometric mean of the phase and group velocity is the speed of light: in fact,

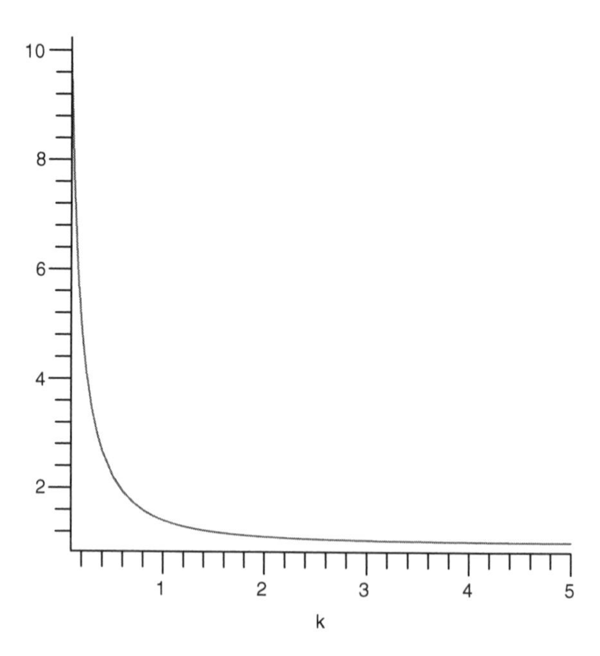

Fig. A.1 The phase velocity $v_p(k)$ as a function of the wave vector

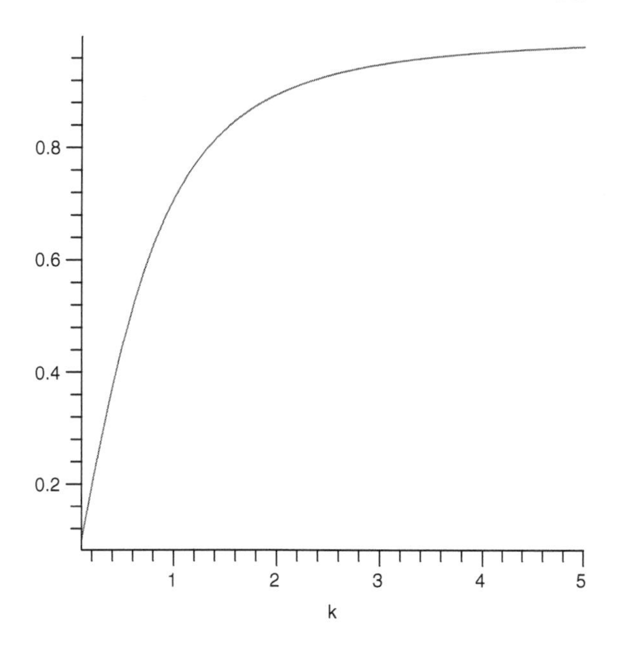

Fig. A.2 The group velocity $v_g(k)$ as a function of the wave vector

$$\sqrt{v_p v_g} = \left\{ c \left[1 + \frac{\omega_p^2}{c^2 k^2} \right]^{1/2} \cdot c \left[1 + \frac{\omega_p^2}{c^2 k^2} \right]^{-1/2} \right\}^{1/2} = c.$$

For a monochromatic plane wave described by the electric field $\mathbf{E} = \mathbf{E_0} \mathrm{e}^{\mathrm{i}(kx-\omega t)}$ with angular frequency $\omega < \omega_p$, the dispersion relation yields

$$\omega^2 = c^2 k^2 + \omega_p^2$$

and, therefore,

$$k = \frac{\sqrt{\omega^2 - \omega_p^2}}{c}.$$

If $\omega < \omega_p$ then the magnitude of the wave vector k is imaginary, $k = \mathrm{i}|k|$, and the electric field of the monochromatic plane wave is

$$\mathbf{E} = \mathbf{E_0} \, \exp\left[\mathrm{i}(kx - \omega t)\right] = \mathbf{E_0} \, \exp\left[\mathrm{i}(i|k|x)\right] \, \exp\left[-i\omega t\right] = \mathbf{E_0} \, \exp\left[-|k|x\right] \, \exp\left[-\mathrm{i}\omega t\right] :$$

the wave cannot propagate through the plasma because its amplitude decreases exponentially fast.

Problems of Chapter 3

3.4 Coordinate time and proper time are related by the time dilation formula

$$\Delta t = \gamma \Delta \tau = \frac{\Delta \tau}{\sqrt{1 - v^2/c^2}}$$

hence, since $v/c = 1/2$,

$$\Delta t = \frac{\Delta \tau}{\sqrt{1 - 1/4}} = \frac{2\Delta \tau}{\sqrt{3}}.$$

The proper times $\tau_{1,2,3} = 1\,\text{s}, 3\,\text{s}, 6\,\text{s}$, correspond to coordinate times

$$t_1 = \frac{2}{\sqrt{3}}\,\text{s}, \quad t_2 = 2\sqrt{3}\,\text{s}, \quad t_3 = 4\sqrt{3}\,\text{s}.$$

The spacetime diagram of the particle and the light cones emitted are sketched in Fig. A.3.

3.5 Refer to Fig. A.4, where the past light cones of the particles are drawn.
The two particles can have interacted at times $t \geq 0$ if there is an overlap between their past light cones in the region of the (x, t) plane above the x-axis. In Fig. A.4 there is no such overlap. In Fig. A.5, instead, the particles have interacted at times $t \geq 0$ and there is such an overlap.
The condition for the particles to have interacted in the past at times $t \geq 0$ is

$$|x_2 - x_1| \leq ct$$

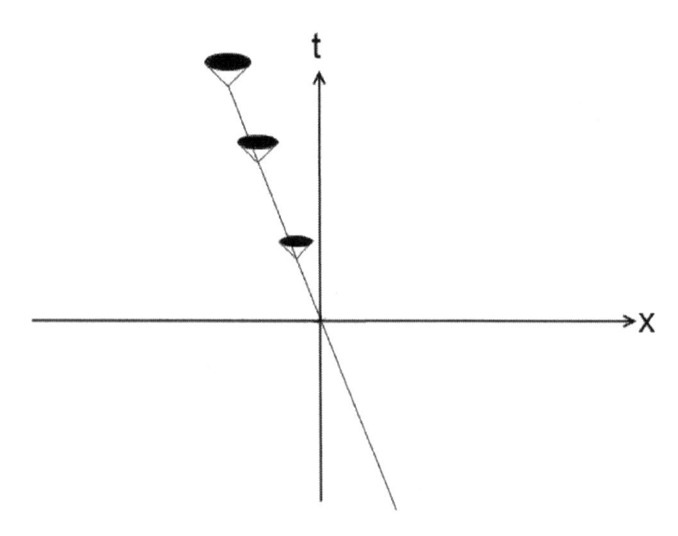

Fig. A.3 The spacetime diagram of the particle of problem 3.4 and three *light cones* emanating from it (not to scale)

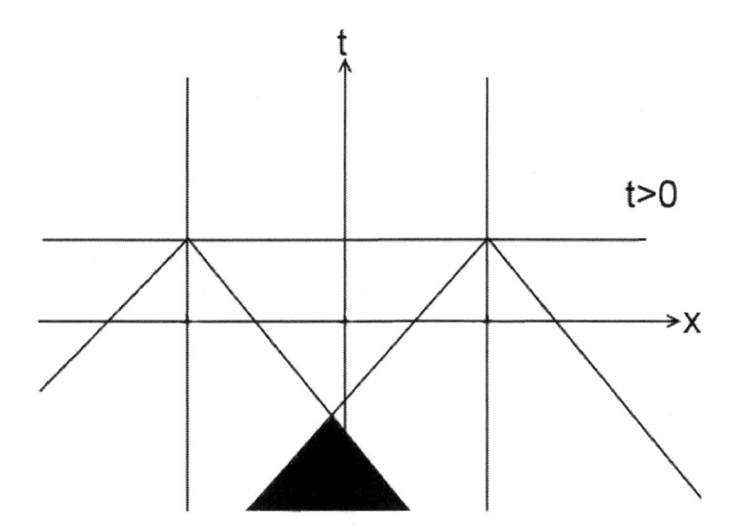

Fig. A.4 The past light cones of the two particles do not overlap in the region $t \geq 0$ above the x-axis. Their overlap (in *black*) lies below the $t = 0$ axis

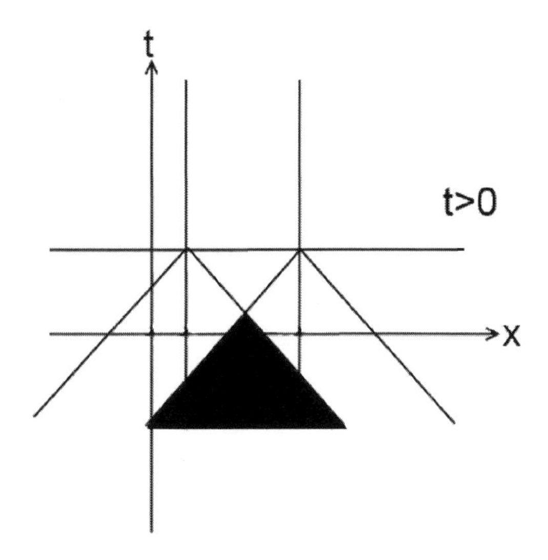

Fig. A.5 Now the past light cones of the two particles do overlap in the region $t \geq 0$ above the x-axis

i.e., the particles must be close enough that in the time interval $(0, t)$ light travels less than the separation $|x_2 - x_1|$.

3.6 Newtonian mechanics corresponds to the limit of Special Relativity in which the speed of light becomes infinite (more rigorously: the limit $v/c \to 0$, where v is the largest speed of particles or signals). In the formal limit $c \to 0$, the light cones in the (x, t) Minkowski space open up and degenerate into planes parallel to the $t = 0$

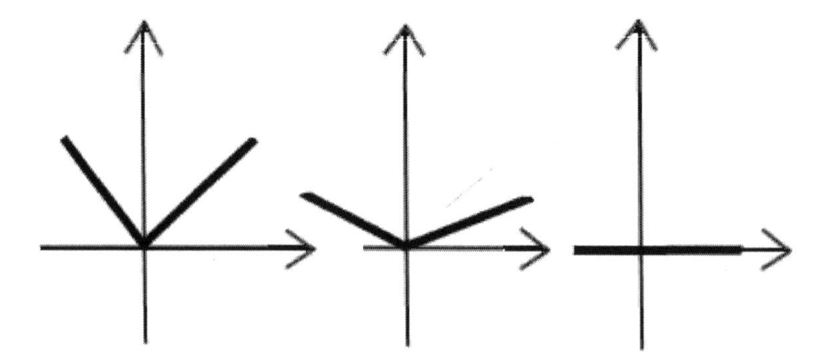

Fig. A.6 As $c \to +\infty$, a light cone opens up more and more until it flattens onto the plane $t = 0$

plane (cf. Fig. A.6). In these planes, interactions propagate instantaneously. A light ray would take zero time to propagate between two spatial points in one of these planes. In Newtonian mechanics there is instantaneous action at distance and there is no time delay between the change of conditions at one point of space and its effects being felt at a distant point.

3.7 Use units in which $c = 1$, then the world trajectory of the particle is given by

$$t(x) = \frac{1}{\omega} \cos^{-1}\left(\frac{x}{x_0}\right).$$

It must be $\left|\frac{dx}{dt}\right| < c$, or $x_0\omega\,|\sin(\omega t)| < c$ and, since $|\sin(\omega t)| \leq 1$ at all times, we obtain $\omega = 2\pi\nu < c/x_0$, where ν is the frequency of the oscillatory motion. The upper bound on the frequency of the motion

$$\nu < \nu_0 = \frac{c}{2\pi x_0}$$

must be satisfied.

3.8 Differentiate the relations

$$cT = x \sinh\left(\frac{at}{c}\right),$$

$$X = x \cosh\left(\frac{at}{c}\right),$$

to obtain

$$cdT = \sinh\left(\frac{at}{c}\right) dx + \frac{ax}{c} \cosh\left(\frac{at}{c}\right) dt,$$

$$dX = \cosh\left(\frac{at}{c}\right) dx + \frac{ax}{c} \sinh\left(\frac{at}{c}\right) dt.$$

Multiply the first equation by $\cosh\left(\frac{at}{c}\right)$ and the second by $\sinh\left(\frac{at}{c}\right)$ and subtract the second equation thus obtained from the first one. This procedure gives

$$\frac{ax}{c^2}\,c\mathrm{d}t = \cosh\left(\frac{at}{c}\right)c\mathrm{d}T - \sinh\left(\frac{at}{c}\right)\mathrm{d}X.$$

Now consider again the same two equations but multiply the first by $\sinh\left(\frac{at}{c}\right)$ and the second by $\cosh\left(\frac{at}{c}\right)$ and subtract the first equation thus obtained from the second. The result is

$$\mathrm{d}x = \cosh\left(\frac{at}{c}\right)\mathrm{d}X - \sinh\left(\frac{at}{c}\right)c\mathrm{d}T.$$

The given line element is

$$\begin{aligned}
\mathrm{d}s^2 &= -\frac{a^2x^2}{c^4}\,c^2\mathrm{d}t^2 + \mathrm{d}x^2 \\
&= -\cosh^2\left(\frac{at}{c}\right)c^2\mathrm{d}T^2 - \sinh^2\left(\frac{at}{c}\right)\mathrm{d}X^2 + 2\sinh\left(\frac{at}{c}\right)\cosh\left(\frac{at}{c}\right)c\mathrm{d}T\mathrm{d}X \\
&\quad + \cosh^2\left(\frac{at}{c}\right)\mathrm{d}X^2 + \sinh^2\left(\frac{at}{c}\right)c^2\mathrm{d}T^2 - 2\sinh\left(\frac{at}{c}\right)\cosh\left(\frac{at}{c}\right)c\mathrm{d}T\mathrm{d}X \\
&= -c^2\mathrm{d}T^2 + \mathrm{d}X^2.
\end{aligned}$$

Therefore, the given line element is nothing but the Minkowski line element in unusual coordinates. Using again the relations

$$\begin{aligned}
cT &= x\sinh\left(\frac{at}{c}\right), \\
X &= x\cosh\left(\frac{at}{c}\right),
\end{aligned}$$

it is easy to see that

$$X^2 - c^2T^2 = x^2\left[\cosh^2\left(\frac{at}{c}\right) - \sinh^2\left(\frac{at}{c}\right)\right] = x^2 \geq 0$$

and, therefore, $|X| \geq c\,|T|$: the coordinates $\{ct, x\}$ cover only the wedge $|X| \geq c\,|T|$ of the 2-dimensional Minkowski space (*Rindler wedge*, see Fig. A.7).

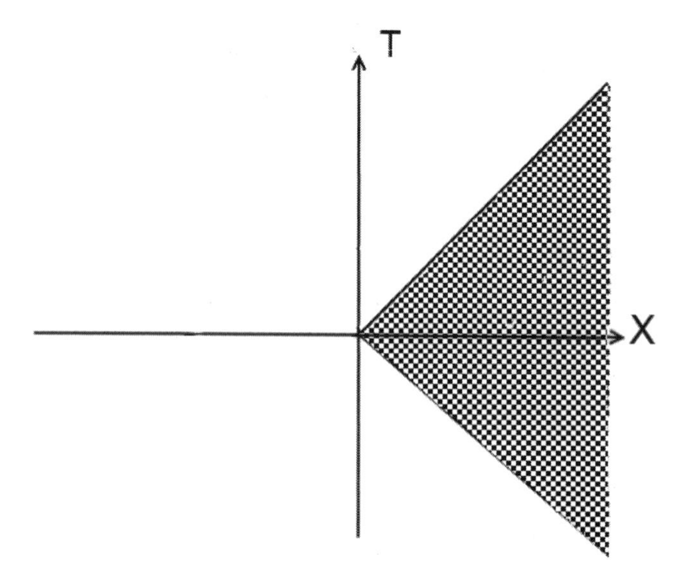

Fig. A.7 The Rindler wedge of Minkowski spacetime

Problems of Chapter 4

4.1 The coordinate transformation

$$x' = x + 2y,$$
$$y' = 3x + 6y,$$

is not an admissible coordinate transformation because it has zero determinant and is not one-to-one. In fact, it is of the form

$$\begin{pmatrix} x' \\ y' \end{pmatrix} = \begin{pmatrix} a & b \\ c & d \end{pmatrix}$$

with $a = 1, b = 2, c = 3, d = 6$ and the determinant is

$$\begin{vmatrix} a & b \\ c & d \end{vmatrix} = ad - bc = 6 - 2 \cdot 3 = 0.$$

The determinant vanishes because the two equations of the transformation are not independent; by substituting the first into the second equation, one obtains $y' = 3x'$ $\forall\,(x, y) \in \mathbb{R}^2$ and, therefore, the entire (x, y) plane is mapped onto the line $y' = 3x'$.

4.2 We have

$$\left(\frac{\partial x^{\mu'}}{\partial x^{\nu}}\right) = \begin{pmatrix} \frac{\partial x'}{\partial x} & \frac{\partial x'}{\partial y} & \frac{\partial x'}{\partial z} \\ \frac{\partial y'}{\partial x} & \frac{\partial y'}{\partial y} & \frac{\partial y'}{\partial z} \\ \frac{\partial z'}{\partial x} & \frac{\partial z'}{\partial y} & \frac{\partial z'}{\partial z} \end{pmatrix} = \begin{pmatrix} a & b & 1 \\ 3 & 1 & 5 \\ 0 & 1/3 & 1 \end{pmatrix}.$$

In order for this transformation to be an admissible coordinate transformation it must be

$$\mathrm{Det}\left(\frac{\partial x^{\mu'}}{\partial x^{\nu}}\right) = -\frac{2a}{3} + 1 - 3b \neq 0$$

or

$$2a + 9b - 3 \neq 0.$$

The components of the vector v^{μ} obey the transformation law $v^{\mu'} = \frac{\partial x^{\mu'}}{\partial x^{\nu}} v^{\nu}$. We have

$$\begin{aligned} v^{1'} &= \frac{\partial x'}{\partial x^{\nu}} v^{\nu} = av^1 + bv^2 + v^3 = a + 11 + 5b, \\ v^{2'} &= \frac{\partial y'}{\partial x^{\nu}} v^{\nu} = 3v^1 + v^2 + 5v^3 = 3 + 5 + 11 = 63, \\ v^{3'} &= \frac{\partial z'}{\partial x^{\nu}} v^{\nu} = 0 \cdot v^1 + \frac{1}{3} v^2 + v^3 = \frac{5}{3} + 11 = \frac{38}{3}, \end{aligned}$$

hence

$$v^{\mu'} = \left(a + 11 + 5b,\ 63,\ \frac{38}{3}\right).$$

Alternatively, in matrix notation, we have

$$\left(v^{\mu'}\right) = \left(\frac{\partial x^{\mu'}}{\partial x^{\nu}}\right)\left(v^{\nu}\right) = \begin{pmatrix} a & b & 1 \\ 3 & 1 & 5 \\ 0 & 1/3 & 1 \end{pmatrix} \begin{pmatrix} 1 \\ 5 \\ 11 \end{pmatrix}$$

$$= \begin{pmatrix} a + 11 + 5b \\ 3 + 5 + 55 \\ 0 + \frac{5}{3} + 11 \end{pmatrix} = \begin{pmatrix} a + 11 + 5b \\ 63 \\ 38/3 \end{pmatrix}.$$

4.4 Use the transformation property

$$A^{\mu'} = \frac{\partial x^{\mu'}}{\partial x^{\nu}} A^{\nu}$$

where $\left\{x^{\mu'}\right\} = \{r, \theta, \varphi\}$ and $\{x^{\mu}\} = \{x, y, z\}$. We have

$$\begin{aligned} x &= r \sin\theta \cos\varphi, \\ y &= r \sin\theta \sin\varphi, \\ z &= r \cos\theta, \end{aligned}$$

and the inverse coordinate transformation

$$\begin{aligned} r &= \sqrt{x^2 + y^2 + z^2}, \\ \theta &= \tan^{-1}\left(\frac{\sqrt{x^2 + y^2}}{z}\right), \\ \varphi &= \tan^{-1}\left(\frac{y}{x}\right), \end{aligned}$$

from which we compute

$$\begin{aligned} \frac{\partial r}{\partial x} &= \frac{x}{\sqrt{x^2 + y^2 + z^2}} = \sin\theta \cos\varphi, \\ \frac{\partial r}{\partial y} &= \frac{y}{\sqrt{x^2 + y^2 + z^2}} = \sin\theta \sin\varphi, \\ \frac{\partial r}{\partial z} &= \frac{z}{\sqrt{x^2 + y^2 + z^2}} = \cos\theta, \\ \frac{\partial \theta}{\partial x} &= \frac{xz}{\left(x^2 + y^2 + z^2\right)\sqrt{x^2 + y^2}} = \frac{\cos\theta \cos\varphi}{r}, \\ \frac{\partial \theta}{\partial y} &= \frac{yz}{\left(x^2 + y^2 + z^2\right)\sqrt{x^2 + y^2}} = \frac{\cos\theta \sin\varphi}{r}, \\ \frac{\partial \theta}{\partial z} &= -\frac{\sqrt{x^2 + y^2}}{x^2 + y^2 + z^2} = -\frac{\sin\theta}{r}, \\ \frac{\partial \varphi}{\partial x} &= -\frac{y}{x^2 + y^2} = -\frac{\sin\varphi}{r \sin\theta}, \\ \frac{\partial \varphi}{\partial y} &= \frac{x}{x^2 + y^2} = -\frac{\cos\varphi}{r \sin\theta}, \\ \frac{\partial \varphi}{\partial z} &= 0. \end{aligned}$$

Therefore, we have

$$A^r = \frac{\partial r}{\partial x} A^x + \frac{\partial r}{\partial y} A^y + \frac{\partial r}{\partial z} A^z = \sin\theta\cos\varphi\, A^x + \sin\theta\sin\varphi\, A^y + \cos\theta\, A^z,$$
$$A^\theta = \frac{\partial \theta}{\partial x} A^x + \frac{\partial \theta}{\partial y} A^y + \frac{\partial \theta}{\partial z} A^z = \frac{\cos\theta\cos\varphi}{r} A^x + \frac{\sin\varphi\cos\varphi}{r} A^y - \frac{\sin\theta}{r} A^z,$$
$$A^\varphi = \frac{\partial \varphi}{\partial x} A^x + \frac{\partial \varphi}{\partial y} A^y + \frac{\partial \varphi}{\partial z} A^z = -\frac{\sin\varphi}{r\sin\theta} A^x + \frac{\cos\varphi}{r\sin\theta} A^y,$$

where the components A^x, A^y, and A^z are now seen as functions of r, θ, and φ.

4.5 The Einstein summation convention means that

$$\delta^\alpha_\alpha \equiv \sum_{\alpha=1}^{n} \delta^\alpha_\alpha = \underbrace{1+1+\cdots+1}_{n \text{ times}} = n.$$

4.6 We have $V^{\mu'} = R^{\mu'}{}_\nu V^\nu$ and

$$V^{\mu'} V_{\mu'} = R^{\mu'}{}_\nu V^\nu R_{\mu'}{}^\alpha V_\alpha = V_\mu V^\mu,$$

which implies that

$$R^{\mu'}{}_\nu R_{\mu'}{}^\alpha = \delta^\alpha_\nu.$$

This equation can be written explicitly as

$$\left(\delta^\mu_\nu + \varepsilon\, l^\mu{}_\nu\right)\left(\delta^\alpha_\mu + \varepsilon\, l_\mu{}^\alpha\right) = \delta^\alpha_\nu.$$

Expanding the left hand side yields

$$\underbrace{\delta^\mu_\nu \delta^\alpha_\mu}_{\delta^\alpha_\nu} + \varepsilon \left(\underbrace{\delta^\nu_\mu\, l_\mu{}^\alpha}_{l_\nu{}^\alpha} + \underbrace{\delta^\alpha_\mu\, l^\mu{}_\nu}_{l^\alpha{}_\nu} \right) + \mathrm{O}\left(\varepsilon^2\right) = \delta^\alpha_\nu,$$

which gives

$$0 + \varepsilon \left(l_\nu{}^\alpha + l^\alpha{}_\nu\right) = \mathrm{O}\left(\varepsilon^2\right),$$

i.e., $l_\mu{}^\nu$ is antisymmetric, $l_\mu{}^\nu = -l^\nu{}_\mu$. Therefore, in four dimensions, $l_\mu{}^\alpha$ can have only $\dfrac{4(4-1)}{2} = 6$ independent components. Since the transformation $R_\mu{}^\nu$ is completely specified by $l_\mu{}^\nu$, there are only six such transformations. We know that they conserve the Minkowski scalar product $V_\mu V^\mu$ and they are easily identified with three spatial rotations and three pure Lorentz boosts. Reflections cannot be connected continuously with the identity, while all the ${R_\mu}^\nu$ transformations reduce to the identity δ^ν_μ in the limit $\varepsilon \to 0$.

4.7 We have

$$\frac{\partial^2 x^\mu}{\partial x^\alpha \partial x^\beta} = \frac{\partial}{\partial x^\beta}\left(\frac{\partial x^\mu}{\partial x^\alpha}\right) = \frac{\partial}{\partial x^\beta}\,\delta^\mu_\alpha = 0.$$

4.8 Under a coordinate change $x^\mu \longrightarrow x^{\mu'}(x^\alpha)$, the 4-vectors transform as

$$A^{\mu'} = \frac{\partial x^{\mu'}}{\partial x^\alpha}\,A^\alpha, \quad B^{\nu'} = \frac{\partial x^{\nu'}}{\partial x^\beta}\,B^\beta,$$

and it is

$$\begin{aligned} C^{\mu'\nu'} \equiv A^{\mu'}B^{\nu'} &= \left(\frac{\partial x^{\mu'}}{\partial x^\alpha}\,A^\alpha\right)\left(\frac{\partial x^{\nu'}}{\partial x^\beta}\,B^\beta\right) \\ &= \frac{\partial x^{\mu'}}{\partial x^\alpha}\,\frac{\partial x^{\nu'}}{\partial x^\beta}\,A^\alpha B^\beta \\ &\equiv \frac{\partial x^{\mu'}}{\partial x^\alpha}\,\frac{\partial x^{\nu'}}{\partial x^\beta}\,C^{\alpha\beta}. \end{aligned}$$

Since $C^{\mu'\nu'} = \dfrac{\partial x^{\mu'}}{\partial x^\alpha}\,\dfrac{\partial x^{\nu'}}{\partial x^\beta}\,C^{\alpha\beta}$, $C^{\mu\nu}$ transforms as a contravariant 2-tensor.

4.10 Consider a coordinate transformation $x^\alpha \longrightarrow x^{\alpha'}$, then it is

$$\begin{aligned} X^{\alpha'}\nabla_{\alpha'}f' = X^{\alpha'}\,\frac{\partial f}{\partial x^{\alpha'}} &= \frac{\partial x^{\alpha'}}{\partial x^\beta}\,X^\beta\,\frac{\partial f}{\partial x^\sigma}\,\frac{\partial x^\sigma}{\partial x^{\alpha'}} \\ &= \frac{\partial x^{\alpha'}}{\partial x^\beta}\,\frac{\partial x^\sigma}{\partial x^\alpha}\,\frac{\partial f}{\partial x^\sigma}\,X^\beta \\ &= \frac{\partial x^\sigma}{\partial x^{\alpha'}}\,\frac{\partial x^{\alpha'}}{\partial x^\beta}\,\frac{\partial f}{\partial x^\sigma}\,X^\beta \\ &= \frac{\partial x^\sigma}{\partial x^\beta}\,\frac{\partial f}{\partial x^\sigma}\,X^\beta = \delta^\sigma_\beta\,\frac{\partial f}{\partial x^\sigma}\,X^\beta = X^\beta\partial_\beta f, \end{aligned}$$

hence

$$X^{\alpha'}\nabla_{\alpha'}f = X^\alpha\nabla_\alpha f.$$

4.11 The Kronecker delta is

$$\delta^\mu_\nu \equiv \begin{cases} 1 \text{ if } \mu = \nu, \\ 0 \text{ if } \mu \neq \nu, \end{cases}$$

in any coordinate system. Let us show that it transforms as a tensor under coordinate changes $x^{\alpha} \longrightarrow x^{\alpha'}(x^{\mu})$. We have

$$\delta^{\mu'}_{\nu'} = \frac{\partial x^{\mu'}}{\partial x^{\nu'}} = \frac{\partial x^{\mu'}}{\partial x^{\alpha}} \frac{\partial x^{\alpha}}{\partial x^{\nu'}} = \frac{\partial x^{\mu'}}{\partial x^{\alpha}} \frac{\partial x^{\beta}}{\partial x^{\nu'}} \delta^{\beta}_{\alpha},$$

which is the transformation law of a $\binom{1}{1}$ tensor.

4.12 By definition, it is $A_{\mu} B^{\mu} \equiv g_{\mu\nu} A^{\mu} B^{\nu}$ and we have

$$A^{\mu} B_{\mu} \equiv g_{\mu\nu} B^{\mu} A^{\nu} = g_{(\mu\nu)} B^{\mu} A^{\nu} = g_{(\mu\nu)} B^{\nu} A^{\mu} = g_{\mu\nu} B^{\mu} A^{\nu} \equiv B_{\mu} A^{\mu}.$$

4.15 We have

$$F^{\mu\nu} = A^{\mu} B^{\nu\alpha} E_{\alpha\beta} D^{\beta} = B^{\mu\rho} C_{\rho} B^{\nu\alpha} E_{\alpha\beta} D^{\beta}$$

$$= B^{\mu\rho} C_{\rho} B^{\nu\alpha} E_{\alpha\beta} A^{\beta} A^{\sigma} A_{\sigma} = B^{\mu\rho} B^{\nu\alpha} C_{\rho} E_{\alpha\beta} A^{\beta} (A^{\sigma} A_{\sigma})$$

and

$$F^{[\mu\nu]} = \left(A^{\sigma} A_{\sigma}\right) \frac{B^{\mu\rho} B^{\nu\alpha} - B^{\nu\rho} B^{\mu\alpha}}{2} C_{\rho} E_{\alpha\beta} A^{\beta}.$$

4.18 We have

$$\begin{aligned}\left(A_{\mu\nu} + A_{\nu\mu}\right) B^{\mu} B^{\nu} &= A_{\mu\nu} B^{\mu} B^{\nu} + A_{\nu\mu} B^{\mu} B^{\nu} \\ &= A_{\mu\nu} B^{\mu} B^{\nu} + A_{\mu\nu} B^{\nu} B^{\mu}.\end{aligned}$$

Since all the indices are dummy indices their names are not important and we can change them, obtaining

$$\left(A_{\mu\nu} + A_{\nu\mu}\right) B^{\mu} B^{\nu} = A_{\rho\sigma} B^{\rho} B^{\sigma} + A_{\alpha\beta} B^{\alpha} B^{\beta} = 2A_{\rho\sigma} B^{\rho} B^{\sigma} = 2A_{\mu\nu} B^{\mu} B^{\nu}.$$

An alternative writing is

$$\begin{aligned}\left(A_{\mu\nu} + A_{\nu\mu}\right) B^{\mu} B^{\nu} &= 2A_{(\mu\nu)} B^{\mu} B^{\nu} = 2A_{(\mu\nu)} B^{(\mu} B^{\nu)} \\ &= 2\left(A_{\mu\nu} - A_{[\mu\nu]}\right) B^{\mu} B^{\nu} = 2A_{\mu\nu} B^{\mu} B^{\nu}\end{aligned}$$

because $A_{[\mu\nu]} B^{\mu} B^{\nu} = 0$.

4.19 We have

$$\begin{aligned} x^{0'} &= ct' = \gamma\left(ct - \frac{v}{c}x\right) = \gamma\left(x^0 - \frac{v}{c}x^1\right), \\ x^{1'} &= x' = \gamma(x - vt) = \gamma\left(x^1 - \frac{v}{c}x^0\right), \\ x^{2'} &= y' = y = x^2, \\ x^{3'} &= z' = z = x^3, \end{aligned}$$

and we compute

$$\begin{aligned} &\frac{\partial x^{0'}}{\partial x^0} = \gamma, \quad \frac{\partial x^{0'}}{\partial x^1} = -\gamma v/c, \quad \frac{\partial x^{0'}}{\partial x^2} = \frac{\partial x^{0'}}{\partial x^3} = 0, \\ &\frac{\partial x^{1'}}{\partial x^0} = -\gamma v/c, \quad \frac{\partial x^{1'}}{\partial x^1} = \gamma, \quad \frac{\partial x^{1'}}{\partial x^2} = \frac{\partial x^{1'}}{\partial x^3} = 0, \\ &\frac{\partial x^{2'}}{\partial x^0} = \frac{\partial x^{2'}}{\partial x^1} = 0, \quad \frac{\partial x^{2'}}{\partial x^2} = 1, \quad \frac{\partial x^{2'}}{\partial x^3} = 0, \\ &\frac{\partial x^{3'}}{\partial x^0} = \frac{\partial x^{3'}}{\partial x^1} = \frac{\partial x^{3'}}{\partial x^2} = 0, \quad \frac{\partial x^{3'}}{\partial x^3} = 1. \end{aligned}$$

Since $A^{\mu\nu}$ is antisymmetric, also $A^{\mu'\nu'}$ is and $A^{\mu'\mu'} = 0$. The other components transform according to $A^{\mu'\nu'} = \frac{\partial x^{\mu'}}{\partial x^{\mu}} \frac{\partial x^{\nu'}}{\partial x^{\nu}} A^{\mu\nu}$, hence

$$\begin{aligned} A^{0'0'} &= 0, \\ A^{0'1'} &= -A^{1'0'} = \frac{\partial x^{0'}}{\partial x^{\mu}} \frac{\partial x^{1'}}{\partial x^{\nu}} A^{\mu\nu} \\ &= \frac{\partial x^{0'}}{\partial x^0} \frac{\partial x^{1'}}{\partial x^{\nu}} A^{0\nu} + \frac{\partial x^{0'}}{\partial x^1} \frac{\partial x^{1'}}{\partial x^{\nu}} A^{1\nu} + \frac{\partial x^{0'}}{\partial x^2} \frac{\partial x^{1'}}{\partial x^{\nu}} A^{2\nu} + \frac{\partial x^{0'}}{\partial x^3} \frac{\partial x^{1'}}{\partial x^{\nu}} A^{3\nu} \\ &= \frac{\partial x^{0'}}{\partial x^0} \frac{\partial x^{1'}}{\partial x^0} A^{00} + \frac{\partial x^{0'}}{\partial x^0} \frac{\partial x^{1'}}{\partial x^1} A^{01} + 0 + \frac{\partial x^{0'}}{\partial x^1} \frac{\partial x^{1'}}{\partial x^0} A^{10} + \frac{\partial x^{0'}}{\partial x^1} \frac{\partial x^{1'}}{\partial x^1} A^{11} + 0 \\ &= \gamma^2 A^{01} + (-\gamma\beta)\gamma A^{10} = \gamma^2(1+\beta)A^{01}, \end{aligned}$$

$$\begin{aligned} A^{0'2'} &= -A^{2'0'} = \frac{\partial x^{0'}}{\partial x^{\mu}} \frac{\partial x^{2'}}{\partial x^{\nu}} A^{\mu\nu} \\ &= \frac{\partial x^{0'}}{\partial x^0} \frac{\partial x^{2'}}{\partial x^{\nu}} A^{0\nu} + \frac{\partial x^{0'}}{\partial x^1} \frac{\partial x^{2'}}{\partial x^{\nu}} A^{1\nu} + \frac{\partial x^{0'}}{\partial x^2} \frac{\partial x^{2'}}{\partial x^{\nu}} A^{2\nu} + \frac{\partial x^{0'}}{\partial x^3} \frac{\partial x^{2'}}{\partial x^{\nu}} A^{3\nu} \\ &= \frac{\partial x^{0'}}{\partial x^0} \frac{\partial x^{2'}}{\partial x^2} A^{02} + \frac{\partial x^{0'}}{\partial x^1} \frac{\partial x^{2'}}{\partial x^2} A^{12} + \frac{\partial x^{0'}}{\partial x^2} \frac{\partial x^{2'}}{\partial x^2} A^{22} + \frac{\partial x^{0'}}{\partial x^3} \frac{\partial x^{2'}}{\partial x^2} A^{32} \\ &= \gamma A^{02} - \gamma\beta A^{12}, \end{aligned}$$

$$
\begin{aligned}
A^{0'3'} &= -A^{3'0'} = \frac{\partial x^{0'}}{\partial x^{\mu}} \frac{\partial x^{3'}}{\partial x^{\nu}} A^{\mu\nu} \\
&= \frac{\partial x^{0'}}{\partial x^{0}} \frac{\partial x^{3'}}{\partial x^{\nu}} A^{0\nu} + \frac{\partial x^{0'}}{\partial x^{1}} \frac{\partial x^{3'}}{\partial x^{\nu}} A^{1\nu} + \frac{\partial x^{0'}}{\partial x^{2}} \frac{\partial x^{3'}}{\partial x^{\nu}} A^{2\nu} + \frac{\partial x^{0'}}{\partial x^{3}} \frac{\partial x^{3'}}{\partial x^{\nu}} A^{3\nu} \\
&= \gamma A^{03} - \gamma\beta A^{13}, \\
A^{1'0'} &= -A^{0'1'}, \\
A^{1'1'} &= 0, \\
A^{1'2'} &= -A^{2'1'} = \frac{\partial x^{1'}}{\partial x^{\mu}} \frac{\partial x^{2'}}{\partial x^{\mu}} A^{\mu\nu} = \frac{\partial x^{1'}}{\partial x^{\mu}} A^{\mu 2} \\
&= \frac{\partial x^{1'}}{\partial x^{0}} A^{02} + \frac{\partial x^{1'}}{\partial x^{1}} A^{12} = -\gamma\beta A^{02} + \gamma A^{12}, \\
A^{1'3'} &= -A^{3'1'} = \frac{\partial x^{1'}}{\partial x^{\mu}} \frac{\partial x^{3'}}{\partial x^{\nu}} A^{\mu\nu} = \frac{\partial x^{1'}}{\partial x^{\mu}} A^{\mu 3} \\
&= \frac{\partial x^{1'}}{\partial x^{0}} A^{03} + \frac{\partial x^{1'}}{\partial x^{1}} A^{13} = -\gamma\beta A^{03} + \gamma A^{13}, \\
A^{2'2'} &= 0, \\
A^{2'3'} &= -A^{3'2'} = \frac{\partial x^{2'}}{\partial x^{\mu}} \frac{\partial x^{3'}}{\partial x^{\nu}} A^{\mu\nu} = A^{23}, \\
A^{3'3'} &= 0.
\end{aligned}
$$

4.20 Let A^{μ} and B^{μ} be 4-vectors with $A_{\mu}B^{\mu} = 0$ and let $x^{\mu} \longrightarrow x^{\mu'}(x^{\alpha})$ be a coordinate change. Then we have

$$
\begin{aligned}
A_{\mu'}B^{\mu'} &= \left(\frac{\partial x^{\alpha}}{\partial x^{\mu'}} A_{\alpha}\right)\left(\frac{\partial x^{\mu'}}{\partial x^{\beta}} B^{\beta}\right) \\
&= \frac{\partial x^{\alpha}}{\partial x^{\mu'}} \frac{\partial x^{\mu'}}{\partial x^{\beta}} A^{\alpha} B^{\beta} = \frac{\partial x^{\alpha}}{\partial x^{\beta}} A^{\alpha} B^{\beta} \\
&= \delta^{\alpha}_{\beta} A^{\alpha} B^{\beta} = A_{\alpha} B^{\alpha}.
\end{aligned}
$$

4.21 Let $T_{\nu\mu} = \pm T_{\mu\nu}$, then we have

$$
T^{\nu\mu} = g^{\nu\alpha} g^{\mu\beta} T_{\alpha\beta} = g^{\nu\alpha} g^{\mu\beta} \left(\pm T_{\beta\alpha}\right) = \pm g^{\nu\alpha} g^{\mu\beta} T_{\beta\alpha} = \pm T^{\mu\nu}.
$$

4.23 On the one hand we have

$$
\begin{aligned}
A_{(\mu\nu)\rho} + A_{[\mu\nu]\rho} &= \frac{1}{2}\left(A_{\mu\nu\rho} + A_{\nu\mu\rho} + A_{\mu\nu\rho} - A_{\nu\mu\rho}\right) \\
&= \frac{1}{2} \cdot 2A_{\mu\nu\rho} = A_{\mu\nu\rho}.
\end{aligned}
$$

On the other hand, it is

$$\begin{aligned} A_{(\mu\nu\rho)} + A_{[\mu\nu\rho]} &= \frac{1}{6}\left(A_{\mu\nu\rho} + A_{\mu\rho\nu} + A_{\rho\mu\nu} + A_{\rho\nu\mu} + A_{\nu\rho\mu} + A_{\nu\mu\rho}\right. \\ &\quad \left. + A_{\mu\nu\rho} - A_{\mu\rho\nu} + A_{\rho\mu\nu} - A_{\rho\nu\mu} + A_{\nu\rho\mu} - A_{\nu\mu\rho}\right) \\ &= \frac{1}{6}\left(2A_{\mu\nu\rho} + 2A_{\rho\mu\nu} + 2A_{\nu\rho\mu}\right) = \frac{1}{3}\left(A_{\mu\nu\rho} + A_{\rho\mu\nu} + A_{\nu\rho\mu}\right), \end{aligned}$$

which is different from $A_{\mu\nu\rho}$.

4.24 Let $A_{\mu\nu} = A_{(\mu\nu)}$ and $B_{\mu\nu} = B_{[\mu\nu]}$, then

$$\begin{aligned} A_{\mu\nu}B^{\mu\nu} &= \left(\frac{A_{\mu\nu} + A_{\nu\mu}}{2}\right)\left(\frac{B_{\mu\nu} - B_{\nu\mu}}{2}\right) \\ &= \frac{1}{4}\left(A_{\mu\nu}B^{\mu\nu} - A_{\mu\nu}B^{\nu\mu} + A_{\nu\mu}B^{\mu\nu} - A_{\nu\mu}B^{\nu\mu}\right) \\ &= \frac{1}{4}\left(A_{\mu\nu}B^{\mu\nu} - A_{\nu\mu}B^{\nu\mu} + A_{\mu\nu}B^{\mu\nu} - A_{\nu\mu}B^{\nu\mu}\right) \\ &= \frac{1}{4}\left(A_{\mu\nu}B^{\mu\nu} - A_{\rho\sigma}B^{\rho\sigma} + A_{\rho\sigma}B^{\rho\sigma} - A_{\rho\sigma}B^{\rho\sigma}\right) \\ &= 0. \end{aligned}$$

4.25 We have

$$\begin{aligned} A_{\mu\nu}B^{\mu}C^{\nu} &= \left(A_{(\mu\nu)} + A_{[\mu\nu]}\right)B^{\mu}C^{\nu} \\ &= A_{(\mu\nu)}B^{\mu}C^{\nu} + \underbrace{A_{[\mu\nu]}}_{\text{antisymmetric}} \overbrace{B^{\mu}C^{\nu}}^{\text{symmetric}} \\ &= A_{(\mu\nu)}B^{\mu}C^{\nu}. \end{aligned}$$

4.27 Using the Levi-Civita symbol ε_{ijk} in three dimensions we have that, for the vector $\mathbf{C} = \mathbf{A} \times \mathbf{B}$, it is $C_i = \varepsilon_{ijk}\, A^j B^k$ and

$$\mathbf{C} \cdot \mathbf{A} = C_i A^i = \varepsilon_{ijk} A^j B^k A^i = \left(\underbrace{\varepsilon_{ijk}}_{\text{antisymmetric in } i,j} \overbrace{A^i A^j}^{\text{symmetric in } i,j} \right) B^k = 0.$$

In a similar fashion we have

$$\mathbf{C} \cdot \mathbf{B} = C_i B^i = \varepsilon_{ijk}\, A^j B^k B^i = \left(\underbrace{\varepsilon_{ijk}}_{\text{antisymmetric in } i,k} \overbrace{B^i B^k}^{\text{symmetric in } i,k} \right) A^j = 0.$$

4.28 The first two properties are trivial:

$$[\mathsf{X},\mathsf{X}] \equiv \mathsf{XX}-\mathsf{XX}=0,$$
$$[\mathsf{Y},\mathsf{X}] \equiv \mathsf{YX}-\mathsf{XY}=-(\mathsf{XY}-\mathsf{YX}) \equiv -[\mathsf{X},\mathsf{Y}].$$

The Jacobi identity is verified by expanding it:

$$\begin{aligned}
&[\mathsf{X},[\mathsf{Y},\mathsf{Z}]]+[\mathsf{Y},[\mathsf{Z},\mathsf{X}]]+[\mathsf{Z},[\mathsf{X},\mathsf{Y}]]\\
&\quad\equiv [\mathsf{X},\mathsf{YZ}-\mathsf{ZY}]+[\mathsf{Y},\mathsf{ZX}-\mathsf{XZ}]+[\mathsf{Z},\mathsf{XY}-\mathsf{YX}]\\
&\quad= [\mathsf{X},\mathsf{YZ}]-[\mathsf{X},\mathsf{ZY}]+[\mathsf{Y},\mathsf{ZX}]-[\mathsf{Y},\mathsf{XZ}]+[\mathsf{Z},\mathsf{XY}]-[\mathsf{Z},\mathsf{YX}]\\
&\quad= \mathsf{XYZ}-\mathsf{YZX}-\mathsf{XZY}+\mathsf{ZYX}+\mathsf{YZX}-\mathsf{ZXY}-\mathsf{YXZ}\\
&\qquad+\mathsf{XZY}+\mathsf{ZXY}-\mathsf{XYZ}-\mathsf{ZYX}+\mathsf{YXZ}=0.
\end{aligned}$$

Problems of Chapter 5

5.1 The null vectors l^μ must satisfy $\eta_{\mu\nu}l^\mu l^\nu = 0$. If $l^\mu = \left(l^0, l^1\right)$, this means that

$$-(l^0)^2 + (l^1)^2 = 0$$

and $l^1 = \pm l^0$. The vectors $\left(l^0, \pm l^0\right)$ are *all* the null vectors of the 2-dimensional Minkowski spacetime.

- If $l^0 > 0$ then the vector $\left(l^0, \pm l^0\right)$ is future-oriented.
- If $l^0 < 0$ then the vector $\left(l^0, \pm l^0\right)$ is past-oriented.
- If $l^0 = 0$ we have the trivial vector $(0, 0)$.

In practice, a null vector is defined up to a multiplicative constant and we can set $l^\mu = (1, \pm 1)$ for future-oriented null vectors and $l^\mu = (-1, \pm 1)$ for past-oriented null vectors.

5.2 We have

$$\begin{aligned}
g_{\mu'\nu'}A^{\mu'}A^{\nu'} &= \frac{\partial x^\alpha}{\partial x^{\mu'}}\frac{\partial x^\beta}{\partial x^{\nu'}}\, g_{\alpha\beta}\, \frac{\partial x^{\mu'}}{\partial x^\rho}A^\rho \frac{\partial x^{\nu'}}{\partial x^\sigma}A^\sigma\\
&= \left(\frac{\partial x^\alpha}{\partial x^{\mu'}}\frac{\partial x^{\mu'}}{\partial x^\rho}\right)\left(\frac{\partial x^\beta}{\partial x^{\nu'}}\frac{\partial x^{\nu'}}{\partial x^\sigma}\right) g_{\alpha\beta}A^\rho A^\sigma\\
&= \frac{\partial x^\alpha}{\partial x^\rho}\frac{\partial x^\beta}{\partial x^\sigma}\, g_{\alpha\beta}A^\rho A^\sigma = \delta^\alpha_\rho \delta^\beta_\sigma\, g_{\alpha\beta}A^\rho A^\sigma = g_{\rho\sigma}A^\rho A^\sigma.
\end{aligned}$$

5.4 We have

$$u_\mu u^\mu = -(u^0)^2 + (u^1)^2 + (u^2)^2 + (u^3)^2 = -1^2 + 0^2 + 0^2 + 0^2 = -1 < 0 \quad \text{timelike}$$

$$
\begin{aligned}
v_\mu v^\mu &= -1^2 + 1^2 + 1^2 + 1^2 = 2 > 0 \quad \text{spacelike} \\
w_\mu w^\mu &= -1^2 + 0^2 + 0^2 + 1^2 = 0 \quad \text{null} \\
x_\mu x^\mu &= -1^2 + 0^2 + 3^2 + (\sqrt{3})^2 = 11 > 0 \quad \text{spacelike} \\
y_\mu y^\mu &= -1^2 + 0^2 + (-1)^2 + 1^2 = 1 > 0 \quad \text{spacelike} \\
z_\mu z^\mu &= -100^2 + 3^2 + 4^2 + 17^2 = -9686 < 0 \quad \text{timelike} \\
q_\mu q^\mu &= -0^2 + 1^2 + 0^2 + 0^2 = 1 > 0 \quad \text{spacelike} \\
t_\mu t^\mu &= -5^2 + 0^2 + 0^2 + (\sqrt{7})^2 = -18 < 0 \quad \text{timelike} \\
r_\mu r^\mu &= -0^2 + 1^2 + 0^2 + 1^2 = 2 > 0 \quad \text{spacelike} \\
s_\mu s^\mu &= -(\sqrt{5})^2 + \pi^2 + (\sqrt{11})^2 + e^2 = \pi^2 + e^2 + 6 > 0 \quad \text{spacelike.}
\end{aligned}
$$

5.6 No. As a counterexample take the 4-vectors $u^\mu = (1, 0, 0, 0)$ (which is timelike since $u_\mu u^\mu = -1$) and $s^\mu = (1, 2, 0, 0)$ (which is spacelike since $s_\mu s^\mu = 3$). We have

$$u_\mu s^\mu = -1 \neq 0.$$

5.11 Without loss of generality, we rotate the spatial coordinate axes so that $l^\mu = \left(l^0, l, 0, 0\right)$ The normalization $l_\mu l^\mu = 0$ yields $|\mathbf{l}| = \pm l^0$, so $l^\mu = \left(l^0, \pm l^0, 0, 0\right)$. Physical vectors are future-oriented, so we assume that $l^0 > 0$. Under a Lorentz transformation we have

$$l^{0'} = \gamma\left(l^0 - \frac{v}{c}\, l^1\right) = \gamma\left(l^0 \mp \frac{v}{c}\, l^0\right) = \gamma\, l^0\left(1 \mp \frac{v}{c}\right).$$

Since $|v|/c < 1$, it is $1 \mp v/c > 0$ and since $\gamma > 0$ and $l^0 > 0$, it is also $l^{0'} > 0$. The proof can be repeated for $l^0 < 0$ and, in general,

$$\text{sign}\left(l^{0'}\right) = \text{sign}\left(l^0\right).$$

5.12 Assume $A^1 = 0$ and A^2 or $A^3 \neq 0$; then, without loss of generality, we can rotate the spatial axes to achieve $A^1 \neq 0$, but this contradicts the assumption that $A^1 = 0$ in *all* frames. Hence, it must be $A^1 = A^2 = A^3 = 0$ in all frames and A^μ has the form $A^\mu = \left(A^0, 0, 0, 0\right)$. By repeating the reasoning if $A^2 = 0$ or $A^3 = 0$ in all frames, we obtain the same result. Now consider a Lorentz transformation $x^\mu \longrightarrow x^{\mu'}$; in the "new" inertial frame it is

$$A^{1'} = \gamma\left(A^1 - \frac{v}{c}\, A^0\right) = -\gamma\, \frac{v}{c}\, A^0 \neq 0,$$

which contradicts the assumption $A^1 = 0$ in *all* inertial frames unless $A^0 = 0$. Then, the only possibility left is that $A^\mu = (0, 0, 0, 0)$.
Similarly, if $A^\mu = (0, \mathbf{A})$ with $A^1 \neq 0$ (otherwise we can rotate the spatial axes so that $A^1 \neq 0$), then we have

$$A^{0'} = \gamma\left(A^0 - \frac{v}{c}A^1\right) = -\gamma\frac{v}{c}A^1 \neq 0,$$

which contradicts the assumption.

5.14 First, consider $X^\mu \equiv \delta^{0\mu}$ in some frame; we have

$$A_{\mu\nu}X^\mu X^\nu = A_{\mu\nu}\delta^{0\mu}\delta^{0\nu} = A_{00} = 0;$$

then consider $X^\mu \equiv \delta^{1\mu}$ and by the same argument we obtain

$$A_{\mu\nu}\delta^{1\mu}\delta^{1\nu} = A_{11} = 0.$$

Repeat this argument for $X^\mu \equiv \delta^{2\mu}$ and for $X^\mu \equiv \delta^{3\mu}$, obtaining $A_{22} = A_{33} = 0$, respectively.

Now consider $X^\mu = \left(X^0, X^1, 0, 0\right)$ with both X^0 and X^1 different from zero. We have

$$\begin{aligned}
A_{\mu\nu}X^\mu X^\nu &= A_{\mu\nu}\left(X^0\delta^{0\mu} + X^1\delta^{1\mu}\right)\left(X^0\delta^{0\nu} + X^1\delta^{1\nu}\right)\\
&= A_{\mu\nu}\left[\left(X^0\right)^2\delta^{0\mu}\delta^{0\nu} + X^0X^1\left(\delta^{0\mu}\delta^{1\nu} + \delta^{1\mu}\delta^{0\nu}\right) + \left(X^1\right)^2\delta^{1\mu}\delta^{1\nu}\right]\\
&= \left(X^0\right)^2 A_{00} + X^0X^1\left(A_{01} + A_{10}\right) + \left(X^1\right)^2 A_{11} = 0
\end{aligned}$$

we then obtain $A_{10} = -A_{01}$.

Next, consider $X^\mu = \left(X^0, 0, X^2, 0\right)$ with both X^0 and X^2 non-vanishing; repeating the reasoning we obtain $A_{20} = -A_{02}$.

Then, consider $X^\mu = \left(X^0, 0, 0, X^3\right)$ with both X^0 and X^3 non-vanishing; with the same reasoning one obtains $A_{30} = -A_{03}$.

Then, consider $X^\mu = \left(X^0, X^1, X^2, 0\right)$ with all of X^0, X^1, and X^2 non-vanishing. We have

$$\begin{aligned}
A_{\mu\nu}X^\mu X^\nu &= A_{\mu\nu}\left[\left(X^0\delta^{0\mu} + X^1\delta^{1\mu} + X^2\delta^{2\mu}\right)\left(X^0\delta^{0\nu} + X^1\delta^{1\nu} + X^2\delta^{2\nu}\right)\right]\\
&= A_{\mu\nu}\left[\left(X^0\right)^2\delta^{0\mu}\delta^{0\nu} + X^0X^1\left(\delta^{0\mu}\delta^{1\nu} + \delta^{1\mu}\delta^{0\nu}\right)\right.\\
&\quad + X^0X^2\left(\delta^{0\mu}\delta^{2\nu} + \delta^{2\mu}\delta^{0\nu}\right) + \left(X^1\right)^2\delta^{1\mu}\delta^{1\nu}\\
&\quad \left. + X^1X^2\left(\delta^{1\mu}\delta^{2\nu} + \delta^{2\mu}\delta^{1\nu}\right) + \left(X^2\right)^2\delta^{2\mu}\delta^{2\nu}\right]\\
&= \left(X^0\right)^2 A_{00} + X^0X^1\left(A_{01} + A_{10}\right)\\
&\quad + X^0X^2\left(A_{02} + A_{20}\right) + \left(X^1\right)^2 A_{11} + X^1X^2\left(A_{12} + A_{21}\right) + \left(X^2\right)^2 A_{22}\\
&= X^1X^2\left(A_{12} + A_{21}\right) = 0,
\end{aligned}$$

from which we obtain that $A_{21} = -A_{12}$.

Next, taking $X^\mu = \left(X^0, X^1, 0, X^3\right)$ one obtains $A_{31} = -A_{13}$ and by taking $X^\mu = \left(X^0, X^1, X^2, X^3\right)$ with all the components non-vanishing, one obtains $A_{32} = -A_{23}$. Therefore, we have $A_{\mu\nu} = -A_{\nu\mu}$ for all values of the indices μ and ν.

5.15 Let $A_\mu A^\mu < 0$, $B_\mu B^\mu < 0$, and

$$A^\mu = \left(A^0, \mathbf{A}\right), \quad B^\mu = \left(B^0, \mathbf{B}\right),$$

with $A^0 B^0 > 0$. Then consider the 4-vector

$$S^\mu = A^\mu + B^\mu,$$

which has norm squared

$$S_\mu S^\mu = \left(A_\mu + B_\mu\right)\left(A^\mu + B^\mu\right) = \underbrace{A_\mu A^\mu}_{<0} + 2A_\mu B^\mu + \underbrace{B_\mu B^\mu}_{<0}.$$

Let us bound the product $A_\mu B^\mu$: we have

$$A_\mu B^\mu = \underbrace{-A^0 B^0}_{<0} + \mathbf{A} \cdot \mathbf{B};$$

since A^μ and B^μ are both timelike, it is

$$-\left(A^0\right)^2 + \mathbf{A}^2 < 0, \quad -\left(B^0\right)^2 + \mathbf{B}^2 < 0,$$

hence $\left(A^0\right)^2 > \mathbf{A}^2$, $\left(B^0\right)^2 > \mathbf{B}^2$ and

$$\left|A^0\right| > |\mathbf{A}|, \quad \left|B^0\right| > |\mathbf{B}|,$$

and finally

$$A^0 B^0 = \left|A^0 B^0\right| \geq |\mathbf{A} \cdot \mathbf{B}| \geq \mathbf{A} \cdot \mathbf{B}.$$

In the end we can write

$$A_\mu B^\mu = -A^0 B^0 + \mathbf{A} \cdot \mathbf{B} \leq -A^0 B^0 + A^0 B^0 = 0.$$

Here the useful inequality is $A_\mu B^\mu \leq 0$, using which we evaluate the norm squared of S^μ as

$$S_\mu S^\mu = \underbrace{A_\mu A^\mu}_{<0} + \underbrace{2A_\mu B^\mu}_{\leq 0} + \underbrace{B_\mu B^\mu}_{<0} < 0$$

and S^μ is timelike.

5.16 Let $A^\mu = \left(A^0, \mathbf{A}\right)$ be timelike, $B^\mu = \left(B^0, \mathbf{B}\right)$ be null, and A^μ and B^μ be isochronous. We have

$$A_\mu A^\mu = -\left(A^0\right)^2 + \mathbf{A}^2 < 0, \quad B_\mu B^\mu = -\left(B^0\right)^2 + \mathbf{B}^2 < 0,$$

and $\left|A^0\right| > |\mathbf{A}|, \left|B^0\right| > |\mathbf{B}|$ with $A^0 B^0 > 0$. Consider the sum vector

$$S^\mu = A^\mu + B^\mu,$$

which has norm squared

$$S_\mu S^\mu = \left(A_\mu + B_\mu\right)\left(A^\mu + B^\mu\right) = \underbrace{A_\mu A^\mu}_{<0} + 2A_\mu B^\mu + \underbrace{B_\mu B^\mu}_{=0}.$$

Let us study the product $A_\mu B^\mu$: we have

$$A_\mu B^\mu = \underbrace{-A^0 B^0}_{<0} + \mathbf{A}\cdot\mathbf{B}$$

and

$$A^0 B^0 = \left|A^0 B^0\right| = \left|A^0\right| |\mathbf{B}| > |\mathbf{A}|\,|\mathbf{B}|$$

while

$$\mathbf{A}\cdot\mathbf{B} \leq |\mathbf{A}\cdot\mathbf{B}| \leq |\mathbf{A}|\,|\mathbf{B}| < A^0 B^0.$$

Then

$$A_\mu B^\mu = -A^0 B^0 + \mathbf{A}\cdot\mathbf{B} < -A^0 B^0 + A^0 B^0 = 0$$

and $A_\mu B^\mu < 0$. The norm squared of S^μ is

$$S_\mu S^\mu = A_\mu A^\mu + 2A_\mu B^\mu < 0$$

and S^μ is timelike. Then we have

$$S^0 A^0 = \left(A^0 + B^0\right) A^0 = \left(A^0\right)^2 + A^0 B^0 > 0$$

and $S^\mu \equiv A^\mu + B^\mu$ is timelike and isochronous with A^μ (and therefore also with B^μ).

5.21 Let $A^\mu = \left(A^0, \mathbf{A}\right)$ be spacelike, $A_\mu A^\mu > 0$. Then there exists an inertial frame in which $A^\mu \doteq (0, \mathbf{A})$. In this frame, define the timelike 4-vector B^μ such that

$$B^\mu \doteq \left(B^0, \mathbf{0}\right) \quad \text{with } B^0 = |\mathbf{A}|.$$

The 4-vectors A^μ and B^μ are orthogonal, $A_\mu B^\mu = 0$, and

$$B_\mu B^\mu \doteq -\left(B^0\right)^2 \doteq -\mathbf{A}^2 \doteq -A_\mu A^\mu.$$

Because this is an equality between world scalars, if it is satisfied in one frame it is satisfied in any frame and we can write

$$B_\mu B^\mu = -A_\mu A^\mu.$$

Now consider the 4-vectors

$$l^\mu \equiv \frac{A^\mu + B^\mu}{2}, \quad m^\mu \equiv \frac{B^\mu - A^\mu}{2};$$

clearly $A^\mu = l^\mu - m^\mu$ and

$$l_\mu l^\mu = \frac{1}{4}\left(A_\mu + B_\mu\right)\left(A^\mu + B^\mu\right) = \frac{1}{4}\left(A_\mu A^\mu + 2A_\mu B^\mu + B_\mu B^\mu\right) = 0,$$
$$m_\mu m^\mu = \frac{1}{4}\left(B_\mu - A_\mu\right)\left(B^\mu - A^\mu\right) = \frac{1}{4}\left(B_\mu B^\mu - 2A_\mu B^\mu + A_\mu A^\mu\right) = 0.$$

In the frame in which $A^\mu \doteq (0, \mathbf{A})$, $B^\mu \doteq \left(B^0, \mathbf{0}\right)$ it is

$$m^\mu \doteq \frac{1}{2}\left(B^0, -\mathbf{A}\right), \quad l^\mu \doteq \frac{1}{2}\left(B^0, \mathbf{A}\right),$$

and m^μ and l^μ are isochronous in this frame, hence they are isochronous in *any* frame.

5.24 Let $A^\mu = \left(A^0, \mathbf{A}\right)$ and $B^\mu = \left(B^0, \mathbf{B}\right)$ be two timelike isochronous 4-vectors, then

$$A_\mu A^\mu = -\left(A^0\right)^2 + \mathbf{A}^2 < 0, \quad B_\mu B^\mu = -\left(B^0\right)^2 + \mathbf{B}^2 < 0, \quad A^0 B^0 > 0.$$

We have $A_\mu B^\mu = -A^0 B^0 + \mathbf{A}\cdot\mathbf{B}$ and, since $A_\mu A^\mu < 0$ and $B_\mu B^\mu < 0$, it is $\left|A^0\right| > |\mathbf{A}|$ and $\left|B^0\right| > |\mathbf{B}|$. Then we have

$$\mathbf{A}\cdot\mathbf{B} \le |\mathbf{A}\cdot\mathbf{B}| \le |\mathbf{A}|\,|\mathbf{B}| < \left|A^0\right|\left|B^0\right| = \left|A^0 B^0\right| = A^0 B^0$$

and

$$A_\mu B^\mu = -A^0 B^0 + \mathbf{A}\cdot\mathbf{B} < -A^0 B^0 + A^0 B^0 = 0,$$

therefore $A_\mu B^\mu < 0$.

5.30 Let $A^\mu = \left(A^0, \mathbf{A}\right)$ and $B^\mu = \left(B^0, \mathbf{B}\right)$ be two 4-vectors. The desired quantity in the new frame is

$$\begin{aligned}
\mathscr{I}' &\equiv \frac{\left(A^{0'} - A^{1'}\right)\left(B^{0'} + B^{1'}\right)}{\left(A^{0'} + A^{1'}\right)\left(B^{0'} - B^{1'}\right)} \\
&= \frac{\gamma^2\left(A^0 - \beta A^1 - A^1 + \beta A^0\right)\left(B^0 - \beta B^1 + B^1 - \beta B^0\right)}{\gamma^2\left(A^0 - \beta A^1 + A^1 - \beta A^0\right)\left(B^0 - \beta B^1 - B^1 + \beta B^0\right)} \\
&= \frac{(1+\beta)\left(A^0 - A^1\right)(1-\beta)\left(B^0 + B^1\right)}{(1-\beta)\left(A^0 + A^1\right)(1+\beta)\left(B^0 - B^1\right)} \\
&= \frac{\left(A^0 - A^1\right)\left(B^0 + B^1\right)}{\left(A^0 + A^1\right)\left(B^0 - B^1\right)} \equiv \mathscr{I}.
\end{aligned}$$

5.32 We have

$$x^2 + y^2 = \left(r^2 + a^2\right)\sin^2\theta\left(\sin^2\varphi + \cos^2\varphi\right) = \left(r^2 + a^2\right)\sin^2\theta$$

and $z^2 = r^2\cos^2\theta$, hence

$$\frac{x^2 + y^2}{r^2 + a^2} + \frac{z^2}{r^2} = 1:$$

the surfaces $r = \text{const.}$ are oblate ellipsoids. Differentiation of the coordinate transformation yields

$$\begin{aligned}
\mathrm{d}x &= \frac{r\mathrm{d}r}{\sqrt{r^2 + a^2}}\sin\theta\cos\varphi + \sqrt{r^2 + a^2}\left(\cos\theta\cos\varphi\mathrm{d}\theta - \sin\theta\sin\varphi\mathrm{d}\varphi\right), \\
\mathrm{d}y &= \frac{r\mathrm{d}r}{\sqrt{r^2 + a^2}}\sin\theta\sin\varphi + \sqrt{r^2 + a^2}\left(\cos\theta\sin\varphi\mathrm{d}\theta + \sin\theta\cos\varphi\mathrm{d}\varphi\right), \\
\mathrm{d}z &= \mathrm{d}r\cos\theta - r\sin\theta\mathrm{d}\theta.
\end{aligned}$$

The Minkowskian line element is

$$\begin{aligned}
\mathrm{d}s^2 &= -c^2\mathrm{d}t^2 + \mathrm{d}x^2 + \mathrm{d}y^2 + \mathrm{d}z^2 \\
&= -c^2\mathrm{d}t^2 + \frac{r^2\mathrm{d}r^2}{r^2 + a^2}\sin^2\theta\cos^2\varphi + \left(r^2\mathrm{d}r^2\right)\left(\cos\theta\cos\varphi\mathrm{d}\theta - \sin\theta\sin\varphi\mathrm{d}\varphi\right)^2 \\
&\quad +2r\sin\theta\cos\varphi\mathrm{d}r\left(\cos\theta\cos\varphi\mathrm{d}\theta - \sin\theta\sin\varphi\mathrm{d}\varphi\right) \\
&\quad + \frac{r^2\mathrm{d}r^2}{r^2 + a^2}\sin^2\theta\sin^2\varphi + \left(r^2 + a^2\right)\left(\cos\theta\sin\varphi\mathrm{d}\theta + \sin\theta\cos\varphi\mathrm{d}\varphi\right)^2 \\
&\quad +2r\sin\theta\sin\varphi\mathrm{d}r\left(\cos\theta\sin\varphi\mathrm{d}\theta + \sin\theta\cos\varphi\mathrm{d}\varphi\right) + \mathrm{d}r^2\cos^2\theta + \mathrm{d}r^2\sin^2\theta\mathrm{d}\theta^2 \\
&\quad -2r\sin\theta\cos\theta\mathrm{d}r\mathrm{d}\theta
\end{aligned}$$

$$
\begin{aligned}
&= -c^2 \mathrm{d}t^2 + \frac{r^2 \mathrm{d}r^2}{r^2+a^2} \sin^2\theta \left(\sin^2\varphi + \cos^2\varphi\right) \\
&\quad + \left(r^2+a^2\right)\left[\mathrm{d}\theta^2 \left(\cos^2\theta\cos^2\varphi + \cos^2\theta\sin^2\varphi\right)\right. \\
&\quad + \mathrm{d}\varphi^2 \left(\sin^2\theta\sin^2\varphi + \sin^2\theta\cos^2\varphi\right) \\
&\quad \left. + 2\mathrm{d}\theta\mathrm{d}\varphi\left(-\sin\theta\cos\theta\sin\varphi\cos\varphi + \sin\theta\cos\theta\sin\varphi\cos\varphi\right)\right] \\
&\quad + 2r\mathrm{d}r\sin\theta\left[\left(\cos\theta\cos^2\varphi + \cos\theta\sin^2\varphi\right)\mathrm{d}\theta\right. \\
&\quad \left. + \left(-\sin\theta\sin\varphi\cos\varphi + \sin\theta\sin\varphi\cos\varphi\right)\mathrm{d}\varphi\right] \\
&\quad + \mathrm{d}r^2\cos^2\theta + r^2\sin^2\theta\mathrm{d}\theta^2 - 2r\sin\theta\cos\theta\mathrm{d}r\mathrm{d}\theta \\
&= -c^2\mathrm{d}t^2 + \frac{r^2\mathrm{d}r^2}{r^2+a^2}\sin^2\theta\mathrm{d}r^2 + \left(r^2+a^2\right)\cos^2\theta\mathrm{d}\theta^2 + \left(r^2+a^2\right)\sin^2\theta\mathrm{d}\varphi^2 \\
&\quad + 2r\sin\theta\cos\theta\mathrm{d}r\mathrm{d}\theta + \mathrm{d}r^2\cos^2\theta + r^2\sin^2\theta\mathrm{d}\theta^2 - 2r\sin\theta\cos\theta\mathrm{d}r\mathrm{d}\theta \\
&= -c^2\mathrm{d}t^2 + \left(\frac{r^2}{r^2+a^2}\sin^2\theta + \cos^2\theta\right)\mathrm{d}r^2 + \left[\left(r^2+a^2\right)\cos^2\theta + r^2\sin^2\theta\right]\mathrm{d}\theta^2 \\
&\quad + \left(r^2+a^2\right)\sin^2\theta\mathrm{d}\varphi^2 .
\end{aligned}
$$

Now we have

$$
\frac{r^2}{r^2+a^2}\sin^2\theta + \cos^2\theta = \frac{r^2+a^2\cos^2\theta}{r^2+a^2}
$$

and the line element becomes

$$
\mathrm{d}s^2 = -c^2\mathrm{d}t^2 + \frac{r^2+a^2\cos^2\theta}{r^2+a^2}\mathrm{d}r^2 + \left(r^2+a^2\cos^2\theta\right)\mathrm{d}\theta^2 + \left(r^2+a^2\right)\sin^2\theta\mathrm{d}\varphi^2 .
$$

The Kerr solution of General Relativity representing the spacetime outside a rotating stationary black hole of mass M and angular moment per unit mass $a = J/Mc$ (in units in which $G = 1$) is

$$
\begin{aligned}
\mathrm{d}s^2 = &-\left(1 - \frac{2Mr}{\Sigma}\right)c^2\mathrm{d}t^2 - \frac{4aMr\sin^2\theta}{\Sigma}\mathrm{d}\theta\mathrm{d}\varphi + \frac{\Sigma}{\Delta}\mathrm{d}r^2 \\
&+ \Sigma\mathrm{d}\theta^2 + \left(r^2 + a^2 + \frac{2Mra^2\sin^2\theta}{\Sigma}\right)\sin^2\theta\mathrm{d}\varphi^2 ,
\end{aligned}
$$

where

$$
\Sigma = r^2 + a^2\cos^2\theta, \quad \Delta = r^2 - 2Mr + a^2,
$$

(Kerr metric in Boyer-Lindquist coordinates). In the limit $M \to 0$ in which gravity disappears, General Relativity reduces to Special Relativity and this spacetime must reduce to the Minkowski spacetime. This is indeed the case because, in this limit,

$$\mathrm{d}s^2 \to \mathrm{d}s_0^2 = -c^2\mathrm{d}t^2 + \frac{\left(r^2 + a^2\cos^2\theta\right)}{r^2 + a^2}\,\mathrm{d}r^2 + \left(r^2 + a^2\cos^2\theta\right)\mathrm{d}\theta^2$$
$$+ \left(r^2 + a^2\right)\sin^2\theta\mathrm{d}\varphi^2,$$

which is the Minkowskian line element in rotating coordinates found above.

Problems of Chapter 6

6.1 We have

$$A_\mu A^\mu = -\left(A^0\right)^2 + \left(A^1\right)^2 + \left(A^2\right)^2 + \left(A^3\right)^2 = -3^2 + 1^2 = -8 < 0.$$

The particle 4-velocity is proportional to A^μ, which is a timelike 4-vector, hence u^μ is timelike and can be associated with a massive particle. To find u^μ, we normalize A^μ because the 4-velocity of a massive particle must satisfy $u_\mu u^\mu = -c^2$. Hence, setting $u^\mu = \alpha\, A^\mu$, it must be

$$-c^2 = u_\mu u^\mu = \alpha^2 A_\mu A^\mu = -8\alpha^2,$$

which implies that $\alpha = \frac{c}{2\sqrt{2}}$. Therefore, the particle 4-velocity is

$$u^\mu = \frac{c}{2\sqrt{2}}\,A^\mu = \left(\frac{3c}{2\sqrt{2}}, 0, 0, \frac{c}{2\sqrt{2}}\right).$$

6.6 Working in Cartesian coordinates, we have

$$u^\mu = (\gamma\, c, \gamma\mathbf{v}) = (\gamma\, c, \gamma\, v, 0, 0)\,.$$

Since $v = \frac{dx}{dt} = \frac{dx}{d\tau}\frac{d\tau}{dt} = v(\tau)$ where τ is the proper time of the particle and

$$u^\mu\left(\tau\right) = \left(\frac{c}{\sqrt{1 - \frac{v^2(\tau)}{c^2}}}, \frac{v(\tau)}{\sqrt{1 - \frac{v^2(\tau)}{c^2}}}, 0, 0\right)$$

while $u^\mu = dx^\mu/d\tau$. Therefore, it is

$$x^\mu\left(\tau\right) = \int_0^\tau \mathrm{d}\tau' u^\mu\left(\tau'\right),$$
$$t\left(\tau\right) = c\int_0^\tau \frac{\mathrm{d}\tau'}{\sqrt{1 - \frac{v^2(\tau')}{c^2}}},$$
$$x\left(\tau\right) = \int_0^\tau \frac{\mathrm{d}\tau' v(\tau')}{\sqrt{1 - \frac{v^2(\tau')}{c^2}}},$$

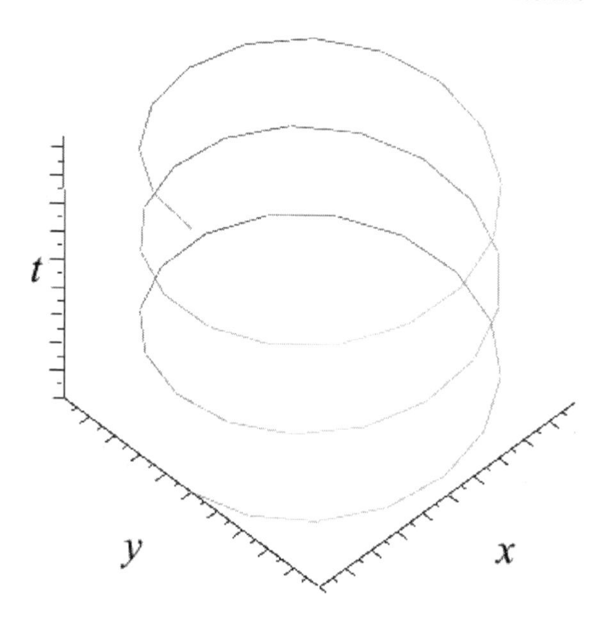

Fig. A.8 The worldline of the particle of problem 6.7 is a helix

$$y(\tau) = \text{const.} = 0,$$
$$z(\tau) = \text{const.} = 0.$$

6.7 We have

$$x = R\cos(\omega t),$$
$$y = R\sin(\omega t),$$

and $v = \omega R$ is the linear velocity of the particle. The worldline is a helix with the time axis as the symmetry axis (Fig. A.8).
Since $v = \omega R$ must be smaller than c, the bound

$$\omega < \frac{R}{c}$$

must be obeyed.
The proper time τ is related to the coordinate time t by the time dilation formula $\mathrm{d}t = \gamma\,\mathrm{d}\tau$, where

$$\gamma = \frac{1}{\sqrt{1 - v^2/c^2}} = \frac{1}{\sqrt{1 - \omega^2 R^2/c^2}}$$

is the Lorentz factor. Since γ is constant, it is

$$t(\tau) = \gamma\tau = \frac{\tau}{\sqrt{1 - \omega^2 R^2/c^2}}.$$

A parametric representation of the particle worldline is

$$x(\tau) = R\cos(\gamma\,\omega\,\tau),$$
$$y(\tau) = R\sin(\gamma\,\omega\,\tau),$$
$$t(\tau) = \frac{\tau}{\sqrt{1 - \omega^2 R^2/c^2}}.$$

6.8 In the instantaneous rest frame of the particle (which is an inertial frame) it is

$$a_\mu u^\mu = \eta_{\mu\nu} a^\mu u^\nu = -a^0 u^0 + a^1 u^1 + a^2 u^2 + a^3 u^3 = 0 \cdot c + \mathbf{0} \cdot \frac{d\mathbf{v}}{dt} = 0.$$

Since $a_\mu u^\mu$ is a world scalar, it assumes the same value in all inertial frames, therefore, it vanishes in all frames.

6.12 Let

$$f(ct, x, y, z) \equiv -(ct - a)^2 + 2(x - b)^2 + y^2 + z^2,$$

then $f = 0$ is the equation of the hypersurface $\mathscr{S}$. The normal to $\mathscr{S}$ is parallel to the gradient of f:

$$N_\mu = \nabla_\mu f = -2(ct - a)\,\delta_{\mu 0} + 4(x - b)\,\delta_{\mu 1} + 2y\,\delta_{\mu 2} + 2z\,\delta_{\mu 3}\Big|_{\mathscr{S}}$$
$$= 2\Big(-ct + a, 2(x - b), y, z\Big)\Big|_{\mathscr{S}}$$

and

$$N^\mu = 2\Big(ct - a, 2(x - b), y, z\Big)\Big|_{\mathscr{S}}.$$

The length squared of N^μ is

$$N_\mu N^\mu = 4\Big[-(ct - a)^2 + 4(x - b)^2 + y^2 + z^2\Big]_{f=0}$$
$$= 4\Big[f(ct, x, y, z) + 2(x - b)^2\Big]_{f=0} = 8(x - b)^2.$$

It is $N_\mu N^\mu > 0$ (and N^μ is spacelike) if $x \neq b$, while $N_\mu N^\mu = 0$ (and N^μ is null) if $x = b$. Therefore, the hypersurface $\mathscr{S}$ is

- *timelike* if $x \neq b$;
- *null* if $x = b$.

When $\mathscr{S}$ is timelike, the unit normal is

$$n_\mu = \frac{N_\mu}{\sqrt{N_\alpha N^\alpha}} = \frac{1}{\sqrt{2}\,|x-b|}\Big(-ct+a, 2(x-b), y, z\Big)\Big|_{f=0}$$

and

$$n^\mu = \frac{1}{\sqrt{2}\,|x-b|}\Big(ct-a, 2(x-b), y, z\Big)\Big|_{f=0}.$$

When $\mathscr{S}$ is null (for $x=b$), the unit normal lies along $\mathscr{S}$ and has arbitrary magnitude. In this case it is

$$f\,(ct, b, y, z) = -\,(ct-a)^2 + y^2 + z^2 = 0$$

and, dropping an irrelevant factor 2,

$$N_\mu = \Big(ct-a, 0, y, z\Big)\Big|_{f=0} = \Big(\pm\sqrt{y^2+z^2}\,, 0, y, z\Big).$$

Problems of Chapter 7

7.4 (a) We have

$$\begin{aligned} x' &= \gamma\,(x-vt) = \gamma\,(x_0 - vt)\,,\\ y' &= y = y_0 - ut,\\ z' &= z. \end{aligned}$$

We eliminate the time t using

$$t = \frac{1}{v}\left(x_0 - \frac{x'}{\gamma}\right) = \frac{1}{\gamma\,v}\left(\gamma\,x_0 - x'\right).$$

Then we have

$$y' = y_0 - ut = y_0 - \frac{u}{\gamma\,v}\left(\gamma\,x_0 - x'\right)$$

and

$$y'\left(x'\right) = \frac{u}{\gamma\,v}\,x' + \left(y_0 - \frac{u x_0}{v}\right).$$

This is still the equation of a straight line but now it has slope $\frac{u}{\gamma v}$. In the moving frame, rain droplets make an angle with the horizontal

$$\theta' = \tan^{-1}\left(\frac{u}{\gamma\,v}\right) = \tan^{-1}\left(\frac{u}{v}\sqrt{1-\frac{v^2}{c^2}}\right);$$

this angle is smaller in the relativistic case than in the non-relativistic case.

- If $v > u$, then $\tan\theta' = \frac{u}{v}\sqrt{1-\frac{v^2}{c^2}} < 1$ and $\theta' < 45°$.
- If $v \leq u$, it can still be $\theta' < 45°$. We have $\theta' = 45°$ at $\tan\theta' = 1 = \frac{u}{v}\sqrt{1-\frac{v^2}{c^2}}$, or

$$v = \frac{u}{\sqrt{1+\frac{u^2}{c^2}}} < u.$$

In the limit $v \to c$ we have $\tan\theta' = \frac{u}{v}\sqrt{1-\frac{v^2}{c^2}} \to 0$ and $\theta' \to 0$ (relativistic beaming becomes extreme).

(b) Now, using the relativistic law of composition of velocities

$$u^{x'} = \frac{u^x - v}{1-\frac{vu^x}{c^2}},$$
$$u^{y'} = \frac{u^y}{\gamma\left(1-\frac{vu^x}{c^2}\right)},$$
$$u^{z'} = \frac{u^z}{\gamma\left(1-\frac{vu^x}{c^2}\right)},$$

we have

$$\mathbf{u} = \left(u^x, u^y, u^z\right) = (0, -u, 0)$$

for O and

$$\mathbf{u}' = \left(u^{x'}, u^{y'}, u^{z'}\right) = \left(-v, -\frac{u}{\gamma}, 0\right)$$

and

$$\tan\theta' = \frac{u^{y'}}{u^{x'}} = \frac{u}{\gamma v}.$$

(c) In (a) we obtain, using $dt = dt'/\gamma$,

$$u^{x'} = \frac{dx'}{dt'} = \frac{dx'}{dt}\frac{dt}{dt'} = \frac{1}{\gamma}\frac{d}{dt}\left[\gamma\left(x_0 - vt\right)\right] = -v,$$
$$u^{y'} = \frac{dy'}{dt'} = \frac{d}{dt'}\left[\frac{u}{\gamma v}x' + \left(y_0 - \frac{ux_0}{v}\right)\right] = \frac{uu^{x'}}{\gamma v} = -\frac{uv}{\gamma v} = -\frac{u}{\gamma},$$
$$u^{z'} = \frac{dz'}{dt'} = 0,$$

hence

$$\mathbf{u}' = \left(u^{x'}, u^{y'}, u^{z'}\right) = \left(-v, -\frac{u}{\gamma}, 0\right),$$

which coincides with the answer obtained in part (b).

Problems of Chapter 8

8.1 We have

$$\begin{aligned}
p_\alpha p^\alpha &= \frac{1}{c^2}\left(Ev_\alpha + \sqrt{E^2c^2 - m^2c^4}e_\alpha\right)\left(Ev^\alpha + \sqrt{E^2c^2 - m^2c^4}e^\alpha\right) \\
&= \frac{1}{c^2}\left[E^2 \underbrace{v_\alpha v^\alpha}_{-c^2} + 2E\sqrt{E^2c^2 - m^2c^4}\underbrace{e_\alpha v^\alpha}_{0} + \left(E^2c^2 - m^2c^4\right)\underbrace{e_\alpha e^\alpha}_{+1}\right] \\
&= -m^2c^2.
\end{aligned}$$

The vector v^α can be interpreted as the 4-velocity of a timelike observer, giving the direction of time for this observer, i.e., there exists a coordinate system $\{x^\mu\} = \{ct, \mathbf{x}\}$ such that $v^\mu = (c, 0, 0, 0)$ in these coordinates. e^α is a spatial vector lying in the hypersurfaces $t =$ const. orthogonal to v^μ. e^α is normalized to unity for convenience: $e_\alpha e^\alpha = 1$ (and $e_\alpha v^\alpha = 0$). e^α has the direction of the 3-momentum of the particle. We have, therefore, a (local) splitting of spacetime into 3-space and time based on the choice of (local) time t (Fig. A.9).
In the rest frame of the observer with 4-velocity v^α, the quantity E can be interpreted as the particle energy with respect to that local frame, with $E \geq mc^2$ (it is $E = mc^2$ only if the particle is at rest in that local frame).

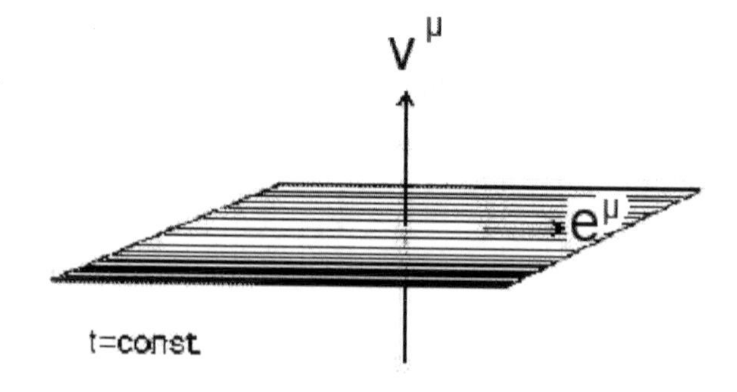

Fig. A.9 The 3 + 1 splitting of spacetime into space and time according to the observer with 4-velocity v^μ.

8.2 Projecting along the "time" direction (i.e., the direction of u^μ) one has

$$u^\alpha \nabla_\alpha \phi = -\frac{\dot{\phi}}{c^2} \underbrace{u_\alpha u^\alpha}_{-c^2} + u^\alpha \tilde{\nabla}_\alpha \phi,$$

where $u^\alpha \tilde{\nabla}_\alpha \phi = 0$, hence

$$\dot{\phi} = u^\alpha \nabla_\alpha \phi.$$

Projecting now onto the 3-space orthogonal to u^α with the projection operator $h_\alpha{}^\sigma$, one has

$$h_\alpha{}^\sigma \nabla_\sigma \phi = -\dot{\phi}\, h_\alpha{}^\sigma \frac{u_\sigma}{c^2} + h_\alpha{}^\sigma \tilde{\nabla}_\sigma \phi$$

but, since $\tilde{\nabla}_\alpha \phi$ is purely spatial, we have

$$\tilde{\nabla}_\alpha \phi = g_\alpha^\sigma \tilde{\nabla}_\sigma \phi = -u_\alpha \underbrace{u^\sigma \tilde{\nabla}_\sigma \phi}_{0} + h^{\alpha\sigma} \tilde{\nabla}_\sigma \phi,$$

hence

$$\tilde{\nabla}_\alpha \phi = h^{\alpha\sigma} \nabla_\sigma \phi$$

is the spatial gradient of the function ϕ.

8.4 Use units in which $c = 1$. We have

$$k_\mu k^\mu = \frac{(x_\mu + t_\mu)}{\sqrt{2}} \frac{(x^\mu + t^\mu)}{\sqrt{2}} = \frac{x_\mu x^\mu + 2x_\mu t^\mu + t_\mu t^\mu}{2}.$$

Using now $x_\mu x^\mu = +1$, $t_\mu t^\mu = -1$, $x_\mu t^\mu = 0$, we obtain

$$k_\mu k^\mu = \frac{1-1}{2} = 0.$$

Proceeding, we have

$$\begin{aligned}
l_\mu l^\mu &= \frac{(x_\mu - t_\mu)}{\sqrt{2}} \frac{(x^\mu - t^\mu)}{\sqrt{2}} = \frac{x_\mu x^\mu - 2x_\mu t^\mu + t_\mu t^\mu}{2} = \frac{1-1}{2} = 0, \\
m_\mu m^\mu &= \frac{(y_\mu + \mathrm{i} z_\mu)}{\sqrt{2}} \frac{(y^\mu + \mathrm{i} z^\mu)}{\sqrt{2}} = \frac{y_\mu y^\mu + 2\mathrm{i} y_\mu z^\mu - z_\mu z^\mu}{2} = \frac{1+0-1}{2} = 0, \\
\bar{m}_\mu \bar{m}^\mu &= \frac{(y_\mu - \mathrm{i} z_\mu)}{\sqrt{2}} \frac{(y^\mu - \mathrm{i} z^\mu)}{\sqrt{2}} = \frac{y_\mu y^\mu - 2\mathrm{i} y_\mu z^\mu - z_\mu z^\mu}{2} = \frac{1+0-1}{2} = 0, \\
k_\mu l^\mu &= \frac{(x_\mu + t_\mu)}{\sqrt{2}} \frac{(x^\mu - t^\mu)}{\sqrt{2}} = \frac{x_\mu x^\mu - t_\mu t^\mu}{2} = \frac{1-(-1)}{2} = 1,
\end{aligned}$$

$$k_\mu m^\mu = \frac{(x_\mu + t_\mu)}{\sqrt{2}} \frac{(y^\mu + \mathrm{i}z^\mu)}{\sqrt{2}} = \frac{x_\mu y^\mu + \mathrm{i}x_\mu z^\mu + t_\mu y^\mu + \mathrm{i}t_\mu z^\mu}{2} = 0,$$

$$k_\mu \bar{m}^\mu = \frac{(x_\mu + t_\mu)}{\sqrt{2}} \frac{(y^\mu - \mathrm{i}z^\mu)}{\sqrt{2}} = \frac{x_\mu y^\mu - \mathrm{i}x_\mu z^\mu + t_\mu y^\mu - \mathrm{i}t_\mu z^\mu}{2} = 0,$$

$$l_\mu \bar{m}^\mu = \frac{(x_\mu - t_\mu)}{\sqrt{2}} \frac{(y^\mu - \mathrm{i}z^\mu)}{\sqrt{2}} = \frac{x_\mu y^\mu - \mathrm{i}x_\mu z^\mu - t_\mu y^\mu + \mathrm{i}t_\mu z^\mu}{2} = 0,$$

$$m_\mu \bar{m}^\mu = \frac{(y_\mu + \mathrm{i}z_\mu)}{\sqrt{2}} \frac{(y^\mu - \mathrm{i}z^\mu)}{\sqrt{2}} = \frac{y_\mu y^\mu - \mathrm{i}y_\mu z^\mu + \mathrm{i}y_\mu z^\mu + z_\mu z^\mu}{2} = \frac{1+1}{2} = 1.$$

To summarize, the null tetrad satisfies the relations

$$\begin{aligned} l_\mu l^\mu &= k_\mu k^\mu = m_\mu m^\mu = \bar{m}_\mu \bar{m}^\mu = 0, \\ k_\mu l^\mu &= m_\mu \bar{m}^\mu = 1, \end{aligned}$$

and all the other products vanish.
We also have

$$\begin{aligned} g_{\mu\nu} &= g_{(\alpha)(\beta)} e^{(\alpha)}{}_\mu e^{(\beta)}{}_\nu \\ &= g_{(0)(0)} e^{(0)}{}_\mu e^{(0)}{}_\nu + g_{(0)(1)} e_\mu{}^{(0)} e_\nu{}^{(1)} + g_{(0)(1)} e_\nu{}^{(0)} e_\mu{}^{(1)} \\ &\quad + g_{(1)(1)} e^{(1)}{}_\mu e^{(1)}{}_\nu + g_{(1)(2)} e^{(1)}{}_\mu e^{(2)}{}_\nu + g_{(1)(2)} e^{(1)}{}_\nu e^{(2)}{}_\mu \\ &\quad + g_{(2)(2)} e^{(2)}{}_\mu e^{(2)}{}_\nu + g_{(2)(3)} e^{(2)}{}_\mu e^{(3)}{}_\nu + g_{(2)(3)} e^{(3)}{}_\mu e^{(2)}{}_\nu \\ &\quad + g_{(3)(3)} e^{(3)}{}_\mu e^{(3)}{}_\nu + g_{(0)(2)} e^{(0)}{}_\mu e^{(2)}{}_\nu + g_{(0)(2)} e^{(0)}{}_\nu e^{(2)}{}_\mu \\ &\quad + g_{(0)(3)} e^{(0)}{}_\mu e^{(3)}{}_\nu + g_{(0)(2)} e^{(0)}{}_\nu e^{(3)}{}_\mu + g_{(1)(3)} e^{(1)}{}_\mu e^{(3)}{}_\nu \\ &\quad + g_{(1)(3)} e^{(1)}{}_\nu e^{(3)}{}_\mu . \end{aligned}$$

Let us compute the coefficients $g_{(\alpha)(\beta)}$:

$$\begin{aligned} g_{(0)(0)} &= g_{\mu\nu} e^\mu{}_{(0)} e^\nu{}_{(0)} = g_{\mu\nu} k^\mu k^\nu = 0, \\ g_{(0)(1)} &= g_{(1)(0)} = g_{\mu\nu} e^\mu{}_{(0)} e^\nu{}_{(1)} = g_{\mu\nu} k^\mu l^\nu = 1, \\ g_{(0)(2)} &= g_{(2)(0)} = g_{\mu\nu} e^\mu{}_{(0)} e^\nu{}_{(2)} = g_{\mu\nu} k^\mu m^\nu = 0, \\ g_{(0)(3)} &= g_{(3)(0)} = g_{\mu\nu} e^\mu{}_{(0)} e^\nu{}_{(3)} = g_{\mu\nu} k^\mu \bar{m}^\nu = 0, \\ g_{(1)(1)} &= g_{\mu\nu} e^\mu{}_{(1)} e^\nu{}_{(1)} = g_{\mu\nu} l^\mu l^\nu = 0, \\ g_{(1)(2)} &= g_{(2)(1)} = g_{\mu\nu} e^\mu{}_{(1)} e^\nu{}_{(2)} = g_{\mu\nu} l^\mu m^\nu = 0, \\ g_{(1)(3)} &= g_{(3)(1)} = g_{\mu\nu} e^\mu{}_{(1)} e^\nu{}_{(3)} = g_{\mu\nu} l^\mu \bar{m}^\nu = 0, \\ g_{(2)(2)} &= g_{\mu\nu} e^\mu{}_{(2)} e^\nu{}_{(2)} = g_{\mu\nu} m^\mu m^\nu = 0, \\ g_{(2)(3)} &= g_{(3)(2)} = g_{\mu\nu} e^\mu{}_{(2)} e^\nu{}_{(3)} = g_{\mu\nu} m^\mu \bar{m}^\nu = 1, \\ g_{(3)(3)} &= g_{\mu\nu} e^\mu{}_{(3)} e^\nu{}_{(3)} = g_{\mu\nu} \bar{m}^\mu \bar{m}^\nu = 0. \end{aligned}$$

Substituting these coefficients into the expression of $g_{\mu\nu}$ yields

$$g_{\mu\nu} = k_\mu l_\nu + k_\nu l_\mu + m_\mu \bar{m}_\nu + m_\nu \bar{m}_\mu.$$

Problems of Chapter 9

9.1 Since $T_{\mu\nu}$ is symmetric, $T_{\nu\mu} = T_{\mu\nu}$, we have

$$0 = \partial^\nu T_{\mu\nu} = \partial^\nu T_{\nu\mu} = \partial^\alpha T_{\alpha\mu}$$

The equation $\partial^\alpha T_{\alpha\mu} = 0$ is rewritten changing the names of the dummy indices as $\partial^\alpha T_{\alpha\nu} = 0$ or as $\partial^\mu T_{\mu\nu} = 0$.

9.2 In the $3+1$ splitting, the spatial 3-metric $h_{\mu\nu}$ is defined by

$$g_{\mu\nu} = -\frac{u_\mu u_\nu}{c^2} + h_{\mu\nu}$$

using the 4-velocity of the fluid u^μ. We have then

$$T_{\mu\nu} = \left(P + \rho c^2\right) \frac{u_\mu u_\nu}{c^2} - P \frac{u_\mu u_\nu}{c^2} + P h_{\mu\nu} = \rho u_\mu u_\nu + P h_{\mu\nu}.$$

9.5 Assume that $T_{\mu\nu} t^\mu t^\nu \geq 0$ for all timelike vectors t^μ. Then, using the form of the perfect fluid stress-energy tensor

$$T_{\mu\nu} = \left(\frac{P}{c^2} + \rho\right) u_\mu u_\nu + P g_{\mu\nu},$$

one obtains

$$T_{\mu\nu} t^\mu t^\nu = \left(\frac{P}{c^2} + \rho\right) \left(u_\mu t^\mu\right)^2 + P t_\mu t^\mu \geq 0$$

or, since $t_\mu t^\mu = -\left|t_\mu t^\mu\right|$,

$$\left(\frac{P}{c^2} + \rho\right) \left(u_\mu t^\mu\right)^2 \geq P \left|t_\mu t^\mu\right|.$$

Now choose $t^\mu = u^\mu$, which implies that $\left(\frac{P}{c^2} + \rho\right) c^4 - P c^2 \geq 0$, or

$$\rho \geq 0$$

and, if $P \geq 0$, then also $\rho + \frac{P}{c^2} \geq 0$. If the pressure P is negative, it must still be

$$\left(\frac{P}{c^2} + \rho\right) \left(u_\mu t^\mu\right)^2 - P \left|t_\mu t^\mu\right| \geq 0;$$

now

$$0 \leq P \left[\frac{\left(u_\mu t^\mu\right)^2}{c^2} - \left|t_\mu t^\mu\right|\right] + \rho \left(u_\mu t^\mu\right)^2 \leq \frac{P}{c^2} \left(u_\mu t^\mu\right)^2 + \rho \left(u_\mu t^\mu\right)^2$$

or

$$0 \leq \left(\frac{P}{c^2} + \rho\right) \left(u_\mu t^\mu\right)^2$$

which implies that

$$\frac{P}{c^2} + \rho \geq 0.$$

9.6 (a) We have

$$\begin{aligned} T_{\mu\nu} u^\mu u^\nu &= \left(P + \rho c^2\right) \frac{\left(u_\mu u^\mu\right)^2}{c^2} + P g_{\mu\nu} u^\mu u^\nu + q_\mu \frac{u_\nu}{c} u^\mu u^\nu + q_\nu \frac{u_\mu}{c} u^\mu u^\nu \\ &= P + \rho c^2 - P \\ &= \rho c^2. \end{aligned}$$

(b) We have also

$$\begin{aligned} h^{\mu\nu} T_{\mu\nu} &= h^{\mu\nu} \left[\left(P + \rho c^2\right) \frac{\left(u_\mu u^\mu\right)^2}{c^2} + P g_{\mu\nu} + q_\mu \frac{u_\nu}{c} + q_\nu \frac{u_\mu}{c}\right] \\ &= P h^{\mu\nu} h_{\mu\nu} + \left(g^{\mu\nu} + u^\mu u^\nu\right) \left(q_\mu \frac{u_\nu}{c} + q_\nu \frac{u_\mu}{c}\right) \\ &= 3P + \underbrace{\frac{q^\mu u_\mu}{c}}_{0} + \underbrace{\frac{q^\nu u_\nu}{c}}_{0} + \underbrace{u^\mu q_\mu}_{0} \frac{u^\nu u_\nu}{c} + \underbrace{u^\nu q_\nu}_{0} \frac{u^\mu u_\mu}{c} \\ &= 3P. \end{aligned}$$

(c) Finally, it is

$$\begin{aligned} T &= \left(P + \rho c^2\right) \frac{\left(u_\mu u^\mu\right)}{c^2} + 4P + 2 \frac{q^\mu u_\mu}{c^2} = -P - \rho c^2 + 4P \\ &= 3P - \rho c^2. \end{aligned}$$

9.8 We have

$$\begin{aligned} T_{\alpha\beta}\,\frac{u^\alpha u^\beta}{c} &= \left(\rho u_\alpha u_\beta + P h_{\alpha\beta} + \frac{q_\alpha u_\beta}{c} + \frac{q_\beta u_\alpha}{c^2} + \pi_{\alpha\beta}\right)\frac{u^\alpha u^\beta}{c^2} \\ &= \Bigg[\rho\left(u^\alpha u_\alpha\right)^2 + P h_{\alpha\beta}u^\alpha u^\beta + q_\alpha\,\frac{u^\alpha}{c}\left(u_\beta u^\beta\right) + q_\beta\,\frac{u^\beta}{c}\left(u_\alpha u^\alpha\right) \\ &\quad + \pi_{\alpha\beta}u^\alpha u^\beta\Bigg]\frac{1}{c^2} \\ &= \left[\rho c^4 - 2cq_\alpha u^\alpha + \pi_{\alpha\beta}u^\alpha u^\beta\right]\frac{1}{c^2} \\ &= \rho c^2. \end{aligned}$$

Then we have

$$\begin{aligned} \frac{1}{3}\,h^{\alpha\beta}T_{\alpha\beta} &= \frac{1}{3}\,h^{\alpha\beta}\left(\rho u_\alpha u_\beta + P h_{\alpha\beta} + q_\alpha u_\beta + q_\beta u_\alpha + \pi_{\alpha\beta}\right) \\ &= \frac{1}{3}\left(P h^{\alpha\beta}h_{\alpha\beta} + q_\alpha h^{\alpha\beta}u_\beta + q_\beta h^{\alpha\beta}u_\alpha + h^{\alpha\beta}\pi_{\alpha\beta}\right). \end{aligned}$$

Using the fact that $h^{\alpha\beta}h_{\alpha\beta} = g^{\alpha\beta}h_{\alpha\beta} = h^\alpha{}_\alpha = 3$, one obtains

$$\frac{1}{3}\,h^{\alpha\beta}T_{\alpha\beta} = \frac{1}{3}\left(3P + \pi^\alpha{}_\alpha\right) = P.$$

Finally, it is

$$\begin{aligned} -h_\alpha{}^\beta u^\sigma T_{\beta\sigma} &= -h^\beta_\alpha u^\sigma\left(\rho u_\beta u_\sigma + P h_{\beta\sigma} + q_\beta\,\frac{u_\sigma}{c} + q_\sigma\,\frac{u_\beta}{c} + \pi_{\beta\sigma}\right) \\ &= -h_\alpha{}^\beta u_\beta\left(u^\sigma u_\sigma\right)\rho - P h_\alpha{}^\beta u^\sigma h_{\beta\sigma} - h_\alpha{}^\beta q_\beta\,\frac{\left(u^\sigma u_\sigma\right)}{c} \\ &\quad - h_\alpha{}^\beta\left(u^\sigma q_\sigma\right)\frac{u_\beta}{c} - h_\alpha{}^\beta\pi_{\beta\sigma}u^\sigma \\ &= cq_\alpha. \end{aligned}$$

In addition, the trace-free part of $h_\alpha{}^\sigma h_\beta{}^\delta T_{\sigma\delta}$ is

$$\begin{aligned} T_{\langle\alpha\beta\rangle} &= h_\alpha{}^\sigma h_\beta{}^\delta T_{\sigma\delta} - \frac{1}{3}\,h_{\alpha\beta}\left(h_\sigma{}^\rho h^{\sigma\delta}T_{\rho\delta}\right) \\ &= h_\alpha{}^\sigma h_\beta{}^\delta\left(\rho u_\sigma u_\delta + P h_{\sigma\delta} + q_\sigma\,\frac{u_\delta}{c} + q_\delta\,\frac{u_\sigma}{c} + \pi_{\sigma\delta}\right) \\ &\quad - \frac{1}{3}\,h_{\alpha\beta}h^{\rho\delta}\left(\rho u_\sigma u_\delta + P h_{\rho\delta} + q_\rho\,\frac{u_\delta}{c} + q_\delta\,\frac{u_\rho}{c} + \pi_{\rho\delta}\right) \\ &= P h_\alpha{}^\sigma h_\beta{}^\delta h_{\sigma\delta} + h_\alpha{}^\sigma h_\beta{}^\delta\,\frac{u_\delta}{c}\,q_\sigma + h_\alpha{}^\sigma h_\beta{}^\delta\,\frac{u_\sigma}{c}\,q_\delta + h_\alpha{}^\sigma h_\beta{}^\delta\pi_{\sigma\delta} \end{aligned}$$

$$
\begin{aligned}
&-\frac{1}{3}\,h_{\alpha\beta}\left(\rho h^{\rho\delta}u_\delta u_\rho + P h^{\rho\delta}h_{\rho\delta} + h^{\rho\delta}\,\frac{u_\delta}{c}\,q_\rho + h^{\rho\delta}\,\frac{u_\rho}{c}\,q_\delta + \pi^\alpha{}_\alpha\right)\\
&= P h_{\alpha\beta} + \pi_{\alpha\beta} - \frac{1}{3}\,h_{\alpha\beta}\left(3P + \pi^\alpha{}_\alpha\right)\\
&= \pi_{\alpha\beta} - \frac{1}{3}\,\pi^\alpha{}_\alpha = \pi_{\alpha\beta}
\end{aligned}
$$

because the trace $\pi^\alpha{}_\alpha$ vanishes.

9.9 Let us calculate the effective quantities using the fact that

$$
\begin{aligned}
\tilde{u}_\alpha &= \gamma\left(u_\alpha + v_\alpha\right),\\
\tilde{h}_{\alpha\beta} &= g_{\alpha\beta} + \frac{\tilde{u}_\alpha\tilde{u}_\beta}{c^2} = -\frac{u_\alpha u_\beta}{c^2} + h_{\alpha\beta} + \frac{\tilde{u}_\alpha\tilde{u}_\beta}{c^2}.
\end{aligned}
$$

We have

$$
\begin{aligned}
T_{\alpha\beta} &= \rho_1 u_\alpha u_\beta + P_1 h_{\alpha\beta} + \rho_2\tilde{u}_\alpha\tilde{u}_\beta + P_2\tilde{h}_{\alpha\beta}\\
&= \rho_1 u_\alpha u_\beta + P_1 h_{\alpha\beta} + \rho_2\gamma^2\left(u_\alpha + v_\alpha\right)\left(u_\beta + v_\beta\right)\\
&\quad + P_2\left[\underbrace{g_{\alpha\beta}}_{-\frac{u_\alpha u_\beta}{c^2}+h_{\alpha\beta}} + \frac{\gamma^2}{c^2}\left(u_\alpha + v_\alpha\right)\left(u_\beta + v_\beta\right)\right]\\
&= \rho_1 u_\alpha u_\beta + P_1 h_{\alpha\beta} + \rho_2\gamma^2 u_\alpha u_\beta + \rho_2\gamma^2\left(u_\alpha v_\beta + u_\beta v_\alpha\right) + \rho_2\gamma^2 v_\alpha v_\beta\\
&\quad - P_2\frac{u_\alpha u_\beta}{c^2} + P_2 h_{\alpha\beta} + \gamma^2 P_2\frac{u_\alpha u_\beta}{c^2} + \gamma^2\frac{P_2}{c^2}\left(u_\alpha v_\beta + u_\beta v_\alpha\right) + P_2\gamma^2 v_\alpha v_\beta\\
&= \left[\rho_1 + \rho_2\gamma^2 + P_2\frac{\left(\gamma^2 - 1\right)}{c^2}\right]u_\alpha u_\beta + \left(\rho_2 + \frac{P_2}{c^2}\right)\gamma^2 v_\alpha v_\beta + \frac{\left(P_1 + P_2\right)}{c^2}h_{\alpha\beta}\\
&\quad + \gamma^2\left(u_\alpha v_\beta + u_\beta v_\alpha\right)\left(\rho_2 + \frac{P_2}{c^2}\right),
\end{aligned}
$$

Now use

$$
\gamma^2 - 1 \equiv \frac{1}{1 - v^2/c^2} - 1 = \frac{v^2/c^2}{1 - v^2/c^2}
$$

where $v^2 \equiv v_\alpha v^\alpha$ to obtain

$$
\begin{aligned}
\rho_1 + \rho_2\gamma^2 + P_2\frac{\left(\gamma^2 - 1\right)}{c^2} &= \rho_1 + \frac{\rho_2}{1 - v^2/c^2} + \frac{P_2 v^2}{c^2\left(1 - v^2/c^2\right)}\\
&= \rho_1 + \rho_2\,\frac{\left(1 - v^2/c^2 + v^2/c^2\right)}{1 - v^2/c^2} + \frac{P_2 v^2/c^2}{1 - v^2/c^2}
\end{aligned}
$$

$$= \rho_1 + \rho_2 + \frac{v^2}{c^2 v^2}\left(\rho_2 + \frac{P_2}{c^2}\right)$$
$$= \rho_1 + \rho_2 + \frac{\gamma^2 v^2}{c^2}\left(\rho_2 + \frac{P_2}{c^2}\right).$$

Substituting back into the expression of $T_{\alpha\beta}$ we obtain

$$T_{\alpha\beta} = \left[\rho_1 + \rho_2 + \frac{\gamma^2 v^2}{c^2}\left(\rho_2 + \frac{P_2}{c^2}\right)\right] u_\alpha u_\beta + \left(\rho_2 + \frac{P_2}{c^2}\right)\gamma^2 v_\alpha v_\beta + (P_1 + P_2)\, h_{\alpha\beta}$$
$$+\gamma^2 \left(\rho_2 + \frac{P_2}{c^2}\right)\left(u_\alpha v_\beta + u_\beta v_\alpha\right).$$

It follows immediately that the effective energy density is given by

$$\bar{\rho} c^2 = T_{\alpha\beta}\frac{u_\alpha u_\beta}{c^2} = \left[\rho_1 + \rho_2 + \frac{\gamma^2 v^2}{c^2}\left(\rho_2 + \frac{P_2}{c^2}\right)\right] c^2.$$

The purely spatial part of $T_{\alpha\beta}$ is

$$\bar{P} h_{\alpha\beta} + q_\alpha \frac{u_\beta}{c} + q_\beta \frac{u_\alpha}{c} + \pi_{\alpha\beta} = \left(\rho_2 + \frac{P_2}{c^2}\right)\gamma^2 \frac{v_\alpha v_\beta}{c^2} + \gamma^2 \left(\rho_2 + \frac{P_2}{c^2}\right) h_{\alpha\beta}$$

and it is straightforward to identify the energy flux density

$$q_\alpha = \gamma^2 \left(\rho_2 + \frac{P_2}{c^2}\right) c v_\alpha.$$

This identification can also be checked by using the definition

$$q_\alpha \equiv -h_\alpha{}^\beta \frac{u^\sigma}{c} T_{\beta\sigma} = -h_\alpha{}^\beta \frac{u^\sigma}{c}\left\{\left[\rho_1 + \rho_2 + \frac{\gamma^2 v^2}{c^2}\left(\rho_2 + \frac{P_2}{c^2}\right)\right] u_\beta u_\sigma\right.$$
$$\left. + \left(\rho_2 + \frac{P_2}{c^2}\right)\gamma^2 v_\beta v_\sigma + \frac{(P_1 + P_2)}{c^2} h_{\beta\sigma} + \gamma^2 \left(\rho_2 + \frac{P_2}{c^2}\right)\left(u_\beta v_\sigma + u_\sigma v_\beta\right)\right\}$$
$$= -\gamma^2 \left(\rho_2 + \frac{P_2}{c^2}\right)\left(h_\alpha{}^\beta v_\beta u^\sigma v_\sigma + h_\alpha{}^\beta v_\beta \frac{u^\sigma}{c} u_\sigma\right)$$
$$= \gamma^2 \left(\rho_2 + \frac{P_2}{c^2}\right) c v_\alpha.$$

We are left with

$$\bar{P}\, h_{\alpha\beta} + \pi_{\alpha\beta} = \left(\rho_2 + \frac{P_2}{c^2}\right)\gamma^2 v_\alpha v_\beta + (P_1 + P_2)\, h_{\alpha\beta}.$$

Taking the trace and remembering that $\pi_{\alpha\beta}$ must be trace-free, one obtains

$$3\bar{P} + \pi^{\alpha}{}_{\alpha} = \left(\rho_2 + \frac{P_2}{c^2}\right)\gamma^2 v^2 + 3\,(P_1 + P_2)$$

and, finally,

$$\bar{P} = P_1 + P_2 + \frac{\gamma^2 v^2}{3}\left(\rho_2 + \frac{P_2}{c^2}\right).$$

Therefore, we have

$$\begin{aligned}
\pi_{\alpha\beta} &= \left(\rho_2 + \frac{P_2}{c^2}\right)\gamma^2 v_\alpha v_\beta + (P_1 + P_2)\,h_{\alpha\beta} - \bar{P}\,h_{\alpha\beta} \\
&= \left(\rho_2 + \frac{P_2}{c^2}\right)\gamma^2 v_\alpha v_\beta + (P_1 + P_2)\,h_{\alpha\beta} - (P_1 + P_2)\,h_{\alpha\beta} \\
&\quad - \frac{\gamma^2 v^2}{3}\left(\rho_2 + \frac{P_2}{c^2}\right)h_{\alpha\beta} \\
&= \gamma^2\left(\rho_2 + \frac{P_2}{c^2}\right)\left(v_\alpha v_\beta - \frac{1}{3}\,h_{\alpha\beta} v^\sigma v_\sigma\right) \\
&= \gamma^2\left(\rho_2 + \frac{P_2}{c^2}\right)v_{\langle\alpha} v_{\beta\rangle}.
\end{aligned}$$

9.10 The requested normalization is

$$k_\alpha k^\alpha = -m^2$$

and the dispersion relation is

$$\omega\,(k) = \sqrt{\mathbf{k}^2 + m^2}.$$

9.11 The scalar field energy-momentum tensor is

$$T_{\mu\nu} = \phi_0^2 k_\mu k_\nu \sin^2\left(k_\alpha x^\alpha\right)$$

and the trace is

$$T = 0.$$

9.13 Work in units in which $c = 1$: the 4-vector u^μ must be normalized to $u_\mu u^\mu = -1$, which gives

$$u^\mu = \frac{\nabla^\mu \phi}{\sqrt{|\nabla_\alpha \phi \nabla^\alpha \phi|}}.$$

We now need to compute

$$\rho = T_{\mu\nu}u^{\mu}u^{\nu} = \left(\nabla_{\mu}\phi\nabla_{\nu}\phi - \frac{1}{2}\,g_{\mu\nu}\,\nabla_{\alpha}\phi\nabla^{\alpha}\phi - V g_{\mu\nu}\right)\frac{\nabla^{\mu}\phi\nabla^{\nu}\phi}{|\nabla_{\alpha}\phi\nabla^{\alpha}\phi|}$$

$$= \frac{\left(\nabla_{\mu}\phi\nabla^{\mu}\phi\right)^2}{|\nabla_{\alpha}\phi\nabla^{\alpha}\phi|} - \frac{1}{2}\frac{\left(\nabla_{\mu}\phi\nabla^{\mu}\phi\right)^2}{|\nabla_{\alpha}\phi\nabla^{\alpha}\phi|} - V\,\frac{\nabla^{\mu}\phi\nabla^{\nu}\phi}{|\nabla_{\alpha}\phi\nabla^{\alpha}\phi|}$$

$$= -\nabla^{\mu}\phi\nabla^{\nu}\phi + \frac{1}{2}\,\nabla_{\mu}\phi\nabla^{\mu}\phi + V$$

$$= -\frac{1}{2}\,\nabla_{\mu}\phi\nabla^{\mu}\phi + V(\phi).$$

Let $g_{\mu\nu} = -u_{\mu}u_{\nu} + h_{\mu\nu}$, with $h_{\mu\nu}u^{\nu} = h_{\mu\nu}u^{\mu} = 0$; then it is

$$P = \frac{1}{3}\,T_{\mu\nu}h^{\mu\nu} = \frac{1}{3}\left[\nabla_{\mu}\phi\nabla_{\nu}\phi - \frac{1}{2}\,g_{\mu\nu}\nabla_{\alpha}\phi\nabla^{\alpha}\phi - g_{\mu\nu}V\right]h^{\mu\nu}$$

$$= \frac{1}{3}\left(-\frac{1}{2}\,g_{\mu\nu}h^{\mu\nu}\nabla_{\alpha}\phi\nabla^{\alpha}\phi - g_{\mu\nu}h^{\mu\nu}V\right)$$

$$= \frac{1}{3}\left(-\frac{1}{2}\,\underbrace{h_{\mu\nu}h^{\mu\nu}}_{3}\,\nabla_{\alpha}\phi\nabla^{\alpha}\phi - \underbrace{h_{\mu\nu}h^{\mu\nu}}_{3}\,V\right)$$

$$= -\frac{1}{2}\,\nabla_{\alpha}\phi\nabla^{\alpha}\phi - V.$$

To summarize, we have

$$\rho = -\frac{1}{2}\,\nabla_{\alpha}\phi\nabla^{\alpha}\phi + V(\phi),$$

$$P = -\frac{1}{2}\,\nabla_{\alpha}\phi\nabla^{\alpha}\phi - V(\phi),$$

$$u^{\mu} = \frac{\nabla^{\mu}\phi}{\sqrt{|\nabla_{\alpha}\phi\nabla^{\alpha}\phi|}}.$$

To check that these results are correct, consider the stress-energy tensor of a perfect fluid constructed with these quantities,

$$(P+\rho)\,u_{\mu}u_{\nu} + P g_{\mu\nu} = \left(-\nabla_{\alpha}\phi\nabla^{\alpha}\phi\right)\frac{\nabla_{\mu}\phi\nabla_{\nu}\phi}{|\nabla_{\alpha}\phi\nabla^{\alpha}\phi|} + \left(-\frac{1}{2}\,\nabla_{\alpha}\phi\nabla^{\alpha}\phi - V\right)g_{\mu\nu}$$

$$= \nabla_{\mu}\phi\nabla_{\nu}\phi - \frac{1}{2}\,\nabla_{\alpha}\phi\nabla^{\alpha}\phi\,g_{\mu\nu} - V g_{\mu\nu}$$

$$= T_{\mu\nu}[\phi].$$

9.14 It is simplest to work in Cartesian coordinates in which $T_{\mu\nu}$ is represented by the diagonal matrix

$$(T_{\mu\nu}) = \begin{pmatrix} \rho c^2 & 0 & 0 & 0 \\ 0 & P & 0 & 0 \\ 0 & 0 & P & 0 \\ 0 & 0 & 0 & P \end{pmatrix}$$

The eigenvalues are $-\rho c^2$ and P, with degeneracies 1 and 3, respectively. The associated normalized eigenvectors are $\frac{u^\mu}{c} = \begin{pmatrix} 1 \\ 0 \\ 0 \\ 0 \end{pmatrix}$ and

$$\left\{ \begin{pmatrix} 0 \\ 1 \\ 0 \\ 0 \end{pmatrix}, \begin{pmatrix} 0 \\ 0 \\ 1 \\ 0 \end{pmatrix}, \begin{pmatrix} 0 \\ 0 \\ 0 \\ 1 \end{pmatrix} \right\}.$$

In fact, it is

$$T_{\mu\nu} u^\nu = \left(\rho u_\mu u_\nu + P\, h_{\mu\nu} \right) u^\nu = \rho u_\mu u_\nu u^\nu = -\rho c^2 u_\mu$$

and, if $e^{(i)}_\mu \in \left\{ \begin{pmatrix} 0 \\ 1 \\ 0 \\ 0 \end{pmatrix}, \begin{pmatrix} 0 \\ 0 \\ 1 \\ 0 \end{pmatrix}, \begin{pmatrix} 0 \\ 0 \\ 0 \\ 1 \end{pmatrix} \right\}$ are purely spatial unit vectors with $u^\mu e^{(i)}_\mu = 0$ (for $i = 1, 2, 3$), then it is

$$T_{\mu\nu} e^{(i)}_\nu = (P + \rho)\, u_\mu \underbrace{u_\nu e^\nu_{(i)}}_{=0} + P\, g_{\mu\nu} e^\nu_{(i)} = P\, e^{(i)}_\nu .$$

For a perfect fluid with equation of state $P = -\rho c^2$, the stress-energy tensor degenerates to

$$T_{\mu\nu} = -\rho c^2 g_{\mu\nu}$$

and any 4-vector is an eigenvector associated with the eigenvalue $-\rho c^2$. In fact, in this case we have

$$T_{\mu\nu} A^\nu = -\rho c^2 g_{\mu\nu} A^\nu = -\rho c^2 A_\nu$$

or $T_{\mu\nu} A^\nu = \lambda A_\nu$ with $\lambda = -\rho c^2$ for any 4-vector A^μ.

9.15 Remember that the Maxwell tensor is

$$
(F_{\mu\nu}) = \begin{pmatrix} 0 & -E_x & -E_y & -E_z \\ E_x & 0 & B_z & -B_y \\ E_y & -B_z & 0 & B_x \\ E_z & B_y & -B_x & 0 \end{pmatrix}.
$$

Then, in Cartesian coordinates in which $u^\mu \doteq c\delta^{0\mu}$ (rest frame of the observer with 4-velocity u^μ) it is

$$
F_{\mu\nu}\,\frac{u^\nu}{c} = F_{\mu 0} = \left(0, E_x, E_y, E_z\right) = (0, \mathbf{E})
$$

and

$$
E^\mu = F^\mu{}_\nu u^\nu = F^\mu{}_0 = (0, \mathbf{E})\,.
$$

Now compute $-\frac{1}{2}\,\varepsilon_{\mu\nu}{}^{\rho\sigma}\,F_{\rho\sigma}\,\frac{u^\nu}{c}$ using $u^\mu = c\delta^{0\mu}$:

$$
\begin{aligned}
-\frac{1}{2}\,\varepsilon_{0\nu}{}^{\rho\sigma}\,F_{\rho\sigma}\,\frac{u^\nu}{c} &= -\frac{1}{2}\,\varepsilon_{00}{}^{\rho\sigma}\,F_{\rho\sigma} = B_0 = 0,\\
-\frac{1}{2}\,\varepsilon_{1\nu}{}^{\rho\sigma}\,F_{\rho\sigma}\,\delta^{\nu 0} &= -\frac{1}{2}\,\varepsilon_{10}{}^{23}\,F_{23} - \frac{1}{2}\,\varepsilon_{10}{}^{32}\,F_{32}\\
&= -\frac{1}{2}\left(-\varepsilon_{01}{}^{23}\,F_{23} + \varepsilon_{01}{}^{23}(-F_{23})\right) = \varepsilon_{01}{}^{23}\,F_{23} = B_x\\
-\frac{1}{2}\,\varepsilon_{2\nu}{}^{\rho\sigma}\,F_{\rho\sigma}\,\frac{u^\nu}{c} &= -\frac{1}{2}\,\varepsilon_{20}{}^{\rho\sigma}\,F_{\rho\sigma} = -\frac{1}{2}\,\varepsilon_{20}{}^{13}\,F_{13} - \frac{1}{2}\,\varepsilon_{20}{}^{31}\,F_{31}\\
&= -\frac{1}{2}\left(\varepsilon_0{}^1{}_2{}^3\,F_{13} - \varepsilon_0{}^1{}_2{}^3\,F_{31}\right) = -\frac{1}{2}\left(\varepsilon_0{}^1{}_2{}^3\,F_{13} + \varepsilon_0{}^1{}_2{}^3\,F_{13}\right)\\
&= -\frac{1}{2}\,2(-B_y) = B_y\\
-\frac{1}{2}\,\varepsilon_{3\nu}{}^{\rho\sigma}\,F_{\rho\sigma}\,\frac{u^\nu}{c} &= -\frac{1}{2}\,\varepsilon_{30}{}^{\rho\sigma}\,F_{\rho\sigma} = -\frac{1}{2}\,\varepsilon_{30}{}^{12}\,F_{12} - \frac{1}{2}\,\varepsilon_{30}{}^{21}\,F_{21}\\
&= -\frac{1}{2}\left(\varepsilon_{30}{}^{12}\,F_{12} - \varepsilon_{30}{}^{12}\,F_{21}\right) = -\frac{1}{2}\left(\varepsilon_{30}{}^{12}\,F_{12} + \varepsilon_{30}{}^{12}\,F_{12}\right)\\
&= -\frac{1}{2}\,2\varepsilon_{30}{}^{12}\,F_{12} = -(-1)B_z = B_z,
\end{aligned}
$$

hence

$$
-\frac{1}{2}\,\varepsilon_{0\nu}{}^{\rho\sigma}\,F_{\rho\sigma}\,\frac{u^\sigma}{c},
$$

is $\mathbf{B} = \left(B_x, B_y, B_z\right)$.

9.16 In the frame S the Maxwell tensor is

$$(F_{\mu\nu}) = \begin{pmatrix} 0 & -E_x & 0 & 0 \\ E_x & 0 & 0 & 0 \\ 0 & 0 & 0 & 0 \\ 0 & 0 & 0 & 0 \end{pmatrix}.$$

The transformation property of $F_{\mu\nu}$ is

$$\begin{aligned} F_{\mu'\nu'} &= \frac{\partial x^{\mu}}{\partial x^{\mu'}} \frac{\partial x^{\nu}}{\partial x^{\nu'}} F_{\mu\nu} = \frac{\partial x^{0}}{\partial x^{\mu'}} \frac{\partial x^{1}}{\partial x^{\nu'}} F_{01} + \frac{\partial x^{1}}{\partial x^{\mu'}} \frac{\partial x^{0}}{\partial x^{\nu'}} F_{10} \\ &= \left(\frac{\partial x^{0}}{\partial x^{\mu'}} \frac{\partial x^{1}}{\partial x^{\nu'}} - \frac{\partial x^{1}}{\partial x^{\mu'}} \frac{\partial x^{0}}{\partial x^{\nu'}} \right) F_{01} \\ &= -E_x \left(\frac{\partial (ct)}{\partial x^{\mu'}} \frac{\partial x}{\partial x^{\nu'}} - \frac{\partial x}{\partial x^{\mu'}} \frac{\partial (ct)}{\partial x^{\nu'}} \right). \end{aligned}$$

The inverse Lorentz transformation

$$\begin{aligned} x &= \gamma \left(x' + \frac{v}{c} ct' \right), \\ ct &= \gamma \left(ct' + \frac{v}{c} x' \right), \\ y &= y', \\ z &= z', \end{aligned}$$

yields

$$\begin{aligned} \frac{\partial (ct)}{\partial x^{\mu'}} &= \gamma \left(\delta_{\mu'0} + \frac{v}{c} \delta_{\mu'1} \right), \\ \frac{\partial x}{\partial x^{\nu'}} &= \gamma \left(\delta_{\nu'1} + \frac{v}{c} \delta_{\nu'0} \right), \end{aligned}$$

and

$$\begin{aligned} F_{\mu'\nu'} &= -E_x \gamma^2 \left[\left(\delta_{\mu'0} + \frac{v}{c} \delta_{\mu'1} \right) \left(\delta_{\nu'1} + \frac{v}{c} \delta_{\nu'0} \right) - \left(\delta_{\mu'1} + \frac{v}{c} \delta_{\mu'0} \right) \left(\delta_{\nu'0} + \frac{v}{c} \delta_{\nu'1} \right) \right] \\ &= -\gamma^2 E_x \left[\delta_{\mu'0}\delta_{\nu'1} + \frac{v}{c} \left(\delta_{\mu'0}\delta_{\nu'0} + \delta_{\mu'1}\delta_{\nu'1} \right) + \frac{v^2}{c^2} \delta_{\mu'1}\delta_{\nu'0} - \delta_{\mu'1}\delta_{\nu'0} \right. \\ &\quad \left. + \frac{v}{c} \left(-\delta_{\mu'1}\delta_{\nu'1} - \delta_{\mu'0}\delta_{\nu'0} \right) - \frac{v^2}{c^2} \delta_{\mu'0}\delta_{\nu'1} \right] \end{aligned}$$

$$= -\gamma^2 E_x \left[\left(\delta_{\mu'0}\delta_{\nu'1} - \delta_{\mu'1}\delta_{\nu'0}\right) + \frac{v^2}{c^2}\left(\delta_{\mu'1}\delta_{\nu'0} - \delta_{\mu'0}\delta_{\nu'1}\right) \right]$$

$$= -\gamma^2 E_x \left(\delta_{\mu'0}\delta_{\nu'1} - \delta_{\mu'1}\delta_{\nu'0}\right) \underbrace{\left(1 - \frac{v^2}{c^2}\right)}_{\gamma^{-2}}$$

$$= -E_x \left(\delta_{\mu'0}\delta_{\nu'1} - \delta_{\mu'1}\delta_{\nu'0}\right).$$

In matrix form, we have the Maxwell tensor

$$\left(F_{\mu'\nu'}\right) = \begin{pmatrix} 0 & -E_x & 0 & 0 \\ E_x & 0 & 0 & 0 \\ 0 & 0 & 0 & 0 \\ 0 & 0 & 0 & 0 \end{pmatrix}.$$

In the inertial frame S' there is a pure electric field and no magnetic field, and this electric field is parallel to the x'-axis.

9.17 The required invariant is

$$F_{\mu\nu}F^{\mu\nu} = 2\left(\mathbf{B}^2 - \mathbf{E}^2\right).$$

For the monochromatic electromagnetic wave represented by the 4-potential

$$A_\mu = a_\mu \exp\left(\mathrm{i}\, k_\alpha x^\alpha\right),$$

it is

$$F_{\mu\nu}F^{\mu\nu} = 2\left(A_\alpha k^\alpha\right)^2 = 0$$

in the Lorentz gauge and, therefore, in any gauge because $F_{\mu\nu}F^{\mu\nu}$ is a gauge-invariant quantity. Then, for this wave, the electric and magnetic fields must have equal magnitudes,

$$|\mathbf{B}| = |\mathbf{E}|.$$

The other invariant $\varepsilon_{\mu\nu\rho\sigma}F^{\mu\nu}F^{\rho\sigma} = 4\,\mathbf{E}\cdot\mathbf{B}$ vanishes for this monochromatic wave, implying that $\mathbf{E}$ is perpendicular to $\mathbf{B}$ (in the usual 3-dimensional sense).

9.18 The wave equation is relativistic, as it can be written in covariant form as

$$\Box u = g^{\mu\nu}\nabla_\mu\nabla_\nu u \doteq \eta^{\mu\nu}\partial_\mu\partial_\nu u = 0,$$

where the last equality holds in Cartesian coordinates, or

$$\frac{\partial^2 u}{\partial x^2} + \frac{\partial^2 u}{\partial y^2} + \frac{\partial^2 u}{\partial z^2} - \frac{1}{c^2}\frac{\partial^2 u}{\partial t^2} = 0.$$

The heat/diffusion equation

$$\frac{\partial u}{\partial t} = a\,\nabla^2 u,$$

by contrast, is non-relativistic because the time derivative of the unknown function is of only first order while the space derivatives are of second order. This feature cannot be expressed using a 4-dimensional spacetime view in which time is a dimension analogous to the three spatial dimensions. Formally, heat conduction or diffusion described by the heat equation occurs with infinite speed (a phenomenon called "heat paradox"). This feature is not a problem for ordinary applications of this equation to every day situations, but it makes the equation untenable in the relativistic regime.

Problems of Chapter 10

10.1 We have

$$\Gamma^{\mu}_{\alpha\beta} = \frac{1}{2}\,g^{\mu\sigma}\left(g_{\sigma\alpha,\beta} + g_{\sigma\beta,\alpha} - g_{\alpha\beta,\sigma}\right)$$

and

$$\begin{aligned}
\Gamma^{\mu'}_{\alpha'\beta'} &= \frac{1}{2}\,g^{\mu'\sigma'}\left(\frac{\partial g_{\sigma'\alpha'}}{\partial x^{\beta'}} + \frac{\partial g_{\sigma'\beta'}}{\partial x^{\alpha'}} - \frac{\partial g_{\alpha'\beta'}}{\partial x^{\sigma'}}\right)\\
&= \frac{1}{2}\,g^{\mu'\sigma'}\left[\frac{\partial x^{\gamma}}{\partial x^{\beta'}}\frac{\partial}{\partial x^{\gamma}}\left(\frac{\partial x^{\nu}}{\partial x^{\sigma'}}\frac{\partial x^{\tau}}{\partial x^{\alpha'}}\,g_{\nu\tau}\right)\right.\\
&\quad+\frac{\partial x^{\gamma}}{\partial x^{\alpha'}}\frac{\partial}{\partial x^{\gamma}}\left(\frac{\partial x^{\nu}}{\partial x^{\sigma'}}\frac{\partial x^{\tau}}{\partial x^{\beta'}}\,g_{\nu\tau}\right)\\
&\quad\left.-\frac{\partial x^{\gamma}}{\partial x^{\sigma'}}\frac{\partial}{\partial x^{\gamma}}\left(\frac{\partial x^{\nu}}{\partial x^{\alpha'}}\frac{\partial x^{\tau}}{\partial x^{\beta'}}\,g_{\nu\tau}\right)\right]\\
&= \frac{1}{2}\,g^{\mu'\sigma'}\left\{\left[\frac{\partial^2 x^{\nu}}{\partial x^{\gamma}\partial x^{\sigma'}}\frac{\partial x^{\tau}}{\partial x^{\alpha'}}\,g_{\nu\tau} + \frac{\partial x^{\nu}}{\partial x^{\sigma'}}\frac{\partial^2 x^{\tau}}{\partial x^{\gamma}\partial x^{\alpha'}}\,g_{\nu\tau} + \frac{\partial x^{\nu}}{\partial x^{\sigma'}}\frac{\partial x^{\tau}}{\partial x^{\alpha'}}\,g_{\nu\tau,\gamma}\right]\frac{\partial x^{\gamma}}{\partial x^{\beta'}}\right.\\
&\quad+\left[\frac{\partial^2 x^{\nu}}{\partial x^{\gamma}\partial x^{\sigma'}}\frac{\partial x^{\tau}}{\partial x^{\beta'}}\,g_{\nu\tau} + \frac{\partial x^{\nu}}{\partial x^{\sigma'}}\frac{\partial x^{\tau}}{\partial x^{\beta'}}\,g_{\nu\tau} + \frac{\partial x^{\nu}}{\partial x^{\sigma'}}\frac{\partial x^{\tau}}{\partial x^{\beta'}}\,g_{\nu\tau,\gamma}\right]\frac{\partial x^{\gamma}}{\partial x^{\alpha'}}\\
&\quad\left.-\left[\frac{\partial^2 x^{\nu}}{\partial x^{\gamma}\partial x^{\alpha'}}\frac{\partial x^{\tau}}{\partial x^{\beta'}}\,g_{\nu\tau} + \frac{\partial x^{\nu}}{\partial x^{\alpha'}}\frac{\partial^2 x^{\tau}}{\partial x^{\gamma}\partial x^{\beta'}}\,g_{\nu\tau} + \frac{\partial x^{\nu}}{\partial x^{\alpha'}}\frac{\partial x^{\tau}}{\partial x^{\beta'}}\,g_{\nu\tau,\gamma}\right]\frac{\partial x^{\gamma}}{\partial x^{\sigma'}}\right\}\\
&= \frac{1}{2}\,g^{\mu'\sigma'}\left\{\left[\frac{\partial^2 x^{\nu}}{\partial x^{\gamma}\partial x^{\sigma'}}\frac{\partial x^{\tau}}{\partial x^{\alpha'}} + \frac{\partial^2 x^{\tau}}{\partial x^{\gamma}\partial x^{\alpha'}}\frac{\partial x^{\nu}}{\partial x^{\sigma'}}\right]\frac{\partial x^{\gamma}}{\partial x^{\beta'}}\right.\\
&\quad+\left[\frac{\partial^2 x^{\nu}}{\partial x^{\gamma}\partial x^{\sigma'}}\frac{\partial x^{\tau}}{\partial x^{\beta'}} + \frac{\partial^2 x^{\tau}}{\partial x^{\gamma}\partial x^{\beta'}}\frac{\partial x^{\nu}}{\partial x^{\sigma'}}\right]\frac{\partial x^{\gamma}}{\partial x^{\alpha'}}
\end{aligned}$$

$$-\left[\frac{\partial^2 x^\nu}{\partial x^\gamma \partial x^{\alpha'}}\frac{\partial x^\tau}{\partial x^{\beta'}}+\frac{\partial^2 x^\tau}{\partial x^\gamma \partial x^{\beta'}}\frac{\partial x^\nu}{\partial x^{\alpha'}}\right]\frac{\partial x^\gamma}{\partial x^{\sigma'}}\Bigg] g_{\nu\tau}$$

$$+\frac{1}{2}g^{\mu'\sigma'}\left[\frac{\partial x^\nu}{\partial x^{\sigma'}}\frac{\partial x^\tau}{\partial x^{\alpha'}}\frac{\partial x^\gamma}{\partial x^{\beta'}}+\frac{\partial x^\nu}{\partial x^{\sigma'}}\frac{\partial x^\tau}{\partial x^{\beta'}}\frac{\partial x^\gamma}{\partial x^{\alpha'}}-\frac{\partial x^\nu}{\partial x^{\alpha'}}\frac{\partial x^\tau}{\partial x^{\beta'}}\frac{\partial x^\gamma}{\partial x^{\sigma'}}\right]g_{\nu\tau,\gamma}$$

$$=\frac{1}{2}g^{\mu'\sigma'}\Bigg[\frac{\partial^2 x^\nu}{\partial x^{\sigma'}\partial x^{\beta'}}\frac{\partial x^\tau}{\partial x^{\alpha'}}+\frac{\partial^2 x^\tau}{\partial x^{\alpha'}\partial x^{\beta'}}\frac{\partial x^\nu}{\partial x^{\sigma'}}+\frac{\partial^2 x^\nu}{\partial x^{\sigma'}\partial x^{\alpha'}}\frac{\partial x^\tau}{\partial x^{\beta'}}$$

$$+\frac{\partial^2 x^\tau}{\partial x^{\beta'}\partial x^{\alpha'}}\frac{\partial x^\nu}{\partial x^{\sigma'}}$$

$$-\frac{\partial^2 x^\nu}{\partial x^{\sigma'}\partial x^{\alpha'}}\frac{\partial x^\tau}{\partial x^{\beta'}}-\frac{\partial^2 x^\tau}{\partial x^{\beta'}\partial x^{\sigma'}}\frac{\partial x^\nu}{\partial x^{\alpha'}}\Bigg]g_{\nu\tau}$$

$$+\frac{1}{2}g^{\mu'\sigma'}\left[\frac{\partial x^\nu}{\partial x^{\sigma'}}\frac{\partial x^\tau}{\partial x^{\alpha'}}\frac{\partial x^\gamma}{\partial x^{\beta'}}+\frac{\partial x^\nu}{\partial x^{\sigma'}}\frac{\partial x^\tau}{\partial x^{\beta'}}\frac{\partial x^\gamma}{\partial x^{\alpha'}}-\frac{\partial x^\nu}{\partial x^{\alpha'}}\frac{\partial x^\tau}{\partial x^{\beta'}}\frac{\partial x^\gamma}{\partial x^{\sigma'}}\right]g_{\nu\tau,\gamma}$$

$$=\frac{1}{2}g^{\mu'\sigma'}\left[\frac{\partial^2 x^\nu}{\partial x^{\sigma'}\partial x^{\beta'}}\frac{\partial x^\tau}{\partial x^{\alpha'}}+2\frac{\partial^2 x^\tau}{\partial x^{\alpha'}\partial x^{\beta'}}\frac{\partial x^\nu}{\partial x^{\sigma'}}-\frac{\partial^2 x^\tau}{\partial x^{\beta'}\partial x^{\sigma'}}\frac{\partial x^\nu}{\partial x^{\alpha'}}\right]g_{\nu\tau}$$

$$+\frac{1}{2}g^{\mu'\sigma'}\left[\frac{\partial x^\nu}{\partial x^{\sigma'}}\frac{\partial x^\tau}{\partial x^{\alpha'}}\frac{\partial x^\gamma}{\partial x^{\beta'}}+\frac{\partial x^\nu}{\partial x^{\sigma'}}\frac{\partial x^\tau}{\partial x^{\beta'}}\frac{\partial x^\gamma}{\partial x^{\alpha'}}-\frac{\partial x^\nu}{\partial x^{\alpha'}}\frac{\partial x^\tau}{\partial x^{\beta'}}\frac{\partial x^\gamma}{\partial x^{\sigma'}}\right]g_{\nu\tau,\gamma}$$

or

$$\Gamma^{\mu'}_{\alpha'\beta'}=+\frac{1}{2}g^{\mu'\sigma'}\left[\frac{\partial x^\nu}{\partial x^{\sigma'}}\frac{\partial x^\tau}{\partial x^{\alpha'}}\frac{\partial x^\gamma}{\partial x^{\beta'}}g_{\nu\tau,\gamma}+(\alpha'\leftrightarrow\beta')-(\sigma'\leftrightarrow\alpha')\right]g_{\nu\tau,\gamma}$$

+second derivative terms

$$=\frac{1}{2}g^{\mu'\sigma'}\Bigg[\frac{\partial x^\nu}{\partial x^{\sigma'}}\frac{\partial x^\tau}{\partial x^{\alpha'}}\frac{\partial x^\gamma}{\partial x^{\beta'}}g_{\nu\tau,\gamma}+\frac{\partial x^\nu}{\partial x^{\sigma'}}\frac{\partial x^\gamma}{\partial x^{\alpha'}}\frac{\partial x^\tau}{\partial x^{\beta'}}g_{\nu\tau,\gamma}$$

$$-\frac{\partial x^\gamma}{\partial x^{\sigma'}}\frac{\partial x^\nu}{\partial x^{\alpha'}}\frac{\partial x^\tau}{\partial x^{\beta'}}g_{\nu\tau,\gamma}\Bigg]+\cdots$$

$$=\frac{1}{2}g^{\mu'\sigma'}\left(\frac{\partial x^\nu}{\partial x^{\sigma'}}\frac{\partial x^\tau}{\partial x^{\alpha'}}\frac{\partial x^\gamma}{\partial x^{\beta'}}\right)\left(g_{\nu\tau,\gamma}+g_{\nu\gamma,\tau}-g_{\gamma\tau,\nu}\right)+\cdots$$

$$=\frac{1}{2}\frac{\partial x^{\mu'}}{\partial x^\varepsilon}\frac{\partial x^{\sigma'}}{\partial x^\delta}g^{\varepsilon\delta}\frac{\partial x^\nu}{\partial x^{\sigma'}}\frac{\partial x^\tau}{\partial x^{\alpha'}}\frac{\partial x^\gamma}{\partial x^{\beta'}}\left(g_{\nu\tau,\gamma}+g_{\nu\gamma,\tau}-g_{\gamma\tau,\nu}\right)+\cdots$$

(using the fact that $\frac{\partial x^{\sigma'}}{\partial x^\delta}\frac{\partial x^\nu}{\partial x^{\sigma'}}=\delta^\nu_\delta$)

$$=\frac{1}{2}\frac{\partial x^{\mu'}}{\partial x^\varepsilon}\frac{\partial x^\tau}{\partial x^{\alpha'}}\frac{\partial x^\gamma}{\partial x^{\beta'}}g^{\varepsilon\nu}\left(g_{\nu\tau,\gamma}+g_{\nu\gamma,\tau}-g_{\gamma\tau,\nu}\right)+\cdots$$

$$=\frac{\partial x^{\mu'}}{\partial x^\varepsilon}\frac{\partial x^\tau}{\partial x^{\alpha'}}\frac{\partial x^\gamma}{\partial x^{\beta'}}\Gamma^\varepsilon_{\tau\gamma}+\cdots$$

Now, it is

$$\left(\frac{\partial^2 x^\nu}{\partial x^{\sigma'}\partial x^{\beta'}}\frac{\partial x^\tau}{\partial x^{\alpha'}}+2\frac{\partial^2 x^\tau}{\partial x^{\alpha}\partial x^{\beta'}}\frac{\partial x^\nu}{\partial x^{\sigma'}}-\frac{\partial^2 x^\tau}{\partial x^{\beta'}\partial x^{\sigma'}}\frac{\partial x^\nu}{\partial x^{\alpha'}}\right)g_{\nu\tau}$$

(we can exchange ν and τ)

$$=\left(\frac{\partial^2 x^\nu}{\partial x^{\sigma'}\partial x^{\beta'}}\frac{\partial x^\tau}{\partial x^{\alpha'}}+2\frac{\partial^2 x^\tau}{\partial x^{\alpha'}\partial x^{\beta'}}\frac{\partial x^\nu}{\partial x^{\sigma'}}-\frac{\partial^2 x^\nu}{\partial x^{\beta'}\partial x^{\sigma'}}\frac{\partial x^\tau}{\partial x^{\alpha'}}\right)g_{\nu\tau}$$

$$=2\frac{\partial^2 x^\tau}{\partial x^{\alpha'}\partial x^{\beta'}}\frac{\partial x^\nu}{\partial x^{\sigma'}}g_{\nu\tau}.$$

Therefore, it is

$$\Gamma^{\mu'}_{\alpha'\beta'}=\frac{\partial x^{\mu'}}{\partial x^{\varepsilon}}\frac{\partial x^\tau}{\partial x^{\alpha'}}\frac{\partial x^\gamma}{\partial x^{\beta'}}\Gamma^{\varepsilon}_{\tau\gamma}+\frac{1}{2}\frac{\partial x^{\mu'}}{\partial x^{\varepsilon}}\frac{\partial x^{\sigma'}}{\partial x^{\delta}}g^{\varepsilon\delta}\cdot 2\frac{\partial^2 x^\tau}{\partial x^{\alpha'}\partial x^{\beta'}}\frac{\partial x^\nu}{\partial x^{\sigma'}}g_{\nu\tau}$$

and, using the fact that $\frac{\partial x^{\sigma'}}{\partial x^{\delta}}\frac{\partial x^\nu}{\partial x^{\sigma'}}=\delta^\nu_\delta$, the last term is seen to be

$$\frac{\partial x^{\mu'}}{\partial x^{\varepsilon}}\frac{\partial^2 x^\tau}{\partial x^{\alpha'}\partial x^{\beta'}}g^{\varepsilon\nu}.$$

Putting everything together, the desired transformation property is

$$\Gamma^{\mu'}_{\alpha'\beta'}=\frac{\partial x^{\mu'}}{\partial x^{\varepsilon}}\frac{\partial x^\tau}{\partial x^{\alpha'}}\frac{\partial x^\gamma}{\partial x^{\beta'}}\Gamma^{\varepsilon}_{\tau\gamma}+\frac{\partial x^{\mu'}}{\partial x^{\nu}}\frac{\partial x^\nu}{\partial x^{\alpha'}\partial x^{\beta'}}.$$

10.2 The identity is a trivial consequence of $\nabla_\sigma g_{\alpha\beta}=0$. In fact, we have

$$0=\nabla_\sigma g_{\alpha\beta}=\partial_\sigma g_{\alpha\beta}-\Gamma^{\mu}_{\sigma\alpha}g_{\mu\beta}-\Gamma^{\mu}_{\sigma\beta}g_{\mu\alpha},$$

from which it follows that

$$\partial_\sigma g_{\alpha\beta}=\Gamma^{\mu}_{\sigma\alpha}g_{\mu\beta}+\Gamma^{\mu}_{\sigma\beta}g_{\mu\alpha}.$$

10.3 The geodesic equation is

$$\frac{d^2x^\alpha}{ds^2}+\Gamma^{\alpha}_{\beta\gamma}\frac{dx^\beta}{ds}\frac{dx^\gamma}{ds}=\gamma\frac{dx^\alpha}{ds}.$$

Introduce a new parameter λ, then

$$\frac{dx^\alpha}{d\lambda}=\frac{dx^\alpha}{ds}\frac{ds}{d\lambda}$$

and

$$\frac{d^2x^\alpha}{d\lambda^2} = \frac{d}{d\lambda}\left(\frac{ds}{d\lambda}\frac{dx^\alpha}{ds}\right) = \frac{d^2s}{d\lambda^2}\frac{dx^\alpha}{ds} + \frac{ds}{d\lambda}\frac{d}{ds}\left(\frac{dx^\alpha}{ds}\right)\frac{ds}{d\lambda}$$
$$= \frac{d^2s}{d\lambda^2}\frac{dx^\alpha}{ds} + \left(\frac{ds}{d\lambda}\right)^2\frac{d^2x^\alpha}{ds^2}.$$

Impose now that with the new parameter λ it is

$$\frac{d^2x^\alpha}{d\lambda^2} + \Gamma^\alpha_{\beta\gamma}\frac{dx^\beta}{d\lambda}\frac{dx^\gamma}{d\lambda} = 0;$$

this requirement leads to

$$\frac{d^2s}{d\lambda^2}\frac{dx^\alpha}{ds} + \left(\frac{ds}{d\lambda}\right)^2\frac{d^2x^\alpha}{ds^2} + \Gamma^\alpha_{\beta\gamma}\frac{dx^\beta}{ds}\frac{dx^\gamma}{d\lambda}\left(\frac{ds}{d\lambda}\right)^2 = 0.$$

Call $\frac{ds}{d\lambda} \equiv f$, then

$$s = \int \mathrm{d}\lambda f(\lambda),$$

$$\frac{d^2s}{d\lambda^2} = \frac{df}{d\lambda},$$

and

$$\frac{d^2x^\alpha}{ds^2} + \frac{1}{f^2}\frac{df}{d\lambda}\frac{dx^\alpha}{ds} + \Gamma^\alpha_{\beta\gamma}\frac{dx^\beta}{ds}\frac{dx^\gamma}{ds} = 0.$$

We have now

$$\frac{d^2x^\alpha}{ds^2} + \Gamma^\alpha_{\beta\gamma}\frac{dx^\beta}{ds}\frac{dx^\gamma}{ds} = -\frac{1}{f^2}\frac{df}{d\lambda}\frac{dx^\alpha}{ds} = \left[\frac{d}{d\lambda}\left(\frac{1}{f}\right)\right]\frac{dx^\alpha}{ds},$$

which is to be compared with

$$\frac{d^2x^\alpha}{ds^2} + \Gamma^\alpha_{\beta\gamma}\frac{dx^\beta}{ds}\frac{dx^\gamma}{ds} = \gamma\frac{dx^\alpha}{ds}.$$

Then, it must be

$$\frac{d}{d\lambda}\left(\frac{1}{f}\right) = \gamma$$

or

$$\frac{d}{ds}\left(\frac{1}{f}\right)\frac{ds}{d\lambda} = \gamma$$

or

$$f \frac{d}{ds}\left(\frac{1}{f}\right) = \frac{d}{ds}\left[\ln\left(\frac{1}{f}\right)\right] = \gamma$$

which gives $\ln(1/f) = \int \mathrm{d}s\, \gamma(s)$ and

$$f(s) = \exp\left[-\int_{s_0}^{s} \mathrm{d}s' \gamma(s')\right] = \frac{ds}{d\lambda}.$$

With one more integration we obtain

$$\lambda(s) = \int \mathrm{d}s\, \mathrm{e}^{-\int_{s_0}^{s} \mathrm{d}s' \gamma(s')}.$$

We can now show that, if λ is an affine parameter, then any new parameter τ obtained with a reparameterization $\lambda \longrightarrow \tau$ will be affine if and only if $\tau = a\lambda + b$, where a and b are constants. Use the relation between λ and s already obtained, with the change $s \longrightarrow \lambda$ and $\lambda \longrightarrow \mu$: then we have

$$\mu(\lambda) = \int \mathrm{d}\lambda\, \mathrm{e}^{\int_{\lambda_0}^{\lambda} \mathrm{d}\lambda' \gamma(\lambda')},$$

but since λ is affine, it is $\gamma = 0$ and

$$\mu(\lambda) = \int \mathrm{d}\lambda = a\lambda + b,$$

where a and b are integration constants.

10.4 Using polar coordinates $\{r, \theta, \varphi\}$ related to Cartesian coordinates by

$$\begin{aligned} x &= r \sin\theta \cos\varphi, \\ y &= r \sin\theta \sin\varphi, \\ z &= r \cos\theta, \end{aligned}$$

the 3-dimensional Euclidean line element is written as

$$\mathrm{d}l^2 = \mathrm{d}r^2 + r^2 \left(\mathrm{d}\theta^2 + \sin^2\theta \mathrm{d}\varphi^2\right).$$

On the surface of the sphere Σ of radius R it is $r = R$ =const. and $\mathrm{d}r = 0$, hence the line element reduces to

$$\mathrm{d}l^2\Big|_{\Sigma} = R^2 \left(\mathrm{d}\theta^2 + \sin^2\theta \mathrm{d}\varphi^2\right).$$

The 2-dimensional metric on this sphere is given by

$$(h_{ab}) = \begin{pmatrix} R^2 & 0 \\ 0 & R^2 \sin^2\theta \end{pmatrix}$$

and its inverse by

$$\left(h^{ab}\right) = \begin{pmatrix} \frac{1}{R^2} & 0 \\ 0 & \frac{1}{R^2 \sin^2\theta} \end{pmatrix}.$$

The determinant is $\mathrm{Det}(h_{ab}) = R^4 \sin^2\theta$. Let us compute the connection coefficients

$$\Gamma^a_{bc} = \frac{1}{2} g^{ad} \left(g_{db,c} + g_{dc,b} - g_{bc,d}\right).$$

It is useful to make a complete list of them before computing them:

$$\Gamma^1_{11}, \quad \Gamma^1_{12} = \Gamma^1_{21}, \quad \Gamma^1_{22}, \quad \Gamma^2_{11}, \quad \Gamma^2_{12} = \Gamma^2_{21}, \quad \Gamma^2_{22}.$$

We then apply the previous formula, obtaining

$$\begin{aligned}
\Gamma^1_{1c} &= \frac{1}{2} g^{1d} \left(g_{d1,c} + g_{dc,1} - g_{1c,d}\right) = \frac{1}{2R^2} \left(g_{11,c} + g_{1c,1} - g_{1c,1}\right) = 0 \\
\Gamma^1_{22} &= \frac{1}{2} g^{1d} \left(g_{d2,2} + g_{d2,2} - g_{22,d}\right) = \frac{1}{2R^2} \left(g_{12,2} + g_{12,2} - g_{22,1}\right) \\
&= -\frac{1}{2R^2} \frac{\partial}{\partial\theta} \left(R^2 \sin^2\theta\right) = -\sin\theta\cos\theta, \\
\Gamma^2_{1c} &= \frac{1}{2} g^{2d} \left(g_{d1,c} + g_{dc,1} - g_{1c,d}\right) = \frac{1}{2R^2 \sin^2\theta} \left(g_{21,c} + g_{2c,1} - g_{1c,2}\right) \\
&= \frac{1}{2R^2 \sin^2\theta} \delta^2_c \frac{\partial}{\partial\theta} \left(R^2 \sin^2\theta\right) = \delta^2_c \frac{\sin\theta\cos\theta}{\sin^2\theta} = \delta^2_c \cot\theta, \\
\Gamma^2_{22} &= \frac{1}{2} g^{2d} \left(g_{d2,2} + g_{d2,2} - g_{22,d}\right) = -\frac{1}{2R^2 \sin^2\theta} g_{22,2} = 0.
\end{aligned}$$

To summarize, the only non-vanishing connection coefficients are

$$\begin{aligned}
\Gamma^1_{22} &= -\sin\theta\cos\theta, \\
\Gamma^2_{12} &= \Gamma^2_{21} = \cot\theta.
\end{aligned}$$

The geodesic equation

$$\frac{d^2x^a}{d\lambda^2} + \Gamma^a_{bc} \frac{dx^b}{d\lambda} \frac{dx^c}{d\lambda} = 0$$

yields, on the sphere,

$$\ddot{\theta} - \sin\theta\cos\theta\,(\dot{\varphi})^2 = 0,$$
$$\ddot{\varphi} + 2\cot\theta\,\dot{\theta}\,\dot{\varphi} = 0,$$

where an overdot denotes differentiation with respect to the (affine) parameter λ. The big circles on the sphere are solutions, hence they are geodesic curves.

- *Meridians:* these are curves with

$$\varphi = \text{const.}, \quad \ddot{\theta} = 0, \quad \theta(\lambda) = \alpha\lambda + \beta,$$

where α and β are integration constants. The meridians are geodesic curves with $\varphi = \text{const.}$ and $0 \le \theta < 2\pi$.
- *Parallels:* these curves are characterized by

$$\theta = \text{const.}, \quad \ddot{\varphi} = 0, \quad \varphi(\lambda) = \alpha\lambda + \beta$$

(where α and β are again integration constants) and by $\sin\theta\cos\theta\,(\dot{\varphi})^2 = 0$. Since $\dot{\varphi} \neq 0$, it must be $\sin\theta = 0$ or $\cos\theta = 0$, corresponding to

$$\theta_{1,2,3} = 0, \pi, \frac{\pi}{2}.$$

The values $\theta_{1,2} = 0, \pi$ describe the North and South poles, which are degenerate geodesics, while the value $\theta_3 = \pi/2$ corresponds to the equator.

10.5 In cylindrical coordinates $\{r, \varphi, z\}$ the answer is

$$\Gamma^2_{12} = \Gamma^2_{21} = \frac{1}{r}$$

and all the other connection coefficients vanish.
In spherical coordinates $\{r, \theta, \varphi, \}$ the answer is

$$\Gamma^1_{22} = -r,$$
$$\Gamma^1_{33} = -r\sin^2\theta,$$
$$\Gamma^2_{12} = \Gamma^2_{21} = \Gamma^3_{13} = \Gamma^3_{31} = \frac{1}{r},$$
$$\Gamma^2_{33} = -\sin\theta\cos\theta,$$
$$\Gamma^3_{23} = \Gamma^3_{32} = \cot\theta,$$

and all the other connection coefficients vanish.

10.6 (a) In cylindrical coordinates the line element is

$$ds^2 = -c^2 dt^2 + dr^2 + r^2 d\varphi^2 + dz^2$$

and, using units in which $c = 1$,

$$(g_{\mu\nu}) = \begin{pmatrix} -1 & 0 & 0 & 0 \\ 0 & 1 & 0 & 0 \\ 0 & 0 & r^2 & 0 \\ 0 & 0 & 0 & 1 \end{pmatrix}.$$

The 4-velocity of the fluid is a timelike vector with components $u^\mu = (1, 0, 0, 0) = \delta^{0\mu}$ in the rest frame of the fluid and satisfying the normalization $u_\mu u^\mu = -1$. The stress-energy tensor of a perfect fluid has the form

$$T_{\mu\nu} = (P + \rho)\, u_\mu u_\nu + P g_{\mu\nu}.$$

Here

$$u_\mu = g_{\mu\nu} u^\nu = g_{\mu 0} u^0 = -\delta_{\mu 0}\ ,$$

and

$$T_{\mu\nu} = (P + \rho)\, \delta_{\mu 0} \delta_{\nu 0} + P g_{\mu\nu}.$$

In these coordinates, $T_{\mu\nu}$ is diagonal and

$$T_{00} = (P + \rho) - P = \rho,$$

$$T_{11} = P g_{11} = P,$$

$$T_{22} = P g_{22} = P r^2,$$

$$T_{33} = P g_{33} = P,$$

or

$$(T_{\mu\nu}) = \begin{pmatrix} \rho & 0 & 0 & 0 \\ 0 & P & 0 & 0 \\ 0 & 0 & P & 0 \\ 0 & 0 & 0 & P \end{pmatrix}.$$

(b) In spherical coordinates the line element is

$$ds^2 = -c^2 dt^2 + dr^2 + r^2 \left(d\theta^2 + \sin^2\theta d\varphi^2\right)$$

and (working again in units in which $c = 1$), it is

$$(g_{\mu\nu}) = \begin{pmatrix} -1 & 0 & 0 & 0 \\ 0 & 1 & 0 & 0 \\ 0 & 0 & r^2 & 0 \\ 0 & 0 & 0 & r^2 \sin^2\theta \end{pmatrix}.$$

Since

$$T_{\mu\nu} = (P + \rho)\, u_\mu u_\nu + P g_{\mu\nu},$$

where

$$u_\mu = g_{\mu\nu} u^\nu = g_{\mu 0} u^0 = -\delta_{\mu 0},$$

in these coordinates $T_{\mu\nu}$ is again diagonal and

$$\begin{aligned} T_{00} &= (P + \rho)\, u_0 u_0 + P g_{00} = P + \rho - P = \rho, \\ T_{11} &= P g_{11} = P, \\ T_{22} &= P g_{22} = P r^2, \\ T_{33} &= P g_{33} = P r^2 \sin^2\theta, \end{aligned}$$

or

$$(T_{\mu\nu}) = \begin{pmatrix} \rho & 0 & 0 & 0 \\ 0 & P & 0 & 0 \\ 0 & 0 & P r^2 & 0 \\ 0 & 0 & 0 & P r^2 \sin^2\theta \end{pmatrix}.$$

10.7 No. Since $\nabla^\nu T_{\mu\nu} = 0$, it is also $\nabla^\mu \nabla^\nu T_{\mu\nu} = 0$ identically for all covariantly conserved forms of matter.

10.8 One can use the projection of the equation of motion $\nabla^\mu T_{\mu\nu} = 0$ onto the 3-space orthogonal to u^μ:

$$h^\mu{}_\alpha \, \partial_\mu P + \left(P + \rho c^2\right) a_\alpha = 0,$$

where $h_{\mu\alpha}$ is the Riemannian metric on the 3-space orthogonal to u^μ and a^μ is the 4-acceleration of the fluid particles. For a dust, $P = 0$ and its gradient, which is the only force acting on the fluid, vanishes identically leaving $a_\mu = 0$.
An alternative solution consists of using the equation of motion of the fluid $\nabla^\mu T_{\mu\nu} = 0$ directly:

$$0 = \nabla^\nu \left(\rho u_\mu u_\nu\right) = \left(\nabla^\nu \rho\right) u_\mu u_\nu + \rho \left(\nabla^\nu u_\mu\right) u_\nu + \rho u_\mu \nabla^\nu u_\nu = 0$$

or

$$u_\mu \left(u^\nu \nabla_\nu \rho + \rho \nabla^\nu u_\nu\right) + \rho \underbrace{u^\nu \nabla_\nu u_\mu}_{a_\mu} = 0$$

By projecting onto the direction of the 4-velocity u^μ, one obtains

$$\underbrace{u^\mu u_\mu}_{-c^2} \left(u^\nu \nabla_\nu \rho + \rho \nabla^\nu u_\nu\right) + \rho \underbrace{a_\mu u^\mu}_{0} = 0$$

and one obtains $u^\nu \nabla_\nu \rho + \rho \nabla^\nu u_\nu = 0$ which, substituted back into the full equation of motion, leads to $a_\mu = 0$.

10.10 Since u^μ is null and an overall multiplicative constant (or function) does not affect its normalization, we can use $l^\mu = \sqrt{\rho}\, u^\mu$ instead, obtaining

$$T_{\mu\nu} = l_\mu l_\nu,$$

where l^μ is a null 4-vector: $l_\mu l^\mu = \rho u_\mu u^\mu = 0$. The trace of the null dust stress energy tensor vanishes,

$$T \equiv T^\mu{}_\mu = l^\mu l_\mu = 0,$$

and

$$T^\alpha_\gamma T^{\gamma\beta} = l^\alpha \left(l_\gamma l^\gamma\right) l^\beta = 0$$

as well. The covariant conservation equation $\nabla^\nu T_{\mu\nu} = 0$ yields

$$\nabla^\nu \left(l_\mu l_\nu\right) = l_\mu \nabla^\nu l_\nu + l^\nu \nabla_\nu l_\mu = 0$$

and, finally,

$$l^\nu \nabla_\nu l^\mu = -\left(\nabla^\nu l_\nu\right) l^\mu.$$

This is the equation of a non-affinely parametrized null geodesic with tangent l^μ.

References

1. G. Woan, *The Cambridge Handbook of Physics Formulas* (Cambridge University Press, Cambridge, 2000)
2. C. Amsler et al., Particle Data Group. Phys. Lett. B **667**, 1 (2008)

Index

V. Faraoni, *Special Relativity*, Undergraduate Lecture Notes in Physics,
DOI: 10.1007/978-3-319-01107-3, © Springer International Publishing Switzerland 2013